# A LEVEL & AS

## PHYSICAL GEOGRAPHY

# AQA Geography

Series editors   Simon Ross   Alice Griffiths

Tim Bayliss   Lawrence Collins

Catherine Hurst   Bob Digby

Andy Slater

UPDATED

**OXFORD**

UNIVERSITY PRESS

OXFORD
UNIVERSITY PRESS

Great Clarendon Street, Oxford, OX2 6DP, United Kingdom

Oxford University Press is a department of the University of Oxford. It furthers the University's objective of excellence in research, scholarship, and education by publishing worldwide. Oxford is a registered trade mark of Oxford University Press in the UK and in certain other countries

© Oxford University Press 2016

Series editor: Simon Ross, Alice Griffiths

Authors: Tim Bayliss, Lawrence Collins, Catherine Hurst, Bob Digby, Andy Slater

The moral rights of the authors have been asserted

Database right of Oxford University Press (maker) 2016

First published in 2016

Revised impression 2020

All rights reserved. No part of this publication may be reproduced, stored in a retrieval system, or transmitted, in any form or by any means, without the prior permission in writing of Oxford University Press, or as expressly permitted by law, by licence or under terms agreed with the appropriate reprographics rights organization. Enquiries concerning reproduction outside the scope of the above should be sent to the Rights Department, Oxford University Press, at the address above.

You must not circulate this work in any other form and you must impose this same condition on any acquirer

British Library Cataloguing in Publication Data
Data available

ISBN 978-0-19-836651-5

10 9 8 7 6 5 4

Paper used in the production of this book is a natural, recyclable product made from wood grown in sustainable forests. The manufacturing process conforms to the environmental regulations of the country of origin.

Printed in India by Manipal Technologies Limited.

**Acknowledgements**

The publisher and authors would like the thank the following for permission to use photographs and other copyright material:

**Cover:** David Baker/Trevillion Images; Earth Imaging/Getty Images; **p4:** Cole Burston/AFP/Getty Images; **p6:** Raymond Llewellyn/Shutterstock; **p10:** Ruth Peterkin/Shutterstock; **p10:** karamysh/Shutterstock; **p10:** Daniel Prudek/Shutterstock; p10: Arsenie Krasnevsky/Shutterstock; p10: StudioSmart/Shutterstock; **p19:** © Crown copyright (2016) OS 100043707; **p20:** Citizen of the Planet/Alamy Stock Photo; **p21:** PhotoStock-Israel/Alamy Stock Photo; **p23:** NASA map by Robert Simmon and Reto Stöckli, based on MODIS data.; **p25:** © imageBROKER/Alamy Stock Photo; **p26:** VLADJ55/Shutterstock; **p26:** Rich Carey/Shutterstock; **p27:** © Ashley Cooper/Alamy Stock Photo; **p27:** katewarn images/Alamy Stock Photo; **p28:** courtesy of NOAA; **p29:** Beawiharta/Reuters; **p30:** Siim Sepp/Alamy Stock Photo; **p31:** courtesy of NOAA; **p32:** Figure 7.3 from Climate Change 2007: The Physical Science Basis. Working Group I Contribution to the Fourth Assessment Report of the Intergovernmental Panel on Climate Change [Solomon, S., D. Qin, M. Manning, Z. Chen, M. Marquis, K.B. Averyt, M. Tignor and H.L. Miller (eds.)]. Cambridge University Press, Cambridge, United Kingdom and New York, NY, USA.; **p33:** Mark Thorpe/Creative Commons Attribution-Share Alike 3.0 Unported; **p39:** BigRoloImages/Shutterstock; **p41:** Cecile Girardin; **p42:** courtesy of SaskPower; **p43:** Frontpage/Shutterstock; **p47:** guentermanaus/Shutterstock; **p48:** Dr. Morley Read/Shutterstock; **p49:** Nigel Dickinson/Alamy Stock Photo; **p50:** powderkeg stock/Alamy Stock Photo; **p50:** ABDUL QODIR/Getty Images; **p52:** cpphotoimages/Shutterstock; **p52:** UK National River Flow Archive http://nrfa.ceh.ac.uk Copyright © 2016 NERC; **p53:** Exmoor Mires Partnership; **p54:** South West Water; **p56:** The Environment Agency; **p56:** Data from the UK National River Flow Archive; **p57:** Figure TS.3 from IPCC, 2014: Climate Change 2014: Mitigation of Climate Change. Contribution of Working Group III to the Fifth Assessment Report of the Intergovernmental Panel on Climate Change [Edenhofer, O., R. Pichs-Madruga, Y.Sokona, E. Farahani, S. Kadner, K. Seyboth, A. Adler, I. Baum, S. Brunner, P. Eickemeier, B. Kriemann, J. Savolainen, S. Schloemer, C. von Stechow, T. Zwickel and J.C. Minx (eds.)]. Cambridge University Press, Cambridge, United Kingdom and New York, NY, USA.; **p58:** MARIO RUIZ/EPA; **p60:** Simon Ross; **p61:** VT750/Shutterstock; **p62:** Rovingmagpie@flickr.com/Getty Images; **p65:** Joel Sartore/NGS; **p67:** MARKA/Alamy Stock Photo; **p68:** Simon Ross; **p69:** David South/Alamy Stock Photo; **p70:** Simon Ross; **p71:** Simon Ross; **p72:** Simon Ross; **p72:** Sylvia Kania/Shutterstock; **p73:** PjrNews/Alamy Stock Photo; **p74:** Tom Grundy/Alamy Stock Photo; **p75:** www.sandatlas.org/Shutterstock; **p75:** Mike P Shepherd/Alamy Stock Photo; **p76:** Darren L. Wickham; **p77:** Frans Lemmens/Getty Images; **p78:** imageBROKER/Alamy Stock Photo; **p79:** Tore Kjeilen; **p80:** Robert Harding Picture Library Ltd/Alamy Stock Photo; **p81:** Walter Meayers Edwards/National Geographic Creative; **p82:** Tony Waltham Geophotos; **p82:** Galyna Andrushko/Shutterstock; **p83:** Ken Barber/Alamy Stock Photo; **p84:** Simon Ross; **p89:** Thomas Mukoya/Reuters Photo Agency; **p90:** Photoshot Holdings Ltd/Alamy Stock Photo; **p91:** Andrew McConnell/Alamy Stock Photo; **p91:** imageBROKER/Alamy Stock Photo; **p92:** John Zada/Alamy Stock Photo; **p93:** Simon Ross; **p93:** Simon Ross; **p94:** Kip Evans/Alamy Stock Photo; **p95:** Quickbird/Digital Globe; **p95:** Mike Malaska; **p96:** WaterFrame/Alamy Stock Photo; **p96:** Art Directors & TRIP/Alamy Stock Photo; **p96:** © howard west/Alamy Stock Photo; **p97:** Dr. Nathalie Maria Vriend,Royal Society Dorothy Hodgkin Research Fellow and College Lecturer, Newnham College; **p98:** Infographic developed by iMMAP as part of the Regional Food Security Analysis Network, a project between FAO and iMMAP, funded by USAID. All rights reserved; **p101:** Simon Ross; **p102:** FRANCISCO LEONG/Getty Images; **p104:** Courtesy of Dennis Decker, WCM, NWS Melbourne, FL/National Oceanic and Atmospheric Administration; **p104:** Michigan Sea Grant; **p106:** Ian Murray/Getty Images; **p106:** Don Mennig/Alamy Stock Photo; **p108:** U. S. Geological Survey; **p110:** John McLellan/REX Shutterstock; **p111:** Shutterstock; **p111:** Dorset Media Service/Alamy Stock Photo; **p112:** Peter Smith Photography; **p114:** Matt Cardy/Getty Images; **p117:** imageBROKER/Alamy Stock Photo; Bob Digby; **p118:** Simon Ross; Bob Digby; **p118:** Geography Photos/Contributor/ Getty Images; Bob Digby; **p119:** Simon Ross; **p119:** Simon Ross; **p120:** Robyn Mackenzie/Shutterstock; **p122:** Simon Ross; **p122:** Simon Ross; **p123:** Giaros/Wikipedia Commons; **p125:** MARK BRAZIER/Alamy Stock Photo; **p125:** David Robertson/Alamy Stock Photo; **p126:** Dan Burton Photo/Alamy Stock Photo; **p126:** Les Gibbon/Alamy Stock Photo; **p127:** imageBROKER/Alamy Stock Photo; **p130:** Simon Ross; **p132:** Digital Globe; **p131:** Robert A. Rohde/Global Warming Art project/Creative Commons Attribution-Share Alike 3.0 Unported; **p132:** FerdinandoScafrogha/Wikipedia Commons/CC_BY_SA2.0; **p133:** GEORGE BERNARD/SCIENCE PHOTO LIBRARY; **p133:** Peter Turner Photography/Shutterstock; **p133:** frans lemmens/Alamy Stock Photo; **p134:** NASA Earth Observatory image; **p134:** Philip Game/Alamy Stock Photo; **p135:** © Crown copyright (2016) OS 100043707; **p136:** Simon Ross; **p136:** Simon Ross; **p137:** Shutterstock/Nick Hawkes; **p137:** Shutterstock/Sue Chillingworth; **p137:** Simon Ross; **p137:** Simon Ross; **p137:** Smallbones; **p138:** Eastern Solent Coastal Partnership; **p138:** © Rebecca Grinham/Stockimo/Alamy Stock Photo; **p138:** Simon Ross; **p138:** Simon Ross; **p138:** Graham Catley/Alamy Stock Photo; **p140:** Based on OS 50m resolution DEM (Crown Copyright NC/01/476); **p141:** Ashley Cooper/Alamy Stock Photo; **p141:** steven gillis hd9 imaging/Alamy Stock Photo; **p141:** Loop Images Ltd/Alamy Stock Photo; **p142:** geogphotos/Alamy Stock Photo; **p142:** © A.P.S. (UK)/Alamy Stock Photo; **p143:** © Crown copyright (2016) OS 100043707; **p143:** Cassini/Royal Geographical Society; **p144:** M. Dharma Raj/NCSCM; **p145:** John T.L/Alamy Stock Photo; **p145:** Saikat Paul/Shutterstock; **p146:** tapasbiswasphotograph /Getty Images; **p146:** AP/Press Association Images; **p147:** Ashish Kothari; **p148:** Universal Images Group North America LLC/Alamy Stock Photo; **p148:** Jack Barker/Alamy Stock Photo; **p149:** PCL Map Collection/University of Texas Libraries; **p156:** Jody/Shutterstock; p158: Simon Ross; **p158:** © David Wall/Alamy Stock Photo; **p159:** © David South/Alamy Stock Photo; **p162:** The University of Texas at Austin; **p167:** © StockShot/Alamy Stock Photo; **p169:** Chris Jackson/Getty Images; **p169:** MyLoupe/UIG/Getty Images; **p170:** Simon Ross; **p170:** Simon Ross; **p171:** © Jan Šiřina/Alamy Stock Photo; **p171:** Steve Garufi; **p172:** Michael J Hambrey; **p175:** Simon Ross; **p175:** Simon Ross; **p176:** Isabelle Laurion; **p178:** Photograph courtesy from www.markhewittphotography; **p179:** © Prisma Bildagentur AG/Alamy Stock Photo; **p179:** © Jo Katanigra/Alamy Stock Photo; **p180:** Gregory Perkins/Alamy Stock Photo; **p180:** Simon Ross; **p181:** © Crown copyright (2016) OS 100043707; **p181:** Rick Ellerman; **p182:** funkyfood London - Paul Williams/Alamy Stock Photo; **p183:** George Rose/Getty Images; **p183:** © david speight/Alamy Stock Photo; **p183:** Dr. Marli Miller/Visuals Unlimited, Inc/Science Photo Library; **p186:** © Naturfoto-Online/Alamy Stock Photo; **p186:** © robertharding/Alamy Stock Photo; **p187:** North York Moors National Park Authority.; p187: Tony Waltham Geophotos; **p188:** © Tom Bean/Alamy Stock Photo; **p190:** Canadian Parks Agency; **p191:** Lynda Dedge/ESS Photo Collection; **p192:** Lynda Dedge/ESS Photo Collection; **p192:** EnviroFoto; **p192:** Reportage/Archivel Image/Alamy Stock Photo; **p193:** Ray Beer; **p193:** Gustaf Hugelius; **p194:** Josef Friedhuber/Getty Images; **p195:** AP Photo/Dmitry Lovetsky; **p195:** WWFDCP; **p196:** Image/photo courtesy of the National Snow and Ice Data Center, University of Colorado, Boulder; **p197:** TITLIS, Switzerland; **p198:** Image/photo courtesy of the National Snow and Ice Data Center, University of Colorado, Boulder; **p198:** Image/photo courtesy of the National Snow and Ice Data Center, University of Colorado, Boulder; **p198:** Steve Morgan/Alamy Stock Photo; **p199:** Steven Jay Kazlowski/Alamy Stock Photo; **p199:** Montipaiton/Shutterstock; **p200:** Tony Waltham Geophotos; **p202:** Arterra Picture Library/Alamy Stock Photo; **p202:** Emanuele Lotti/Getty Images; **p203:** Pontresina Tourism; **p204:** Erlend Bjørtvedt (CC-BY-SA); **p205:** The Norwegian Ministry of Trade, Industry and Fisheries/Trond Viken.; **p205:** ArcticPhoto.com; **p206:** with kind permission from Svalbard Expeditions; **p207:** Norsk Polarinstitutt; **p209:** Polarportal.dk/Peter Langen; **p210:** David Ramos/Getty Images; **p213:** stockphoto mania/Shutterstock; **p214:** benedictus/Shutterstock; **p216:** SCIENCE PHOTO LIBRARY; **p218:** © The Protected Art Archive/Alamy Stock Photo; **p219:** WORLDSAT INTERNATIONAL/SCIENCE PHOTO LIBRARY; **p220:** Christophe Hormann; **p225:** Science Source/UGS/SCIENCE PHOTO LIBRARY; **p226:** S-F/Shutterstock; **p226:** Ross Ressmeyer/Getty Images; **p227:** Sipa Press/REX/ Shutterstock; **p228:** club4traveler/Shutterstock; **p229:** Steve Jurvetson; **p229:** AP/Press Association Photos; **p230:** Caribbean Helicopters; **p231:** Stocktrek Images/Richard Roscoe/Getty Images; **p231:** NASA Earth Observatory; **p232:** Shutterstock; **p239:** David Rydevik; **p240:** STR/AFP//Getty Images; **p240:** Paul Kennedy/Getty Images; **p241:** 3777190317/Shutterstock; **p243:** REX/Shutterstock; **p243:** Max McClure/Alamy Stock Photo; **p244:** George Allen Penton/Shutterstock; **p245:** Mark Pearson/Alamy Stock Photo; **p245:** View Pictures Ltd/Alamy Stock Photo; **p246:** AP/Press Association; **p247:** The Asahi Shimbun/Getty Images; **p247:** Digital Globe (licensed under the Creative Commons Attribution-Share Alike 3.0 Unported license); **p248:** epa european pressphoto agency b.v./Alamy Stock Photo; **p248:** Daniel Berehulak/Staff/Getty Images; **p249:** Greg Baker/AP/Press Association; **p250:** Jeff Schmaltz, MODIS Rapid Response Team, NASA/GSFC; **p252:** handout/Getty Images; **p253:** Atsuko Ellie Tera/Alamy Stock Photo; **p253:** LUIS EL VIR/AP/Press Association Images; **p254:** courtesy NOAA; **p255:** courtesy NOAA; **p258:** Jeremy Woodhouse/Masterfile; **p258:** © ton koene/Alamy Stock Photo; **p259:** © epa european pressphoto agency b.v./Alamy Stock Photo; **p259:** Kamira/Shutterstock; **p260:** Simon Ross; **p260:** © Clarence Holmes Photography/Alamy Stock Photo; **p261:** © Joerg Boethling/Alamy Stock Photo; **p262:** Handout/Getty Images; **p263:** David Edwards; **p266:** Steven Sanford; **p267:** © Franck Fotos/Alamy Stock Photo; **p268:** Jerome Garot; **p268:** NASA/EARTH OBSERVATORY/JOSHUA ST; **p270:** DigitalGlobe/ScapeWare3d/Getty Images; **p270:** DigitalGlobe/ScapeWare3d/Getty Images; **p272:** US Force Photo/Alamy Stock Photo; **p273:** Image provided by The Weather Channel; **p274:** Emmanuel Dunand/Getty Images; **p274:** 123 RF; **p276:** UNICEF; **p277:** Steven Saphore/Getty Images; **p280:** kajornyot wildlife photography/Shutterstock; **p282:** Arno van Dulmen/Shutterstock; **p283:** © Custom Life Science Images/Alamy Stock Photo; **p283:** IUCN (Red List Unit); **p284:** Tims Images/Shutterstock; **p286:** Zuma Press Inc./Alamy Stock Photo; **p287:** Friedemann Vogel/Getty Images; **p289:** Aleksey Stemmer/Shutterstock; **p290:** Marvin Dembinsky Photo Associates/Alamy Stock Photo; **p292:** Amy Tseng/Shutterstock; **p294:** racorn/Shutterstock; **p295:** Dr. Morley Read/Shutterstock; **p295:** Protasov AN/Shutterstock; **p295:** robert cicchetti/Shutterstock ; **p297:** Simon Ross; **p300:** Ashok Jain/Naturepl; **p300:** Mark Blinch/Reuters; **p300:** Robert Pedley; **p301:** Fix The Fells; **p302:** Tom Bean/Getty Images; **p306:** Dennis van de Water/Shutterstock; **p306:** sittitap/Shutterstock; **p307:** Marcus VDT/Shutterstock; **p308:** Fish Ho Hong Yun/Shutterstock; **p308:** nattanan726/Shutterstock; **p309:** Thomas Schoch; p309: Denis Van de Water/Shutterstock; **p310:** Front page/Shutterstock; **p310:** kietr/Shutterstock; **p310:** T. Photography/Shutterstock; **p310:** Alfredo Maiquez/Shutterstock; **p310:** Mykola Gomeniuk/Shutterstock; **p310:** Ian Trower/Getty Images; **p312:** Oleg Znamenskiy/Shutterstock; **p313:** EcoPrint/Shutterstock; **p315:** Oliver Reichenauer/Shutterstock; **p315:** Rich Carey/Shutterstock; **p315:** Paulette Sinclair/Alamy Stock Photo; **p315:** John Wollwerth/Shutterstock; **p317:** BlueOrange Studio/Shutterstock; **p318:** © Colin Harris/ era-images/Alamy Stock Photo; **p318:** Ashley Cooper/Alamy Stock Photo; **p319:** David A Eastley/ Alamy Stock Photo; **p321:** Helen Baines; **p322:** Jane McIlroy/Shutterstock; **p322:** Forestry Commission; **p323:** Daniel J. Rao/Shutterstock; **p324:** Rebecca Cole/Alamy Stock Photo; **p325:** Simon Ross; **p326:** Andrey Armyagov/Shutterstock; **p326:** courtesy of NOAA; **p327:** Brian Kinney/Shutterstock; **p328:** LWM/NASA/LANDSAT/Alamy Stock Photo; **p329:** Jad Davenport/Getty Images; **p329:** National Geographic Creative/Alamy Stock Photo; **p330:** National Geographic Creative/Alamy Stock Photo; **p330:** Vilainecrevette/Shutterstock; **p331:** B. Anthony Stewart/Getty Images; **p332:** Jeff Schmaltz, MODIS Rapid Response Team, Goddard Space Flight Center; **p333:** World Resources Institute; **p334:** Axel Bueckert/Shutterstock; **p335:** Keith Burdett/Alamy Stock Photo; **p336:** Construction Photography/Alamy Stock Photo; **p336:** Tim Cuff/Alamy Stock Photo Alamy; **p337:** imageBROKER/Alamy Stock Photo; **p337:** Meibion/Alamy Stock Photo; **p338:** Richard Ash/New Ferry Butterfly Park; **p338:** Richard Ash/New Ferry Butterfly Park; **p339:** prinzessinnengarten; **p339:** prinzessinnengarten; **p340:** David Young/Shutterstock; **p341:** Mark Hannaford/Alamy Stock Photo; **p341:** Dave Gibbeson/Alamy Stock Photo; **p343:** South West Water; **p344:** Holmes Garden Photos/Alamy Stock Photo; **p345:** Geoffrey Robinson/Alamy Stock Photo; **p346:** Board of Trustees of Royal Botanic Gardens, Kew; **p347:** Board of Trustees of Royal Botanic Gardens, Kew; **p348:** Tim Bayliss; p349: Simon Ross; **p351:** Simon Ross; **p352:** Simon Ross; **p353:** Shutterstock; **p355:** © catrinhelen/Stockimo/Alamy Stock Photo; **p355:** Tim Bayliss; **p355:** © Edward Simons/Alamy Stock Photo; **p356:** Tim Bayliss; **p356:** © Bubbles Photolibrary/Alamy Stock Photo; **p356:** Alice Griffiths; **p356:** Alice Griffiths; **p356:** Alice Griffiths; **p360:** © 2013 British Geological Survey, NERC. Boreas published by John Wiley and Sons Ltd., on behalf of The Boreas Collegium.

All artwork by Kamae Design except p26(b) and p38(t) by Giorgio Bacchin.

Third party website addresses referred to in this publication are provided by Oxford University Press in good faith and for information only and Oxford University Press disclaims any responsibility for the material contained therein

Every effort has been made to contact copyright holders of material reproduced in this book. Any omissions will be rectified in subsequent printings if notice is given to the publisher.

**Approval message from AQA**

This textbook has been approved by AQA for use with our qualification. This means that we have checked that it broadly covers the specification and we are satisfied with the overall quality. Full details of our approval process can be found on our website.

We approve textbooks because we know how important it is for teachers and students to have the right resources to support their teaching and learning. However, the publisher is ultimately responsible for the editorial control and quality of this book.

Please note that when teaching the AQA AS and A Level courses, you must refer to AQA's specification as your definitive source of information. While this book has been written to match the specification, it cannot provide complete coverage of every aspect of the course.

A wide range of other useful resources can be found on the relevant subject pages of our website: www.aqa.org.uk.

# Contents

This book has been updated to reflect the changes to A Level Geography questions that AQA issued in September 2019:

**Paper 1 Physical geography Section C: Hazards, Ecosystems under stress**

Questions 05.1 to 0.5.4 and 06.1 to 06.4 have been replaced with a 4-mark question

These 4 mark questions will test AO1 and are the same question style as those found at the beginning of Sections A and B.

All practice questions within this book have been reviewed and many updated in order to reflect more closely the latest AQA exam question format and wording. Some facts and statistics have also been updated.

# Contents

## How to use this book

This is one of two books in this series, written for the AQA GCE in Geography. This particular book (Physical geography) has been written to meet the content requirements of the A Level course, but can equally well be used for the separate AS course.

Skills questions indicated by the Ⓢ icon are aimed at meeting the geographical and statistical skills requirements for both AS and A Level.

Practice questions have been included for both AS and A Level, with marks allocated. Please note that the Practice questions used in this book allow students a genuine attempt at practising exam skills, but are not intended to replicate the exact nature of final exam questions.

At appropriate points, chapters focus on providing fieldwork opportunities. These, plus associated questions, will help to prepare you for the fieldwork requirements for both AS and A Level.

# 1 Water and carbon cycles

*Natural events like the Fort McMurray forest fires in Canada 2016 can impact the carbon cycle*

## Your exam

**AL** 'Water and carbon cycles' is a core topic. You must answer all questions in Section A: Water and carbon cycles as part of Paper 1: Physical geography. Paper 1 makes up 40% of your A Level.

**AS** You must answer one question in Section A of the Paper 1: Physical geography and people and the environment, from a choice of three: Water and carbon cycles or Coastal systems and landscapes or Glacial systems and landscapes. Paper 1 makes up 50% of your AS Level.

## Your key skills in this chapter

**S** In order to become a good geographer you need to develop key geographical skills. In this chapter, whenever you see the skills icon you will practise a range of quantitative and relevant qualitative skills, within the theme of 'water and carbon cycles'. Examples of these skills are:

- Understanding and calculating simple mass balance 1.1, 1.5, 1.12
- Describing and analysing quantitative data, patterns and trends 1.10, 1.11, 1.17, 1.18
- Presenting and analysing field data 1.6
- Drawing and annotating diagrams of physical systems 1.2, 1.3, 1.4, 1.7, 1.9, 1.11, 1.13, 1.15, 1.16

## Fieldwork opportunities

One of the characteristics of these cycles is the way in which they can be applied at a range of scales—so, what you learn during local investigations will inform your understanding of physical processes that also operate at a wider scale. The topics in this chapter provide excellent opportunities for a range of local fieldwork investigations. Examples include:

### 1. Investigations of rates of infiltration

Infiltration is a key process responsible for the partitioning of rainfall between overland flow (runoff) and soil storage or subsurface flow. Infiltration is easy to measure using simple infiltration rings which can be made from plastic pipe. Infiltration rates may be affected by a range of factors such as surface cover, soil moisture, soil texture, angle of slope, and soil compaction allowing different groups of students within your class to conduct a range of related but distinct investigations in a single location.

### 2. Measurement of water balance

Catchment discharge is a fundamental parameter in the drainage basin water balance. You can use secondary rainfall and runoff data to construct a simple water balance for catchments. Practice measuring stream discharge and rainfall in the field will allow you to understand the potential errors associated with these estimates. Understanding errors is central to any budgeting exercise. Measurement of rainfall in multiple simple rain gauges (even around school grounds) will give an opportunity to examine spatial variation in rainfall and its potential impact on creating good estimates of rainfall inputs.

### 3. Estimation of carbon stocks in woodland.

The stock of carbon within woodland can be simply estimated. There are standard equations to estimate living biomass of trees from the diameter of the tree measured at 1.3m height.

Tree biomass is 50% carbon so it is a simple conversion to work out how much carbon is stored in the tree. Where tree age can also be estimated, either from the girth of the tree, knowledge of the site, or from tree ring evidence on similar felled trees then the rate of carbon sequestration as mass of carbon per year can be calculated, in this case students can estimate both the stock of carbon and the flux.

In this section you will learn about the concept of systems in physical geography and their application to the water and carbon cycles

## What is a system?

You will likely be familiar with the term 'ecosystem'. An ecosystem describes the interrelationships between living and non-living components within a particular environment, such as a pond (Figure 1) or a forest.

A simple diagram is often used to show the different components (parts) of an ecosystem and the relationships or links (**flows/transfers**) between them. This is essentially what is meant by a **system**.

**Inputs** include:
• precipitation
• leaf fall during the autumn
• seeds carried by wind and birds

**Stores/components** include:
• water
• soil
• plants

**Outputs** include:
• water soaking through soil and rocks
• evaporation
• seed dispersal

**Flows/transfers** include:
• photosynthesis
• infiltration
• transpiration

**Figure 1**  *Garden pond ecosystem*

In geography, a systems approach can help us to understand the physical and human world around us. It enables us to see the whole picture. We can apply this approach to physical systems such as drainage basins or to human systems such as the operations on a farm or in a factory.

A systems approach helps us to understand how energy is transferred between the components of a system and how those components themselves can change. This approach also helps us to appreciate how both natural change and human activities can impact upon an environment.

**Figure 2**  *Systems approach terminology*

| Systems term | Definition | Examples | |
|---|---|---|---|
| | | **Drainage basin** | **Woodland carbon cycle** |
| Input | Material or energy moving into the system from outside | Precipitation | Precipitation with dissolved carbon dioxide ($CO_2$) |
| Output | Material or energy moving from the system to the outside | Runoff | Dissolved carbon within runoff |
| Energy | Power or driving force | Latent heat associated with changes in the state of water | Production of glucose through the process of photosynthesis |
| Stores/components | The individual elements or parts of a system | Trees, puddles, soil | Trees, soil, rocks |
| Flows/transfers | The links or relationships between the components | Infiltration, groundwater flow, evaporation | Burning, absorption |
| Positive feedback | A cyclical sequence of events that amplifies or increases change. Positive feedback loops exacerbate the outputs of a system, driving it in one direction and promoting environmental instability. | Rising sea levels (due to thermal expansion and melting freshwater ice) can destabilise ice shelves, increasing the rate of calving. This leads to an increase in melting, causing sea levels to rise further. | Increased temperatures due to climate change cause melting of permafrost. Trapped greenhouse gases are released, enhancing the greenhouse effect, raising temperatures further. |
| Negative feedback | A cyclical sequence of events that damps down or neutralises the effects of a system, promoting stability and a state of dynamic equilibrium. | Increased surface temperatures lead to an increase in evaporation from the oceans. This leads to more cloud cover. Clouds reflect radiation from the sun, resulting in a slight cooling of surface temperatures. | Increased atmospheric $CO_2$ leads to increased temperatures, promoting plant growth and rates of photosynthesis. This, in turn, removes more $CO_2$ from the air, counteracting the rise in temperature. |
| Dynamic equilibrium | This represents a state of balance within a constantly changing system | Remote and unaffected drainage basin/woodland where there has been no significant natural or human impacts, or one that has had time to adjust to change | |

## Applying the systems approach to the water and carbon cycles

In this chapter you will learn about the **water** and **carbon cycles** and the complex relationships between their many component parts. You will learn that natural change, and also change due to human activities, often upsets the dynamic equilibrium of the cycle. It is this complexity and the need for us to see the 'whole picture' that is the reason why a systems approach is the ideal mechanism for studying these two vital cycles together with other aspects of physical and human geography.

### The water cycle system

Look at Figure **1** (1.2). It shows a simplified version of the water cycle. In its entirety, the water cycle system is a **closed system** – water is not lost to or gained from space. However, at a local scale, such as a drainage basin, it is an **open system**. Precipitation is an input and runoff to the oceans is an output. There are many components and **stores**, such as trees, built-up areas and soil. Flows and transfers include *throughflow* and *groundwater flow* (see 1.3).

### The carbon cycle system

Figure **3** shows the carbon cycle in the form of a systems diagram. As this is the global carbon system, it is a closed system – there are no inputs to or outputs from the system as a whole. At a local scale, such as a forest, it is an open system with both inputs and outputs.

There are many components and stores, such as rocks, the oceans and the atmosphere. Flows and transfers include *photosynthesis* and *respiration*.

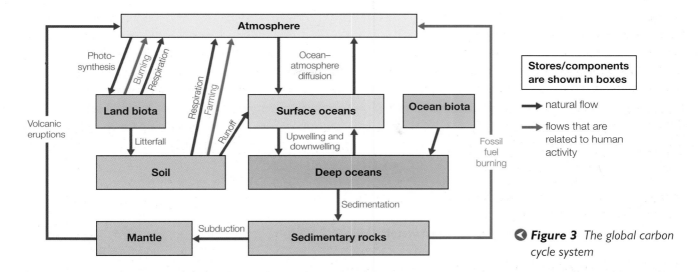

◀ **Figure 3** *The global carbon cycle system*

### ACTIVITIES

1 Study Figure **1**. Work in pairs or small groups to consider how the systems terminology listed in Figure **2** can be applied to the pond ecosystem. Present your answers in the form of a table.

**(S)** 2 Study Figure **1** (1.2). Attempt to represent the information about the water cycle in the form of a systems diagram similar to Figure **3** above, which shows the carbon cycle system. You may need to construct several draft versions before completing your final diagram.

3 Explain why geographers find a systems approach useful in studying physical systems.

### STRETCH YOURSELF

Use the internet to find further examples of positive and negative feedbacks within the water and carbon cycles. You might find it easier to focus on local scale cycles, say involving a drainage basin, a lake or a forest. Analyse the role of feedbacks within and between cycles. Is the ideal scenario to have a dynamic equilibrium?

In this section you will learn about the global water cycle and its stores

## What is the water cycle?

Water is essential for life on Earth. It is constantly being recycled, stored and transferred between the land, oceans and atmosphere. Water is not evenly distributed across the Earth. Some regions enjoy plentiful supplies, others suffer severe shortages which can lead to human misery, migration and famines.

Ownership of water can be a controversial political issue. Some people believe that future wars may be about securing water supplies.

Figure **1** shows the global water cycle. Notice that there are a number of components which fall into two kinds of process.

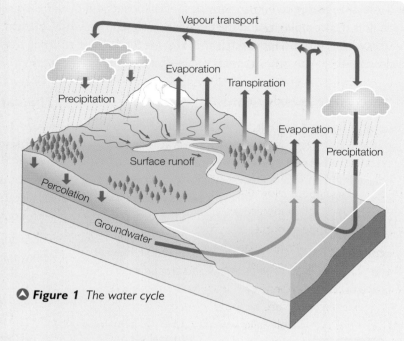

▲ **Figure 1** *The water cycle*

- Stores – most of the Earth's water is stored as saline (salt) water in the oceans. Of the freshwater stores, ice sheets (Antarctica and Greenland) and groundwater are the main stores. Rivers, lakes and the atmosphere contain remarkably small amounts of the global water stores.

- Transfers – these are the processes involved in transferring water between stores. For example, precipitation transfers water from the atmosphere to the Earth's surface. Evaporation moves it back to the atmosphere. Water may infiltrate the ground or percolate slowly through the rocks as groundwater flow.

## The main stores in the water cycle

Water is stored within four major physical systems – the **lithosphere** (land), **hydrosphere** (liquid water), **cryosphere** (frozen water – snow and ice) and **atmosphere** (air).

Look at Figure **2** to see the breakdown of water storage. There are some important, and maybe surprising, facts to notice.

- The vast majority of the Earth's water is saline water (97.5 per cent), most of which is stored in the oceans.

- Only 2.5 per cent of the Earth's water is freshwater, almost all of which is stored as snow and ice (68.7 per cent) and groundwater (30.1 per cent).

- Surface and other freshwater comprises only 1.2 per cent of all freshwater.

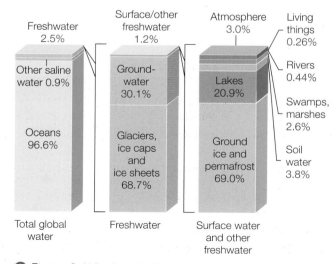

▲ **Figure 2** *Where is the Earth's water?*

## The global distribution of water stores

Water stores have a geographical component in that they are not evenly distributed across the world. Consider the distribution of land, sea and ice sheets as shown in Figure **3**. This has a profound impact on the global distribution of water. On land, there is an uneven distribution of rivers, lakes and groundwater aquifers.

## What is the global distribution of groundwater aquifers?

Just over 30 per cent of all freshwater is stored in rocks deep below the ground surface forming vast underground reservoirs called *aquifers*. These sources are crucial for sustaining civilisations across the world. Figure **3** shows the global distribution of aquifers.

Aquifers most commonly form in rocks such as chalk and sandstone, which are porous (contain pores – air pockets) and permeable (allow water to pass through). Water enters the rocks either directly, where they are exposed on the ground, or very slowly, as water drains through the overlying soil. Soils vary enormously in their capacity to store and transfer water – this is the **soil water budget**. Porous, sandy soils hold little moisture as water is easily transferred through the pore spaces. Clay soils tend to store water, with very limited water transfer.

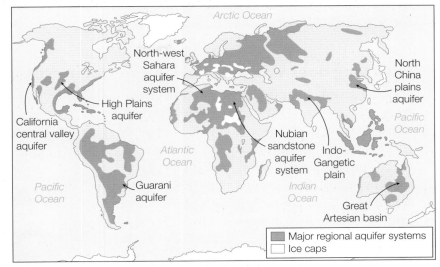

▲ **Figure 3** *Distribution of land, sea, ice sheets and major world aquifers*

The upper level of saturated rock is called the *water table*. This rises and falls in response to groundwater flow, water abstraction by people or by *recharge* (additional water flowing into the rock). Through careful management, the water table needs to be maintained at the same level – a state of equilibrium.

Aquifers in the deserts of Africa, the Middle East and Australia are called *fossil aquifers* and were formed thousands of years ago when the climate in those regions was much wetter. Many aquifers are being exploited unsustainably as more water is extracted. Saline aquifers exist where seawater has infiltrated into the rocks, often due to the over-abstraction.

## ACTIVITIES

 **1** Make a copy of Figure **1**. Use different colours to indicate the stores and transfers (processes) in the water cycle.

**2 a** Study Figure **2**. The oceans are the main store of saline water. Can you account for the 'other saline water' stores?

    **b** Many of the world's glaciers are melting. Where do you think this freshwater is going?

    **c** Calculate the percentages of rivers and also of lakes of the 'total global water'.

    **d** Why is 'groundwater' a more important freshwater source than 'glaciers, ice caps and ice sheets'?

    **e** Why do you think the atmosphere stores such a small amount of the world's water?

**3 a** Study Figure **3**. Describe the distribution of the major regional aquifers.

    **b** Suggest why some aquifers are located in present-day arid regions. What are the issues associated with this?

## STRETCH YOURSELF

Find out more about the distribution of lakes, an important store of freshwater.

- Are lakes distributed evenly across the world?
- What determines the location of lakes?
- How important are lakes in water supply?
- Are there any signs of stress on the world's lakes?

In this section you will learn about the processes driving changes in the magnitude of the water cycle stores over time and space

## How long does water remain in the water cycle stores?

Look at Figure **1**. It lists the typical timescales that water remains in each store. Notice, for example, that water in the soil (soil water or moisture) does not remain very long (1–2 months). It may quickly soak into the underlying soil and be transpired by plants, be transferred into rivers by throughflow or simply evaporated back into the atmosphere. Groundwater replacement, in contrast, can take hundreds or even thousands of years! These varying time scales are extremely important in understanding the complexity of transfers within the water cycle. Also consider how these will vary from place to place.

## What are the processes of change?

In section 1.2 we have seen that the water cycle is made up of stores and transfers. The amount of water held in each store is largely determined by the transfer processes that act as inputs and outputs.

It is important to appreciate that the magnitude of water held within a store will vary over time and space. Consider the seasonal changes that occur in the Arctic, with the annual cycle of freezing and melting of the sea ice. Also consider the passage of frontal systems with their associated bands of rain and the recent trend of shrinking glaciers around the world. Over time and space, the magnitude of these stores has changed (Figure **2**).

## Climate change

At the peak of the last Ice Age (about 18 000 years ago), about a third of the Earth's land area was covered by glaciers and ice sheets . With water 'locked up' as snow and ice, the magnitude of this store increased significantly. With less liquid water reaching the oceans, sea levels fell by over 100 m compared with the present day.

During warmer periods in the past – say, about three million years ago – ocean levels were about 50 m higher than they are today as the amount of water stored as snow and ice declined. You can see why scientists are so concerned about the possible impacts of global warming on sea levels!

🔺 **Figure 1** *Typical residence times of water found in various stores*

| Process (flow/transfer) | Definition |
| --- | --- |
| **Precipitation** | Transfer of water from the atmosphere to the ground. It can take the form of rain, snow, hail, dew. |
| **Evaporation** (evapotranspiration, if combined with transpiration, water loss from plants) | Transfer of water from liquid state to gaseous state (water vapour). The vast majority occurs from the oceans to the atmosphere. |
| **Condensation** | Transfer of water from a gaseous state to a liquid state, for example, the formation of clouds. |
| **Sublimation** | Transfer from a solid state (ice) to a gaseous state (water vapour) and vice versa. |
| **Interception** | Water intercepted and stored on leaves of plants. |
| **Overland flow** | Transfer of water over the land surface. |
| **Infiltration** | Transfer of water from the ground surface into soil where it may then percolate into underlying rocks. |
| **Throughflow** | Water flowing through soil towards a river channel. |
| **Percolation** | Water soaking into rocks. |
| **Groundwater flow** | Transfer of water very slowly through rocks. |

🔺 **Figure 2** *The main global transfer processes*

## Cloud formation and the causes of precipitation

Cloud formation and subsequent precipitation varies considerably with time and space. If you look at a satellite photograph of the Earth, you will see that clouds are very unevenly distributed, as is associated precipitation. The driving force behind cloud formation and precipitation is the global atmospheric circulation model (Figure **3**).

Simplified to suggest the presence of three interconnected cells, the atmospheric circulation model identifies latitudinal zones of rising and falling air.

At the Equator, for example, high temperatures result in high rates of evaporation. The warm, moist air rises, cools and condenses to form towering banks of cloud and heavy rainfall in a low pressure zone called the ITCZ (Inter-Tropical Convergence Zone). Seasonally, this zone moves north and south, with the overhead sun illustrating both the spatial and temporal changes in transfers and store magnitudes that occur within the water cycle.

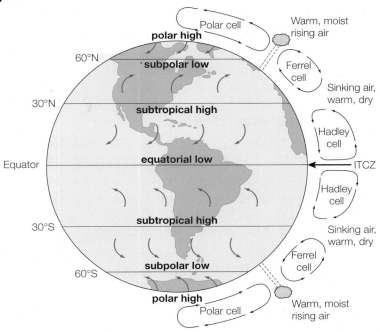

**Figure 3** *Atmospheric circulation model*

In the mid latitudes, cloud formation is mostly driven by the convergence of warm air from the Tropics and cold air from the Arctic. The boundary of these two distinct air masses – the polar front – results in rising air and cloud (and rain) formation. Strong upper-level winds (the jet stream) drive these unstable weather systems across the mid latitudes, establishing the largely changeable conditions experienced in the UK.

Cloud formation can occur on a more localised scale. The formation of thunderstorms from intense convective activity clearly demonstrates the variations in both time and space of water cycle transfer processes.

## Cryospheric processes

After oceanic water, the largest stores of water on Earth take the form of frozen water (ice) – 95 per cent of the ice is locked up in the world's two great ice sheets covering Antarctica and Greenland. While the Earth's ice masses may seem stable and lacking in change, this is far from the case – it is all a matter of the timescale involved.

Snow falling on glaciers and ice sheets becomes compressed and enters long-term storage, forming layers of glacial ice. Scientists in the Antarctic have drilled down into layers of ice over 400 000 years old!

On a shorter timescale, snow accumulated during the winter adds to the mass of a glacier or ice sheet. In the summer, melting occurs or ice calves (breaks away). On a glacier, the equilibrium line marks the altitude where annual accumulation and melting are equal. In recent decades the climate has warmed, causing the equilibrium line to move to ever higher altitudes. Most glaciers in the world are now shrinking and retreating.

The melting of freshwater ice has a profound impact on sea levels – the total melting of all the polar ice sheets could result in a 60 m rise in sea level, adding a great deal of water to the ocean store. We have already identified the **positive feedback** loop whereby rising sea levels destabilise ice shelves, triggering calving and further melting (Figure **2**, 1.1).

## Processes of change at the local scale

So far we have mostly considered processes of change at the global scale. While alluding to local scale changes, say with thunderstorms or individual glaciers, it is worth focusing now on the processes operating on local hillslopes, considered by hydrologists to be the most important local unit of study. River basins will be addressed in 1.4 *The drainage basin system*.

### The hillslope water cycle

Look at Figure **4**. It shows the water cycle stores and transfers on a typical hillslope. See how the components of a hillslope water cycle are affected by a variety of natural and human-related factors. Notice the different places where water can be stored. The amount of water held in each of these stores will depend on many factors operating over relatively short timescales. The magnitude of the stores changes in response to a wide variety of processes. Perhaps the most influential is *infiltration* – the movement of water from the ground surface into the soil. Water that is effectively trapped on the ground surface – either because the soil is saturated or frozen or because the rate of precipitation exceeds the capacity of the soil to absorb it – will either be stored as surface storage, evaporate or start to flow downslope as overland flow. The rapid transfer of water overland is, of course, a major factor leading to flooding.

⬇ **Figure 4** *The hillslope water cycle system*

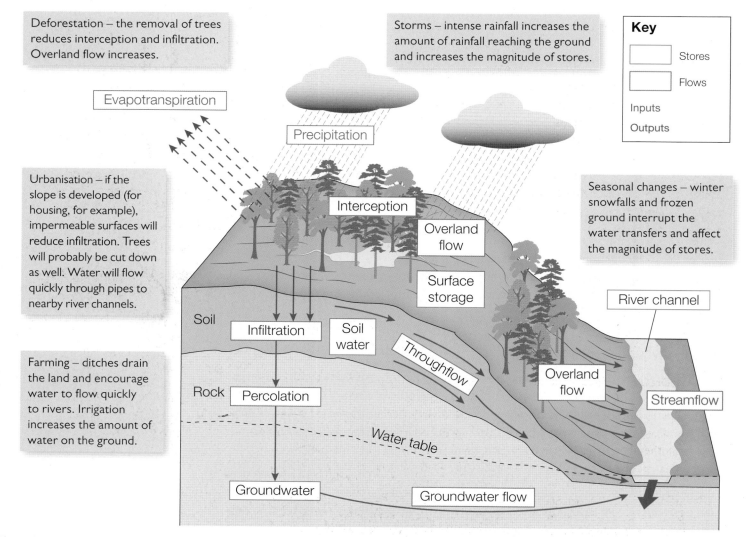

Deforestation – the removal of trees reduces interception and infiltration. Overland flow increases.

Storms – intense rainfall increases the amount of rainfall reaching the ground and increases the magnitude of stores.

**Key**

Stores

Flows

Inputs

Outputs

Evapotranspiration

Precipitation

Urbanisation – if the slope is developed (for housing, for example), impermeable surfaces will reduce infiltration. Trees will probably be cut down as well. Water will flow quickly through pipes to nearby river channels.

Seasonal changes – winter snowfalls and frozen ground interrupt the water transfers and affect the magnitude of stores.

Interception

Overland flow

Surface storage

River channel

Soil

Infiltration

Soil water

Throughflow

Overland flow

Streamflow

Farming – ditches drain the land and encourage water to flow quickly to rivers. Irrigation increases the amount of water on the ground.

Rock

Percolation

Water table

Groundwater

Groundwater flow

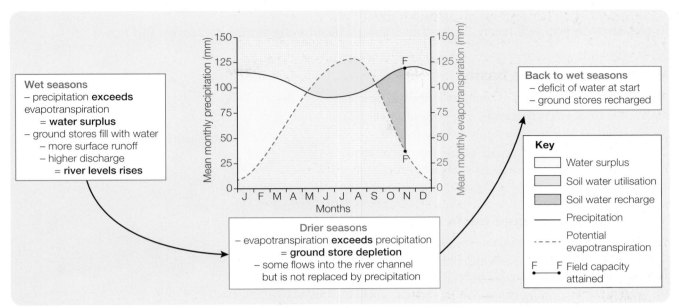

**Figure 5** *The soil water budget*

Water that is able to infiltrate the soil may be stored for very long periods of time either in the soil or deep within the underlying bedrock. The soil water budget describes the changes in the soil water store during the course of a year (Figure **5**).

Notice that during the wetter winter months, precipitation exceeds potential **evapotranspiration**. This leads to a water surplus. Soils will be saturated, surface overland flow encouraged and river levels will be high. In the summer months, when potential evapotranspiration exceeds precipitation, the soil starts to dry out. In the early autumn, rates of evapotranspiration start to fall and the soil water is replenished (recharge).

When the soil holds as much water as it can without any outputs occurring, it is said to have reached its *field capacity*. Thereafter, a water surplus occurs and water transfer processes will become active.

Soil water budgets will vary considerably from place to place depending on the type and depth of the soil, its texture and permeability. Much the same is true of the underlying bedrock, as its capacity to store and transfer water will depend on its lithology and its structure (porosity and permeability). For example, water will move slowly through older, porous rocks such as some sandstones, yet it will move very rapidly through widely jointed limestones. Consider, for example, how potholers can find themselves in danger very quickly in limestone caverns when the water table rises rapidly in response to a rainstorm.

## ACTIVITIES

1  Study Figure **1**.
   a  Which water store has the shortest residence time? Suggest reasons for this.
   b  Why does deep groundwater have the longest residence time?
   **S**  c  Draw a simple diagram showing the inputs and outputs for the glacier store.
   d  Suggest how the glacier store may change in magnitude over time.
   e  Rivers have a very short residence time. What are the implications of this for the other stores and for human uses of rivers?

2  What are the major processes responsible for change in the magnitude of global water cycle stores over time and space?

3  Study Figure **4**. For each of the boxed factors, consider the impact on the magnitudes of the hillslope water stores. You could present your answer in the form of a table or as annotations on photos obtained from the internet to represent each factor.

## STRETCH YOURSELF

Investigate the factors that affect the height of the water table on a hillslope. Consider rock type, soil characteristics, relief, vegetation and water abstraction. To what extent can the height of the water table be used to indicate the relative magnitude of water stores?

In this section you will learn about the drainage basin system and its stores and flows

## What is a drainage basin?

A drainage basin is the area of land that is drained by a river and its tributaries (Figure **1**). The edge of a river basin is marked by a boundary called the *watershed*. Drainage basins vary enormously in size, from small local basins to major river systems such as the Mississippi, Nile and Amazon.

## What is the drainage basin system?

The movement of water within the drainage basin is illustrated by the drainage basin hydrological cycle or the *drainage basin system* (Figure **2**). This is an open system, with inputs (precipitation) and outputs (runoff, evapotranspiration). Notice that part of the diagram in Figure **2** shows the way that water moves down a hillslope towards a river – this is the *hillslope system* that we studied in 1.3.

In a hydrological sense the drainage basin is an *open* system. However, for planning purposes, it is often considered to be a *closed* system – the principles of cause and effect are contained and do not spread outside its area. So, in terms of flood management, water supply and pollution control, the drainage basin is the basic spatial unit used by planners and organisations such as the Environment Agency. What happens in a drainage basin is contained within that basin and will not affect neighbouring basins.

## Precipitation

Water enters the drainage basin system as precipitation. Some of it may be intercepted by plants and trees where it may be stored before being evaporated. It takes time for the water to drip through the leaves or down the stems (**stemflow**) to the ground surface. Here, it is stored as puddles, flows over the ground as overland flow or infiltrates the soil. Some water may be taken up by plants before being transpired.

## Groundwater flow

Groundwater flow feeds rivers through their banks and bed. Being generally a slow method of transfer, it carries on supplying water well after an individual rainfall event has occurred. This explains why rivers continue to flow during long dry periods. Eventually, water moves out of the system as runoff, when the river flows into lakes or the sea, or evapotranspiration.

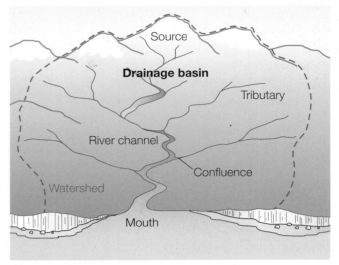

▲ **Figure 1** *The drainage basin*

## Infiltration

The *infiltration capacity* (rate of infiltration) is an extremely important factor and will vary according to soil type and antecedent conditions, i.e. to what extent the soil is already saturated. Infiltration capacity is exceeded when the soil is unable to absorb water at the rate at which it is falling (or melting if it is snow). Thin, frozen or already saturated soils will usually have a low infiltration capacity. Trees may promote infiltration as the roots form pathways for water to percolate underground. Water actually soaks into the soil by a combination of capillary action (the attraction of water molecules to soil particles) and gravity, with the latter usually dominating.

## Overland flow

If water is unable to infiltrate it may run off the surface as overland flow, flowing across a large surface area (*sheetflow*) or concentrated into small channels called *rills*. Overland flow on agricultural land is not common in the UK as much of the land is covered by vegetation, although it can sometimes be seen in winter when the soil is bare. In urban areas, particularly on roads, overland flow is extremely common, often exacerbating flooding.

## Throughflow

Once in the soil, water may be either stored as **soil water** or pass through as throughflow, dependent on the depth and texture of the soil – coarse, sandy soil absorbs and transfers water rapidly, especially through discrete 'pipes' within the soil, caused by animal activity or the growth of plant roots. This contributes significantly to the flood hazard. Such soils are said to have a *low field capacity* (retain little water). Clay soils drain more slowly and have a *high field capacity*. These soils tend to be wet as they have tiny pore spaces, which do not allow water to the transferred readily.

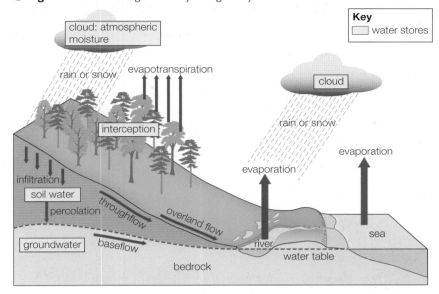

⊘ *Figure 2* The drainage basin hydrological cycle

**Key**
☐ water stores

Water passes through the soil until it reaches the water table (the upper level of saturated ground) or the underlying bedrock. If the bedrock is impermeable, no further downward movement will occur. If permeable, water will seep into the cracks and holes within the rock. Water transfer through rock can take tens or even hundreds of years and rock is, therefore, an important water store.

However, some jointed rocks such as limestone and even granite can transmit water very quickly.

For example, a study of water flow through limestone in Cheddar Gorge, Somerset, calculated rates of 583 cm/hour, which is considerably faster than most rates of throughflow. As a comparison, flow rates through sandstones tend to be about 200 cm/hr and through unconsolidated gravels, up to 20 000 cm/hr.

### EXTENSION

#### The role of vegetation

Most drainage basins are clothed by one or more types of vegetation that will, to some extent, intercept precipitation (scrub and bushes in semi-arid regions, grassland in temperate latitudes or coniferous forest and tundra in high latitudes).

| Type of vegetation biome | Loss of water by interception (average per year) |
|---|---|
| Temperate pine forest | 94% (low-intensity rain); 15% (high-intensity) |
| Brazilian evergreen rainforest | 66% |
| Grass | 30–60% |
| Pasture (clover) | 40% in growing season |
| Coniferous forest | 30–35% |
| Temperate deciduous forest | 20% with leaves; 17% without leaves |
| Cereal crops | 7–15% in growing season |

A study of similar-sized upland catchments of the River Severn (contained 67.5 per cent forest) and the River Wye (contained 1.2 per cent forest) in the 1970s provided interesting comparative data on the role of forest interception in reducing runoff. Having received the same amount of rainfall during the study period, the Severn catchment 'lost' some 38 per cent of the total precipitation (e.g. through evaporation) compared with just 17 per cent for the Wye. The clear inference is that forest cover has an important impact on the drainage basin system.

### ACTIVITIES Ⓢ

Study Figure **2**. Construct a flow diagram, using a series of boxes and arrows, to describe how water is transferred from the atmosphere into river channels.

- Use a colour-coding system to separate stores from flows.
- Refer to the relative transfer speeds, either in the text or by using proportional symbols such as different-sized arrows.
- Use simple diagrams, sketches or thumbnail photos to help you describe what is happening at each stage.

In this section you will learn about the water balance and the causes of variation in runoff

## What is the water balance?

In order to gain a better understanding of the drainage basin system we can use a simple equation called the **water balance**. This helps hydrologists to plan for future water supply and flood control by understanding the unique hydrological characteristics of an individual drainage basin.

## What causes variations in runoff?

An important aspect of the equation is the total runoff (expressed as a percentage of precipitation). This is a measure of the proportion of the total precipitation that makes its way into streams and rivers.

The two river basins in Figure **1** record very different runoff percentages. This is because of the differences in soil water, rock type and vegetation cover. Also think about how the time of year will affect the rates of evapotranspiration and vegetation growth (interception).

The type and intensity of precipitation are also important. Intense rainfall is more likely to pass quickly into rivers, increasing the amount of runoff. Drizzle will be held in the trees and on the grass, much of which will evaporate. Snow will delay any runoff but when frozen soils melt, runoff values might be high.

### The River Wye, Wales

With a total length of 215 km, the River Wye is the fifth-longest river in the UK. From its source in the Plynlimon Hills in mid-Wales, it flows south-eastwards before joining the Severn Estuary at Chepstow (Figure **2**). The river is rich in wildlife, with a variety of habitats. It is an Area of Outstanding Natural Beauty and also has a Site of Special Scientific Interest.

The upper part of the basin is characterised by steep slopes, acidic soils and grassland. Much of this area was originally forested but this has been largely cleared to make way for pasture and sheep grazing. This has reduced interception and increased the potential for overland flow. Ditches have been dug to drain the land to make it more productive, but this has increased the speed of water transfer, making the river more prone to flooding.

The rocks in much of the upper river basin are impermeable mudstones, shales and grits. Further south, the river flows over sandstones before cutting its way through a limestone gorge between Symonds Yat and Chepstow.

**S** The water balance is expressed as:

$$P = Q + E +/- S \quad \text{where}$$

P = precipitation

Q = total runoff (streamflow)

E = evapotranspiration

S = storage (in soil and rock)

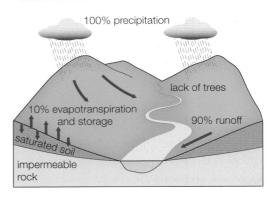

The runoff percentage is high (90%), so most of the precipitation is transferred straight to the river – little is lost or stored on the way. Under conditions like this, flooding is likely.

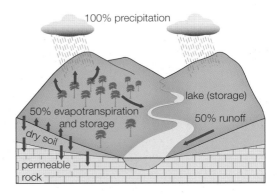

The runoff percentage (50%) is much lower than above (90%), so a higher proportion of the precipitation is lost or stored before it reaches the river channel. Reasons for this might include a heavily forested river basin, or one that has permeable rocks. Under these conditions, flooding is much less likely.

**Figure 1** *Variations in runoff and water balance between drainage basins*

Because the underlying rock is mainly impermeable, groundwater flow is therefore limited throughout the basin: soils quickly become saturated and are unable to absorb excess water. This encourages overland flow, increasing the risk of flooding downstream – Hereford has been affected by flooding on many occasions.

Rainfall totals are highest in the western upland parts of the river basin while higher temperatures and rates of evapotranspiration occur in the east. Runoff tends to be higher in the winter when rainfall totals are high and rates of plant growth and evapotranspiration are low.

Figure **3** provides monthly data for the River Wye's drainage basin system. Notice that there are significant variations in precipitation and runoff during the year.

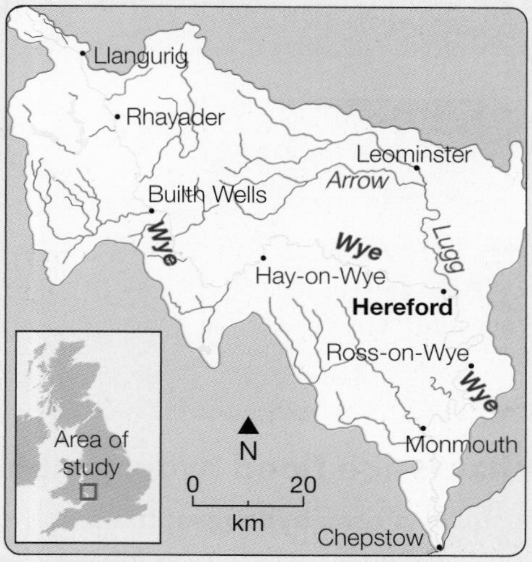

**Figure 2** *The River Wye and its major tributaries*

| Month | Precipitation | Runoff | Evapotranspiration | Storage | Runoff as a % of precipitation |
|---|---|---|---|---|---|
| January | 280.8 | 275.7 | 10.6 | −5.5 | 98.2 |
| February | 191.7 | 145.6 | 12.1 | 34.0 | 76.0 |
| March | 491.0 | 440.2 | 35.9 | | |
| April | 103.8 | 43.7 | 62.2 | | |
| May | 168.9 | 126.4 | 65.3 | | |
| June | 98.7 | 92.8 | 71.0 | | |
| July | 142.2 | 83.0 | 76.8 | | |
| August | 93.8 | 50.8 | 75.6 | | |
| September | 285.1 | 199.5 | 46.6 | | |
| October | 497.9 | 449.8 | 25.5 | | |
| November | 279.4 | 264.8 | 12.1 | | |
| December | 188.4 | 141.2 | 3.7 | | |

**Figure 3** *River Wye water balance*

## ACTIVITIES

1 Study Figure **3**.
   a Use the water balance equation to help you complete the 'storage' column. (January and February have already been completed).
   b Why are there some negative storage values?
   c Do there appear to be any seasonal trends with the positive and negative values? Can you explain these trends?
   d Why do you think there is a high positive storage value in September?
2 Now complete the final column. To do this you need to divide each runoff value by the precipitation value and multiply by 100.
   a In which month is the value of 'runoff as a percentage of precipitation' the highest?
   b Suggest reasons for this very high percentage.
   c Why do you think there was a particularly high percentage runoff value in June?

3 Assess how the following factors cause variations in runoff: type and intensity of precipitation, climate, soil water, rock type, human activities (such as reservoirs, land use change and urbanisation).

## STRETCH YOURSELF

Find out more about the characteristics of the River Wye's drainage basin. Look at the rock type, vegetation and land use and support your study with maps and satellite photos. Use your research to help to explain the water balance data in Figure **3**.

• Is flooding an issue?

• What are the issues of water supply?

In this section you will learn about the flood hydrograph

## What is discharge?

Imagine standing on a bridge looking down on a fast-flowing river. The volume of the water passing beneath you is the river's *discharge*. It is essentially a measurement of runoff at a moment in time.

Values for river discharge are expressed in 'cumecs' (cubic metres per second). Discharge is most commonly calculated using the following equation:

$$\text{Discharge (m}^3\text{ per second)} = \text{cross-sectional area (m}^2\text{)} \times \text{velocity (metres per second)}$$

## What is the flood hydrograph?

The **flood** (storm) **hydrograph** is simply a graph showing the discharge of a river following a particular storm event. Figure **1** shows a typical flood hydrograph and identifies its main features.

Despite the unique nature of river hydrographs, it is possible to identify two models representing polar opposites (Figure **2**).

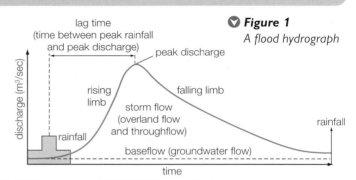

▼ **Figure 1**
A *flood hydrograph*

| Drainage basin and precipitation characteristics | 'Flashy' hydrograph with a short lag time and high peak | | Low, flat hydrograph with a low peak | |
|---|---|---|---|---|
| Basin size | Small basins often lead to a rapid water transfer. | | Large basins result in a relatively slow water transfer. | |
| Drainage density | A high density speeds up water transfer. | | A low density leads to a slower transfer. | |
| Rock type | Impermeable rocks encourage rapid overland flow. | | Permeable rocks encourage a slow transfer by groundwater flow. | |
| Land use | Urbanisation encourages rapid water transfer. | | Forests slow down water transfer because of interception. | |
| Relief | Steep slopes lead to rapid water transfer. | | Gentle slopes slow down water transfer. | |
| Soil water | Saturated soil results in rapid overland flow. | | Dry soil soaks up water and slows down its transfer. | |
| Rainfall intensity | Heavy rain may exceed the infiltration capacity of vegetation, and lead to rapid overland flow. | | Light rain will transfer slowly. | |

▲ **Figure 2** *Characteristics that affect hydrographs*

Rivers tend to have unique responses to rainfall events, as illustrated by their 'typical' hydrograph. Some respond very quickly ('flashy') while others respond much more slowly, with the hydrograph being attenuated (spread out) over a long period of time. In the winter of 2015/16 there were serious floods in parts of northern UK. In the Lake District, rivers responded very quickly to the heavy rainfall events. The already, saturated soil conditions and steep slopes, meant water moved rapidly overland and along river channels to devastate villages such as Glenridding.

York was similarly affected, but it took days for the water to work its way down the tributaries and into the River Ouse, which flows through the city. The need to understand an individual river's response pattern (hydrograph) is essential if hydrologists are to plan effectively for flood control and mitigation.

### STRETCH YOURSELF

Use the National River Flow Archive at www.ceh.ac.uk/data/nrfa/data/search.html to search for hydrological information about a local river or a river that you have studied. Find out about the river flow patterns and try to account for their variations.

| Time from start (hours) | Discharge (m³/sec) |
|---|---|
| 0 | 0.35 |
| 1 | 0.39 |
| 2 | 0.55 |
| 3 | 2.11 |
| 4 | 4.79 |
| 5 | 4.25 |
| 6 | 3.36 |
| 7 | 4.70 |
| 8 | 6.08 |
| 9 | 6.20 |
| 10 | 6.76 |
| 11 | 5.56 |
| 12 | 4.52 |
| 13 | 3.51 |
| 17 | 1.78 |
| 20 | 1.38 |
| 24 | 0.95 |

**Figure 3** *Discharge data for the River Easan Biorach during a 24-hour period in late October*

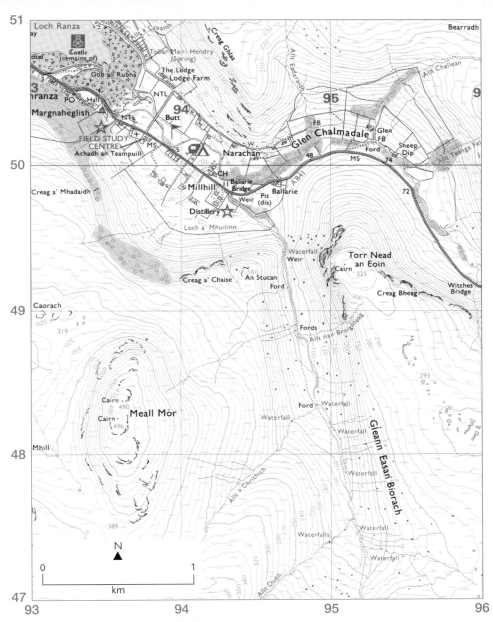

**Figure 4** *OS 1:25 000 map of the drainage basin of the River Easan Biorach on the Isle of Arran in Scotland*

## ACTIVITIES ⓢ

1 Study the discharge data for the River Easan Biorach in Figure **3**. Figure **4** is a map extract showing the drainage basin of the river. The data in Figure **3** was collected at grid reference 938503.

  **a** On a sheet of graph paper, draw the graph axes in Figure **5** (draw the horizontal axis in full, 0–24).

  **b** Draw the rainfall histogram onto your graph.

  **c** Now carefully plot the discharge data in Figure **3** to construct a hydrograph for the river.

  **d** Use Figure **1** to help you add labels to your hydrograph to describe its main features.

  **e** Calculate the lag time to the nearest hour.

  **f** Use the OS map extract in Figure **4** to attempt to explain these features of the hydrograph:
- the steep rising limb
- the short time lag to the first peak
- the presence of two distinct peaks
- the second peak being higher than the first peak
- the gentle falling limb.

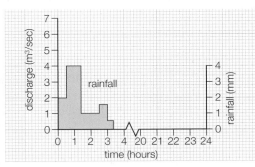

**Figure 5** *Rainfall histogram for the River Easan Biorach*

In this section you will learn about the factors affecting changes in the water cycle

## Natural (physical) variations affecting change

Extreme weather events such as severe storms or periods of drought can have significant impacts on the water cycle. They can affect both stores and transfers.

### Californian drought (2012–16)

California suffered a severe drought between 2012 and 2016 (Figure **1**). Rivers and lakes dried up, agricultural productivity declined and fires raged across tinder-dry forests and grasslands.

- Drought causes reduction in water stores in rivers and lakes.
- Vegetation dies back or is destroyed by fire – it affects processes such as transpiration, interception and infiltration.
- Groundwater flow becomes more important – it is a long-term transfer and not affected by short-term weather extremes.

- Heat and dry air causes initial high rates of evapotranspiration. This declines as water on the ground dries up (less water available to be evaporated) and trees transpire less.
- Soils dry out – the soil water store is reduced and throughflow ceases.

**Figure 1**
Castiac Lake, California at half its usual capacity, 2014

Seasonal changes are quite marked in mid- to high-latitude countries.

**Figure 2** *Some of the effects of summer and winter variations on the UK water cycle*

| Water cycle component | Summer | Winter |
| --- | --- | --- |
| Precipitation | Total rainfall may be less but storms are more frequent. | Greater quantities of rainfall with a likelihood of snow. |
| Vegetation – interception, transpiration, etc | Vegetation grows rapidly increasing interception and transpiration. | Vegetation dies back reducing interception and transpiration. |
| Evaporation | Higher temperatures encourage rapid evaporation (warm air can hold more moisture). | Lower temperatures reduce rates of evaporation. |
| Soil water | Dry soils encourage infiltration. But hard, baked soils encourage overland flow. | Soils may become saturated, leading to overland flow. |
| River channel flow | Low flow conditions are more likely. | High flow conditions are more likely. |

## Human activities affecting change

### Land-use change

The land-use changes that impact the most on the water cycle are urbanisation and deforestation.

- Urbanisation is the replacement of vegetated ground with impermeable concrete and tarmac. Water cannot infiltrate the soil, which increases overland flow and makes flooding more likely. Soil water and groundwater stores are reduced.

- Deforestation is the removal of trees, leading to surface runoff and soil erosion and reducing soil water stores.

### Farming practices

Farmers are able to control the local water cycle through irrigation or land drainage. Soils covered with plants have higher infiltration and soil water rates, and, therefore, reduced runoff.

If **desertification** occurs (see 2.13), the capacity to retain water is much lower. This capacity is lost completely once the soil is sealed.

## Water abstraction

The extraction of water from rivers or groundwater aquifers is referred to as *water abstraction*. Water that is abstracted for irrigation, industry and domestic purposes can have significant effects on the local water cycle.

Aquifers can become depleted. They can also become contaminated by inflowing saltwater if the water table drops below sea level – this has become an issue with the chalk aquifer beneath London. Abstraction can result in low flow conditions in rivers, which can have harmful impacts on ecosystems.

### Irrigation in the Middle East

Irrigation has a significant impact on water stores (aquifers and rivers) and transfer processes (evaporation and infiltration). In parts of the Middle East, water is being abstracted from underground aquifers that were formed thousands of years ago. They are in serious danger of becoming depleted as the rate of recharge is far slower than the rate of use. Figure **3** shows how technology can be used to reduce evaporation in hot environments.

⬆ **Figure 3** *Netting of a banana plantation, the Jordan Valley, Israel*

### ACTIVITIES

1 Study Figure **1**.
   a Draw a simple diagram to show the water cycle. Add annotations to show how a drought can impact on its stores and transfers.
   b Work in pairs to attempt a similar diagram to show the impacts of a severe storm event.
2 Figure **2** describes the influence of seasonal changes on selected components of the water cycle. Suggest impacts for stores and transfers that are not listed in the table.
3 Outline ways in which human activities can lead to changes in the stores and transfers in the water cycle.

### Land drainage in the UK

The low-lying land of the East Anglian Fens and the Somerset Levels were once submerged. Through the construction of deep drains and a network of ditches, which move water quickly through the system, this land is now highly productive farmland, although still vulnerable to occasional floods (Somerset in 2014). Moorland drainage ditches have also been held partly responsible for increasing the flood risk in the city of York.

The drainage of peatlands can have significant impacts on both the water cycle and the carbon cycle – the water table is lowered, changing rates of infiltration and evaporation. Dry peat is friable and vulnerable to erosion. In the past, excessive drainage in the Fens led to clouds of peat being formed – during the infamous 'Fen Blows' peat would be whisked up into the air to create huge black clouds.

Peatlands are essentially thick deposits of partly decomposed vegetation and, as such they act as important carbon stores. English peatlands alone are thought to store some 584 million tonnes of carbon. Vegetation on top of the peat also absorbs carbon dioxide from the atmosphere. As the peatlands are drained, air penetrates deeper, enabling decomposition of the carbon, releasing carbon dioxide. Dry peat can also ignite releasing carbon. It has been estimated that if all the peat in England were to be destroyed, the amount of carbon released to the atmosphere would be equivalent to about five years of England's current carbon dioxide emissions!

### STRETCH YOURSELF

In 2015 floods in South America, drought in East Africa and even the warmest and wettest December on record in the UK, have all been linked to the cyclical phenomenon, *El Niño*. Every six years or so, warmer water replaces the cold water in the Eastern Pacific off the coast of South America. This has direct consequences for the local weather patterns and also global weather conditions. These short-term cycles have implications for the water cycle – rainfall patterns become distorted which, in turn, affect other stores and transfers.

Find out more about the 2015/16 El Niño event. Describe the effects that it had on the world's weather and consider the implications of these effects on stores and transfers within the water cycle. Is there any connection between El Niño and climate change?

# The global carbon cycle: stores

In this section you will learn about the global carbon cycle and its stores

## The global carbon cycle

Carbon (C) is a basic chemical element that, along with nitrogen, phosphorous and sulphur, is needed by all plants and animals in order to survive. The recycling of carbon is essential for life on Earth. It enables food to be provided for plants and animals and energy sources to be created to fuel industrial development. In recent years there has been much concern about the increasing levels of carbon dioxide in the atmosphere and the impact that this is having on climate change.

Figure **1** shows that there are a number of components of the global carbon cycle. These components fall into two kinds of processes:

- *Stores* – the main stores of carbon are the lithosphere (rocks and soil), hydrosphere (oceans), cryosphere (snow and ice), atmosphere and biosphere (plants). The term **carbon sink** is used to describe a store that absorbs more carbon than it releases. A **carbon source** releases more carbon than it absorbs.
- *Transfers* – these are the processes involved in transferring carbon between the stores. For example, the process of photosynthesis takes carbon out of the atmosphere in the form of carbon dioxide and converts it into carbohydrates, such as glucose within plants. Transfers are the inputs and outputs that affect the magnitude of the stores at any one time.

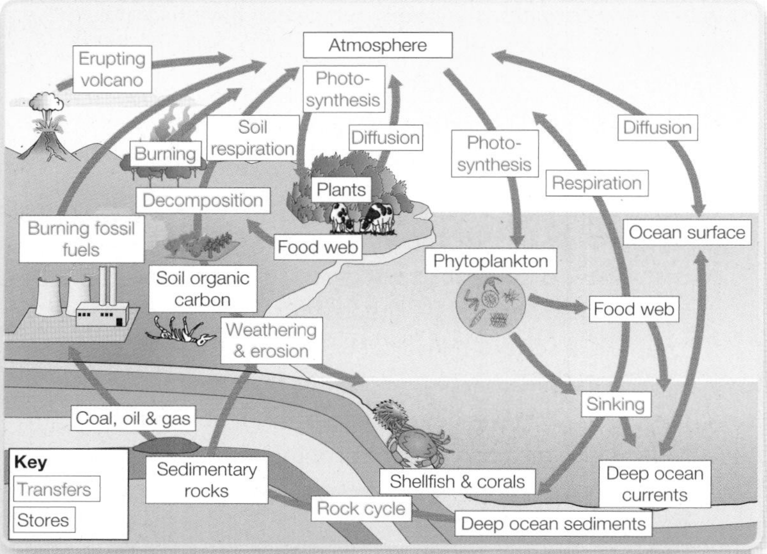


***Figure 1*** *The global carbon cycle*

## What are the main stores in the carbon cycle?

The amount of carbon held in each store is subject to change over timescales ranging from a few minutes to millions of years. These changes are known as *fluxes*.

Carbon stores can be identified at all geographical scales, from individual plants and trees within a forest through to entire vegetation successions (seres), such as woodlands, heaths, coastal sand dunes, and, at the global scale, the world's oceans and continents (see 6.4).

Carbon stores have a geographical component in that they are not evenly distributed across the world. Consider the distribution of land and sea, for example. This has a profound impact on the global distribution of carbon stores such as oceans, soils and rocks. Also, fossil fuels are only found in certain parts of the world. It is in these areas where much of the world's industrial production is concentrated, with a subsequent high level of carbon emissions caused by the burning of fossil fuels (Figure **2**).

**Figure 2** *Carbon stores in the global carbon cycle*

| Carbon store | Amount in billions of metric tons | Description |
|---|---|---|
| Marine sediments and sedimentary rocks | 100 000 | This is by far the largest store of carbon. It is a long-term store, with rocks taking millions of years to form. |
| Ocean | 38 000 | The oceans are a very important carbon store. Carbon dioxide is absorbed directly from the air and river water discharges carbon carried in solution. Since the industrial revolution, the oceans have absorbed more carbon dioxide from the air due to increased carbon emissions. |
| Fossil fuel deposits | 4000 | Hydrocarbons such as coal, oil and gas are important long term stores of carbon. Since the industrial revolution they have been exploited for heat and power. The resulting combustion has pumped huge quantities of carbon dioxide into the atmosphere causing climate change. |
| Soil organic matter | 1500 | Soils contain rotting organic matter and are important carbon stores. Carbon can remain in soils for hundreds of years. Deforestation, land use change and soil erosion can, however, release the stored carbon very rapidly. |
| Atmosphere | 750 | Carbon is held in the atmosphere in the form of carbon dioxide. In recent decades the amount of carbon dioxide has increased due to emissions from power stations, vehicles and deforestation. This has led to the enhanced greenhouse effect and climate change. |
| Terrestrial plants | 560 | Plants are vital for all life on Earth. They convert energy from the sun into carbohydrates that support life. Plants can store carbon for many years and transfer it to the soil. However, through deforestation this carbon can be released back into the atmosphere very rapidly. |

## The global pattern of vegetation carbon storage

Figure **3** is a satellite image showing the global distribution of carbon storage in plants, indicated by photosynthesis.

Some regions of the world (e.g. the Sahara Desert), have virtually no plant storage. Other areas have flourishing vegetation growth and productivity.

Scientists have identified that carbon uptake is increasing in middle and high latitudes of the northern hemisphere, but less carbon is being absorbed in the tropics and southern hemisphere. A major cause of this decrease is thought to be drought (possibly linked to climate change), impacting crop yields, timber production and expanses of natural vegetation.

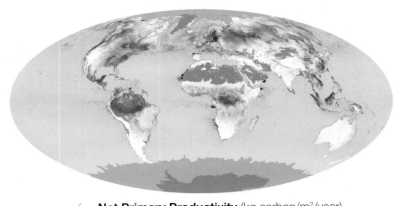

**Net Primary Productivity** (kg carbon/m²/year)

land
ocean

−0.5    0    0.5    1    1.5    2    2.5

**Figure 3** *Global measurements of the carbon stored by plants during photosynthesis*

## ACTIVITIES

1  Study Figure **1**.
   a  What form does carbon take when it is in the atmosphere?
   b  How does carbon become stored in deep ocean sediments?
   c  How is carbon stored on land?
   d  Suggest ways in which people affect the carbon cycle.
2  Study Figure **2**. For each individual store, calculate the proportion of the total carbon stored. Present this information as a pie chart.
3  Study Figure **3**. Describe the global pattern of carbon storage in plants. Suggest reasons why it varies across the world.

## STRETCH YOURSELF

Find out more about the global distribution of fossil fuels. How is the distribution likely to affect local and regional carbon cycles?

Find out about the recent decision to exploit oil in Alaska. Think about how this might impact on the carbon cycle.

In this section you will learn about the transfers in the carbon cycle

## The carbon cycle at the local scale

### The carbon cycle of an individual tree

Look at Figure **1**. Notice that the tree acts as a carbon store (wood is about 50 per cent carbon) and there are several processes that transfer carbon between the atmosphere and the land.

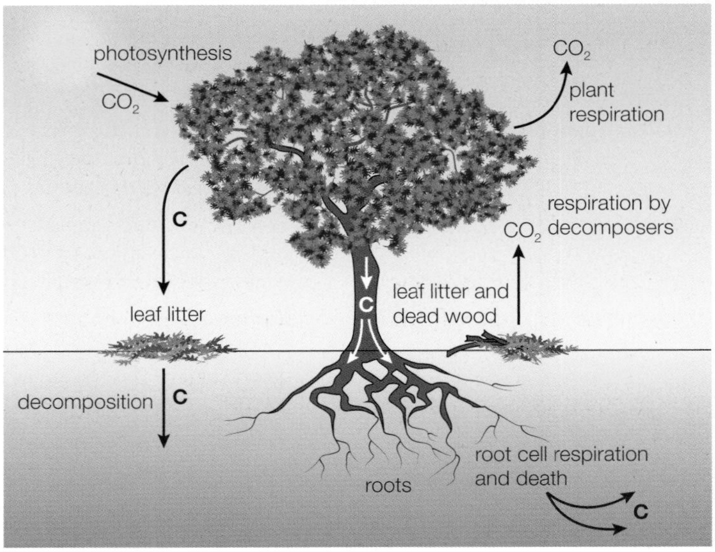

### A terrestrial carbon cycle: the lithosere

When rock is exposed for the first time, say after glacier retreat, it is vulnerable to processes of weathering. As the rock is slowly broken down, carbon may be released, often dissolved in water. Over time, vegetation such as lichen and moss grow on the bare rock and carbon exchange starts to take place involving photosynthesis and respiration. Gradually, as organic matter is added to the broken fragments of rock, a soil develops that can support a wider range of plants.

Soil is an important component of the carbon cycle as it can absorb and store carbon over moderate time periods. Over hundreds of years, the plant species become more diverse, benefiting from the supply of carbon in the soil. A number of different habitats are established and wildlife becomes abundant.

This sequence of changes is called a *vegetation succession* (see 6.9). A succession that relates to a specific environment is called a **sere** and each stage in the succession can be referred to as a **seral stage**. A **lithosere** is a vegetation succession that occurs on bare rock. Other seres include **hydrosere** (water – freshwater pond), **halosere** (salt – coastal salt marsh) and **psammosere** (coastal sand – sand dunes).

Eventually, the final stage of a sere is reached when a state of environmental equilibrium or balance is achieved. This is usually in response to the climate and is therefore termed the *climatic climax* (see 6.6). The climax vegetation for a lithosere in the UK will usually be a deciduous woodland (Figure **2**). As Figure **2** illustrates, the carbon cycle is highly complex, involving a number of different stores and many transfer mechanisms, each varying considerably over time and space.

**Figure 1** *Carbon cycle of a tree*

**Figure 2** *A simple terrestrial (lithosere) carbon cycle*

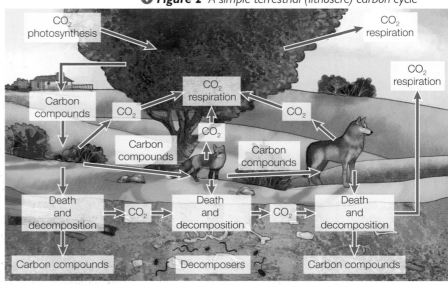

## What are the main transfers operating in the carbon cycle?

Look back to the global carbon cycle (Figure **1**, 1.8) to remind yourself of the main transfers responsible for the flow of carbon between the various carbon stores.

### Photosynthesis

**Photosynthesis** is the process whereby plants use the light energy from the sun to produce carbohydrates in the form of glucose (Figure **3**).

◆ Green plants absorb the light energy using chlorophyll (a green substance found in chloroplasts in plant cells) in their leaves.

◆ The absorbed light energy converts carbon dioxide in the air and water from the soil into glucose. During this process, oxygen is released into the air.

◆ Some glucose is used in respiration, the rest is converted into starch, which is insoluble but can be converted back into glucose for respiration.

The process can be summarised as:

carbon dioxide + water $\xrightarrow{\text{light energy}}$ glucose + oxygen

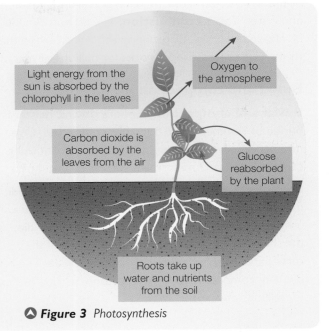

Light energy from the sun is absorbed by the chlorophyll in the leaves

Oxygen to the atmosphere

Carbon dioxide is absorbed by the leaves from the air

Glucose reabsorbed by the plant

Roots take up water and nutrients from the soil

⬆ **Figure 3** *Photosynthesis*

### Respiration

**Respiration** is a chemical process that happens in all cells and is common to both plants and animals. Glucose is converted into energy that can be used for growth and repair, movement and control of body temperature in mammals. Carbon dioxide is then returned to the atmosphere, mostly by exhaled air.

### Decomposition

When organisms die they are consumed by decomposers such as bacteria, fungi and earthworms. During this process of **decomposition**, carbon from their bodies is returned to the atmosphere as carbon dioxide (Figure **4**). Some organic material passes into the soil where it may be stored for hundreds of years.

### Did you know?

Researchers from Leeds University have used elevation data together with satellite data of forest cover to assess the accuracy with which maps measure the area of forests. They discovered that more than 75 per cent of tropical mountain forests are on slopes in excess of 27 degrees. This means that most maps underestimate the area covered by forest. Their research suggested that the global area of tropical mountain forests is 40 per cent greater than the area reported on maps. This means that tropical mountain forests are more important stores, and therefore transfers, of carbon than previously thought.

▶ **Figure 4** *Decomposing elephant, Bouba-Ndjida National Park, Cameroon*

## Combustion

Organic material contains carbon. When it is burned in the presence of oxygen (e.g. coal in a power station, Figure **5**) it is converted into energy, carbon dioxide and water. This is **combustion**. The carbon dioxide is released into the atmosphere, returning carbon that might have been stored in rocks for millions of years.

## Burial and compaction

**Burial and compaction** is where organic matter is buried by sediments and becomes compacted. Over millions of years, these organic sediments containing carbon may form *hydrocarbons* such as coal and oil.

Corals and shelled organisms take up carbon dioxide from the water and convert it to calcium carbonate, used to build their shells (Figure **6**). When they die, the shells accumulate on the seabed. Some of the carbonates dissolve, releasing carbon dioxide. The rest become compacted to form limestone, storing carbon for millions of years.

**Figure 5** *Coal-burning power station, Moscow, Russia*

**Figure 6** *Corals use carbon dioxide to build their exoskeletons*

## Carbon sequestration

**Carbon sequestration** is an umbrella term used to describe the transfer of carbon from the atmosphere to plants, soils, rock formations and oceans. Sequestration is both a natural and human process. *Carbon capture and storage (CCS)* involves the technological 'capturing' of up to 90 per cent of the carbon dioxide ($CO_2$) emissions produced from the use of fossil fuels in electricity generation and industrial processes (Figure **7**). There are several CCS facilities operating in the USA and elsewhere (see 1.14). From 2020, carbon dioxide from chemical plants and oil refineries in Rotterdam will be sequestered in the North Sea seabed. Smaller-scale sequestration can also take place, for example by a change in farming practices (see 1.11).

## Weathering

**Weathering** involves the breakdown or decay of rocks in their original place at or close to the surface. When carbon dioxide is absorbed by rainwater it forms a mildly acidic carbonic acid. Through a series of complex chemical reactions, rocks will slowly dissolve with the carbon being held in solution (Figure **8**). Transported via the water cycle to the oceans, this carbon can then be used to build the shells of marine organisms.

**Figure 7** *The Hazelwood coal-fired power station, Latrobe Valley, Victoria, Australia, which is trialling carbon capture and storage*

**Figure 8** *Chemical weathering of limestone by the process of carbonation*

### ACTIVITIES

1 Study Figure **1**.
   a How is carbon transferred to the soil?
   b How is carbon released to the atmosphere?
   c If the leaf litter was burned in a fire, what impact would this have on the carbon cycle?
   d Suggest why trees are important carbon sinks.

2 Study Figure **2**.
   a What role do the animals play in the carbon cycle?
   b Imagine that foxes were wiped out from this ecosystem. What effect would this have on the carbon cycle?

3 Apart from burning fossil fuels to generate power (Figure **5**), how else can combustion release carbon dioxide into the atmosphere?

**S**
4 Draw a flow diagram in the form of a cycle to show how weathering on the land can result in carbon being stored as deep-sea sediments and then turned into limestone, which can then itself be weathered.

### STRETCH YOURSELF

Find out more about the changes in the carbon cycle that occur during the development of a sere of your choice (not a lithosere!). Identify the different seral stages for your chosen sere and suggest how the carbon cycle develops in its complexity as the succession progresses. In referring to both stores and transfers, consider the changes in magnitude that take place over time and space. Include diagrams similar to Figure **2** to represent a seral stage.

Attempt to produce a simplified sketch with labels or annotations, similar to Figure **2**, that describes the carbon cycle.

Changes in the carbon cycle: physical causes

In this section you will learn how changes due to physical activity take place in the stores and transfers within the carbon cycle

## Why does the carbon cycle experience change?

While people are right to be concerned about recent rapid increases in carbon dioxide emissions into the atmosphere, the carbon cycle has experienced change throughout the billions of years of the Earth's history.

Carbon, like water, is subject to significant levels of change over time and space. Although carbon may be stored deep within the Earth's crust for millions of years, in other contexts it can be transferred in a matter of seconds from one store to another, say during a wildfire. Carbon within the carbon cycle is in a constant state of flux, undergoing changes in magnitudes at all scales and in all time frames.

There are many reasons, both physical and human, why these changes take place.

## Physical causes of changes in the carbon cycle
### Natural climate change

During the Quaternary geological period (from 2.6 million years ago to the present day) global climates fluctuated considerably between warm (interglacial) and cold periods (glacial). Look at Figure **1**. It shows temperature and atmospheric carbon dioxide levels during the last 800 000 years based on data obtained from ice cores in the Antarctic ice sheet. There are several regularly occurring cold (glacial) periods interspersed with warmer interglacial periods.

Notice that the trends for temperature and carbon dioxide mirror one another – higher temperatures are associated with higher levels of carbon dioxide in the air. It is interesting to see the scale of the recent rise in carbon dioxide levels, which have now surpassed 400 ppmv (parts per million by volume).

The causal relationship between temperature and carbon dioxide is an interesting one to explore. An increase in the level of carbon dioxide in the atmosphere leads to enhanced global warming and subsequent temperature increase. Lower levels of carbon dioxide reduce the effectiveness of the greenhouse effect, which leads to global cooling. So, in theory carbon dioxide levels trigger temperature change. However, temperature change also has an impact on levels of carbon dioxide.

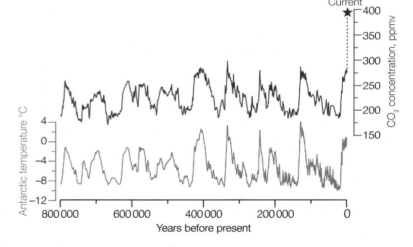

▲ **Figure 1** *Global temperature and carbon dioxide fluctuations during the last 800 000 years (ppmv = parts per million by volume)*

## The impacts of cold conditions on carbon stores and transfers

◆ Chemical weathering processes would have been more active because cold water can hold more carbon dioxide.

◆ Forest coverage would be very different both in total area and in geographical location. This would have affected the significance and distribution of processes such as photosynthesis and respiration.

◆ Decomposers would have been less effective, so carbon transfer to soils would have been reduced.

◆ Less water would have flowed into the oceans as it was locked up as snow and ice on land. There would be less sediment transfer along rivers and less build-up of sediments on the ocean floor.

◆ The soil would have been frozen over vast areas of land. This would have stopped transfers of carbon.

### Wild fires

Wild fires can be started naturally by lightning strikes. However, increasingly they are started deliberately by people. Despite being restricted to tiny parts of the Earth's surface wild fires can have regional impacts. In 1997–8 and again in 2013 there were many huge fires in Indonesia (Figure **2**) that burned out of control for months. Smoke from these fires spread across parts of south-east Asia, affecting the lives of millions of people. The fires released a large quantity of carbon dioxide into the atmosphere, causing a noticeable spike in the rising trend of carbon emissions recorded since the late 1950s (see 1.16). In 2015 an air pollution crisis (named the Southeast Asian haze) affected more than 28 million people in the region, causing widespread respiratory illness and some deaths. The haze resulted from forest fires – mostly caused by illegal slash and burn practices – on the Indonesian islands of Sumatra and Kalimantan.

Wild fires can turn forests from being a carbon sink to being a carbon source, as combustion returns huge quantities of carbon back into the atmosphere (see 5.16).

## The impacts of warm conditions on carbon stores and transfers

As indicated by Figure **1**, in recent years global temperatures have risen and this has been particularly noticeable in the high latitudes. One of the effects of this has been the melting of permafrost in tundra regions, for example Siberia in Russia.

Carbon stored within the permafrost – together with other gases such as methane – is now being released into the atmosphere where it further enhances the greenhouse effect, leading to increased warming. This is an excellent example of a positive feedback leading to further destabilisation of the system.

**▼ Figure 2** *Indonesian wild fires emitting carbon dioxide (2013)*

## Volcanic activity

Volcanic activity returns to the atmosphere carbon that has been trapped for millions of years in rocks deep within the Earth's crust (Figure **3**). During the Palaeozoic era (542–251 million years ago) volcanoes were much more active than they are today. A vast amount of carbon dioxide was emitted into the atmosphere, where it remained for a very long time.

At present, volcanoes emit between 130 and 380 million tonnes of carbon dioxide per year. By comparison, human activities emit about 30 billion tonnes of carbon dioxide per year, mainly as a result of burning fossil fuels. Volcanoes also erupt lava, which contains silicates that will slowly weather. This converts carbon dioxide in the air to carbonates in solution. In this way carbon dioxide is absorbed very, very slowly from the atmosphere.

▼ **Figure 3** *Volcanic gases emitted on Hawaii*

## ACTIVITIES

**(S)**

1  Study Figure **1**. Describe the patterns of temperature and carbon dioxide over the last 800 000 years.

2  Suggest how the carbon cycle was affected by cold glacial periods during the Quaternary period. How did this affect the magnitude of the stores and the operation of the transfers?

3  How do volcanic eruptions contribute to the carbon cycle?

4  How can trees be both a carbon store and a carbon sink at different times or locations?

### EXTENSION

5  Describe in detail the comparative trends between temperature and carbon dioxide according to data obtained from the Vostok ice core.

6  Critically evaluate the theory that orbital cycles triggered the initial temperature rise at the end of the last glacial period but thereafter it was the increase in carbon dioxide that drove increases in temperature.

## STRETCH YOURSELF

Find out about the possible impact on the carbon cycle of a future eruption of the Yellowstone supervolcano.

Explain the effects on the atmosphere and assess the possible consequences for life on Earth.

## EXTENSION

# Is there a causal link between carbon dioxide and temperature in explaining glacial cycles?

Look at Figure **4**. By superimposing the carbon dioxide and temperature curves for data obtained from the Vostok ice core, it can be seen that the two curves do not match perfectly. Assuming that the data are correct (this is an important assumption), there appears to be a time delay between temperature change and changes in the level of carbon dioxide.

Some scientists believe that the trigger for these long-term trends in temperature and carbon dioxide is orbital change – the Milankovitch Cycles.

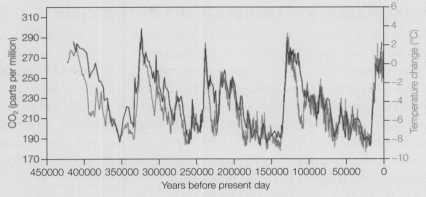

▲ **Figure 4** *Vostok ice core records for carbon dioxide concentration (blue) and temperature change (red)*

These regular cycles of orbital eccentricity cause slight variations in the amount of the sun's radiation that warms up the Earth. So, as temperatures start to rise at the end of a glacial period (triggered by orbital change), there is a surge of carbon dioxide released into the atmosphere by the warming of the oceans and the 'unlocking' of the land surface that had previously been frozen. This surge of carbon dioxide enhances the greenhouse effect, amplifying the warming trend. This is an excellent example of a positive feedback loop.

A study in 2012 by Shakun et al. looked at temperature changes during the transition from the last glacial period to the current warmer period, about 20 000 years ago. They found that:

◆ the Earth's orbital cycles triggered warming in the Arctic approximately 19 000 years ago, causing large amounts of ice to melt, flooding the oceans with fresh water

◆ this influx of fresh water then disrupted ocean current circulation, in turn causing a see-sawing of heat between the hemispheres

◆ the southern hemisphere and its oceans warmed first, starting about 18 000 years ago. As the Southern Ocean warmed, the solubility of carbon dioxide in the water fell and this causes the oceans to give up more carbon dioxide, releasing it into the atmosphere.

So, there is evidence to suggest that orbital cycles triggered the initial warming at the end of the last glacial period leading to a surge in carbon dioxide emissions, which in turn amplified the warming trend. Overall, scientists believe that more than 90 per cent of the post-glacial warming occurred after the rise in atmospheric carbon dioxide.

There is still a considerable debate about the causal connections between temperature and carbon dioxide and the role played by the Milankovitch Cycles. Issues concerning data reliability further complicate an already heated debate. Whatever the causes of climate change in the past, there is little doubt among the scientific community that the current high levels of carbon dioxide are the result of anthropogenic (human) factors and that this is causing the recent rise in global temperatures.

In this section you will learn how changes due to human activity take place in the stores and transfers within the carbon cycle

## Human causes of changes in the carbon cycle

According to the Intergovernmental Panel on Climate Change (IPCC), about 90 per cent of anthropogenic (human related) carbon release comes from the combustion of fossil fuels, primarily coal, but also oil and natural gas. The remaining 10 per cent results from land-use change, such as deforestation, land drainage and agricultural practises. Roughly half of the anthropogenic carbon is absorbed equally by oceans and vegetation and the remainder is absorbed by the atmosphere.

Since the 1960s, global concentrations of carbon dioxide have increased dramatically from about 320 ppm to just over 400 ppm, the highest level ever recorded. In Figure **1** the boxes are stores of carbon and the arrows indicate fluxes and the processes that drive those fluxes. Black figures are estimates of the natural stores and fluxes and red figures indicate anthropogenic effects on the carbon cycle after 1750, the start of the Industrial Revolution.

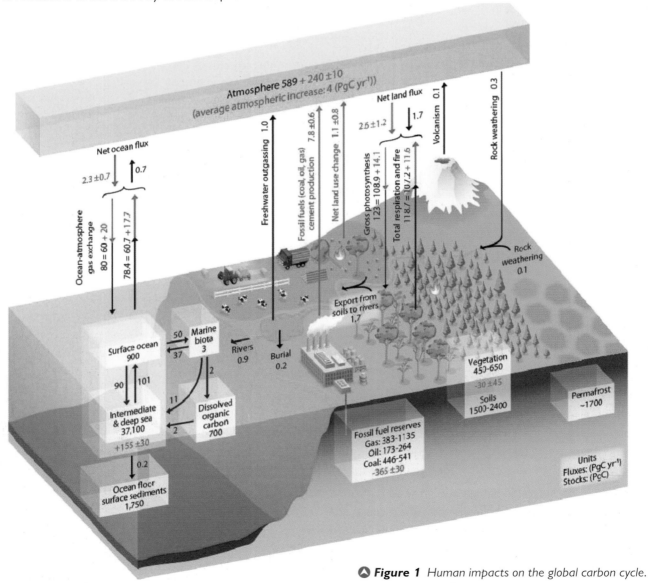

⬆ **Figure 1** *Human impacts on the global carbon cycle.*

## Combustion of fossil fuels

Fossil fuels are natural sources of energy formed from the remains of living organisms, primarily plants. They are extremely important long-term carbon stores comprising carbon locked away within the remains of organic matter. Today, most of the world's gas and oil is extracted from rocks that are 70–100 million years old. The carbon has remained locked up in these deposits for all that time but, when burnt to generate energy and power, the stored carbon is released, primarily as carbon dioxide into the atmosphere, accelerating the cycling of this carbon.

Fossil fuels are mainly composed of carbon and hydrogen, hence the term *hydrocarbons*. Methane, the main component of natural gas, has the chemical formula $CH_4$. Oil (petroleum) is a more complex compound comprising carbon, hydrogen, nitrogen and sulphur along with other impurities. When combustion takes place, reactions occur with oxygen releasing carbon dioxide and water.

$$CH_4 + 2O_2 \longrightarrow CO_2 + 2H_2O$$

Since the Industrial Revolution, fossil fuels have been burnt in increasing quantities, pumping carbon dioxide into the atmosphere. Once in the atmosphere it enhances the natural greenhouse effect, increasing global temperatures – so-called *global warming*. Figure **2** shows this dramatic increase since the 1950s, driven by the rapid industrialisation of developing nations (such as China) as well as the continued demand from the world's industrialised nations (such as the USA).

Since the late 1950s carbon dioxide in the atmosphere has been measured by the Hawaiian Volcanic Observatory, their research showing a huge increase in levels of atmospheric carbon dioxide (Figure **3**). These figures have been used to support the concept of human-induced climate change and global warming.

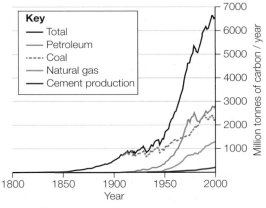

▲ **Figure 2** *Trends in global carbon emissions since 1800*

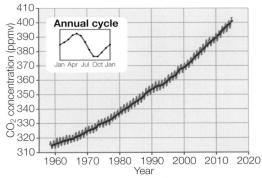

▲ **Figure 3** *Recent increases in atmospheric carbon dioxide*

### Land-use change

Land-use change is responsible for about 10 per cent of carbon release globally, which impacts on relatively short-term stores and has direct links to issues of climate change and global warming. Furthermore, at the local scale, land-use changes can have a very significant impact on small-scale carbon cycles.

### Farming practices

Ploughing and harvesting, rearing livestock, using machinery fuelled by fossil fuels and using fertilisers based on fossil fuels are all farming practices that release carbon. On many farms it is the use of artificial fertilisers that is the main source of carbon emission.

Methane is a potent greenhouse gas, and some farming practices result in high levels of methane emissions. Livestock, especially cattle, ruminate (regurgitate food and masticate a second time – 'chewing the cud'), which produces methane as a by-product. This has raised issues worldwide about the desirability of moving away from such a high dependence on meat and dairy products. Cattle in the USA emit around 5.5 million tonnes of methane per year into the atmosphere – around 20 per cent of the total methane emissions in the USA.

Methane is also produced from the cultivation of rice. Studies indicate that rice may contribute up to 20 per cent of global methane production. Interestingly, research in Asia and North America has found that rice yields have increased by 25 per cent due to increased levels of carbon dioxide in the air. But this in turn has resulted in a 40 per cent increase in methane emissions. Rice is the primary food source for 50 per cent of the world's population, mostly in developing regions, so this trend is likely to continue.

### Shimpling Park Farm, Bury St Edmunds, Suffolk

Look at Figure **4**. It shows the carbon budget for Shimpling Park Farm in Suffolk. The farm is a large (645 ha), organic arable farm specialising in wheat, barley, oats, spelt and quinoa along with some grazing for sheep.

The farm has calculated its carbon emissions as 1150 tonnes $CO_2$, mainly from fuel for machinery and also nitrous oxide emissions (included as carbon emissions) from crop residues and green manures. Fertilisers are usually the major source of emissions on non-organic farms. The flock of 250 sheep contribute 6.6 per cent of emissions, mostly as methane.

About 40 per cent of the carbon emissions are offset by carbon sequestration on the farm. Values for sequestration in the soil do not appear in Figure **4** but this is likely to be a significant carbon sink. In the future, the farm intends to raise the level of organic matter within the soils which will store even more carbon. An increase of just 0.1 per cent in soil organic matter over the entire farm could increase the sequestration of carbon to a level of about 4500 tonnes $CO_2$, four times greater than the emissions!

▼ **Figure 4** *Carbon emissions budget for Shimpling Park Farm, Suffolk*

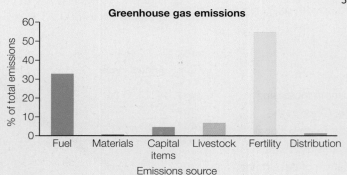

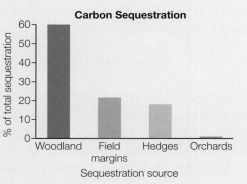

### Deforestation

Trees are removed, either by burning or felling, for building, ranching, mining or the growing of commercial crops such as oil palm and soya. This is *deforestation*. The timber itself is a valuable product in the production of furniture and other wood products. Forests are also harvested for firewood.

Deforestation is widespread across the world but is particularly concentrated in tropical regions, for example, in Indonesia. In total, it accounts for about 20 per cent of all global carbon dioxide emissions.

In a natural system, when a tree dies it decomposes very slowly and releases carbon over a long period of time. During that time, new vegetation starts to grow that quickly compensates for the carbon being released by the dead tree – the system is *carbon neutral*. When deforestation by burning occurs, carbon is immediately released into the atmosphere. If the land is then used for a different purpose, such as grassland for cattle ranching, the future absorption of carbon dioxide will be reduced. The system has now become a source of carbon rather than a sink (Figure **5**). This is extremely significant in terms of the carbon cycle both globally and regionally, as forest ecosystems are limited to certain regions of the world.

### Urbanisation

Replacing open countryside with concrete and tarmac is known as 'urbanisation'. It is a major change in land use (see *AQA Geography for A Level & AS Human Geography*, Contemporary urban enivronments). This has a significant impact on the local carbon cycle – important stores are either replaced (vegetation) or covered up (soils) with impermeable surfaces.

Globally, urban areas occupy about 2 per cent of the total land area. However, these areas account for 97 per cent of all anthropogenic carbon dioxide emissions! The major sources of these emissions are transport, the development of industry, the conversion of land use from natural to urban, and cement production for the building sector.

Cement is an extremely important material used in construction across the world. Carbon dioxide is a byproduct of a chemical conversion process used in the production of clinker, a component of cement, in which limestone ($CaCO_3$) is converted to lime ($CaO$). Carbon dioxide is also emitted during cement production by the use of fossil-fuel combustion. It has been estimated that cement production contributes about 2.4 per cent of global carbon emissions (not including the use of fossil fuels), although this is highly concentrated close to the major cement-producing plants.

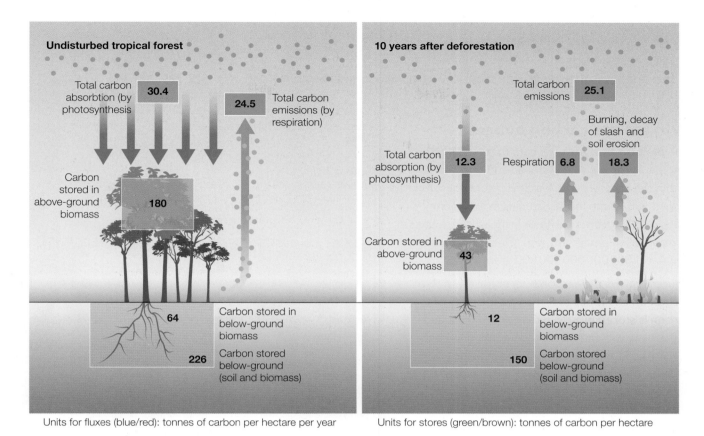

Units for fluxes (blue/red): tonnes of carbon per hectare per year          Units for stores (green/brown): tonnes of carbon per hectare

## ACTIVITIES

**S**

1   Use Figure **1** to help you draw a diagram to show in detail the impact that human activities are having (and will have in the future) on the carbon cycle. Add annotations to show how the various factors discussed in this section can affect transfers and stores in the carbon cycle. Use photos from the internet to illustrate your diagram. What other factors cause changes to the carbon cycle?

2   Study Figure **3**.
   **a**   Describe the pattern of carbon dioxide emissions since the late 1950s.
   **b**   How does the seasonal growth of vegetation cause annual variations in carbon emissions?
   **c**   What are the main causes of the rapid increase in emissions?

3   Assess how farming practices can affect the carbon stores and transfers at the local scale. Use the internet to find out more about the main forms of emission (such as fuel, fertility and livestock) and the options available for sequestration of carbon (such as woodlands, hedges and marginal land).

4   Study Figure **5**. Outline how deforestation affects the stores and transfers of carbon in a tropical rainforest. Use the statistics to provide numerical comparisons to support your answer.

## EXTENSION

5   Carry out your own research and consider how the use of fracking and tar (oil) sands is changing the patterns and geography of carbon emissions. Searching www.energycentral.com is a good place to start.

**▲ Figure 5** *Changes in carbon cycling between tropical forest and areas of deforestation*

## STRETCH YOURSELF

Find out more about the impact of urbanisation on the carbon cycle (the article 'The Role of Urbanization in the Global Carbon Cycle' maybe of use; search for it at www.frontiersin. org). Find some photos from the internet and add labels to describe this impact. In which regions of the world is urbanisation yet to peak? What are the implications for the carbon cycle?

In this section you will learn about:
- the carbon budget
- the impacts of the carbon cycle on land, ocean and atmosphere

## What is the carbon budget?

The **carbon budget** uses data to describe the amount of carbon that is stored and transferred within the carbon cycle. It's similar to a household budget with some money coming in and money going out. Carbon is commonly measured in units of mass called petagrams (Pg). One petagram is equal to 1 000 000 000 000 000 grams!

Figure **1** shows the carbon budget for the global carbon cycle. Notice that the vast majority of carbon is stored mostly in the Earth's crust and in the oceans. Relatively low amounts of carbon are stored in the atmosphere and in plants. This is because the carbon transfers are extremely active, with carbon constantly flowing back and forth between the atmosphere and the land.

Figure **2** shows the carbon budget for a single tree. Notice how the units of carbon move from one store to another. Also notice the significance of the photosynthesis transfer. This is largely responsible for a net loss of carbon in the atmosphere and shows the importance of trees as carbon sinks (they absorb more carbon dioxide than they release).

## The impacts of the carbon cycle

The most important role of the carbon cycle is the release of carbon dioxide and other gases such as methane into the atmosphere. These gases absorb long-wave radiation from the Earth and warming lower atmosphere, enabling life to exist on Earth. This is the *greenhouse effect*.

In recent decades, increased emissions resulting from anthropogenic activities, such as burning fossil fuels and deforestation, have increased the concentration of the greenhouse gases, making them more effective in trapping radiation given off by the Earth. This is called the *enhanced greenhouse effect* (Figure **4**), which explains why average global temperatures have increased by about 0.8 °C since 1880.

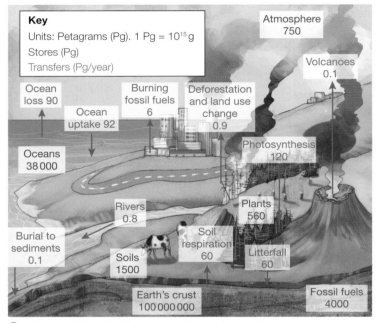

**Key**

Units: Petagrams (Pg). 1 Pg = $10^{15}$ g
Stores (Pg)
Transfers (Pg/year)

Atmosphere 750

Volcanoes 0.1

Ocean loss 90

Ocean uptake 92

Burning fossil fuels 6

Deforestation and land use change 0.9

Photosynthesis 120

Oceans 38 000

Rivers 0.8

Plants 560

Burial to sediments 0.1

Soil respiration 60

Litterfall 60

Soils 1500

Earth's crust 100 000 000

Fossil fuels 4000

▲ **Figure 1** The global carbon budget

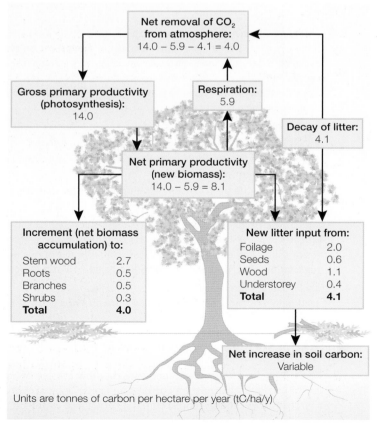

Net removal of $CO_2$ from atmosphere:
14.0 − 5.9 − 4.1 = 4.0

Gross primary productivity (photosynthesis): 14.0

Respiration: 5.9

Decay of litter: 4.1

Net primary productivity (new biomass):
14.0 − 5.9 = 8.1

Increment (net biomass accumulation) to:

| | |
|---|---|
| Stem wood | 2.7 |
| Roots | 0.5 |
| Branches | 0.5 |
| Shrubs | 0.3 |
| **Total** | **4.0** |

New litter input from:

| | |
|---|---|
| Foliage | 2.0 |
| Seeds | 0.6 |
| Wood | 1.1 |
| Understorey | 0.4 |
| **Total** | **4.1** |

Net increase in soil carbon: Variable

Units are tonnes of carbon per hectare per year (tC/ha/y)

▲ **Figure 2** The carbon budget for a forest tree

| Land | Ocean | Atmosphere |
|---|---|---|
| • The carbon cycle is responsible for the formation and development of soil. Carbon in the form of organic matter (litterfall) introduces important nutrients and provides a structure to the soil.<br><br>• Carbon in the form of organic matter is essential for plant growth and the production of food.<br><br>• Carbon stored in grass provides fodder for animals.<br><br>• Carbon provides a valuable source of energy in the form of wood and fossil fuels. | • Carbon can be converted into calcium carbonate, which is used by some marine organisms to build shells.<br><br>• The carbon cycle has an impact on the presence and proliferation of phytoplankton, a basic food for many marine organisms. Phytoplankton consumes carbon dioxide during photosynthesis. The carbon is then passed along the marine food chain. | • Carbon dioxide in the atmosphere helps to warm the Earth through the greenhouse effect. Without this, there would be no life on Earth.<br><br>• Increases in carbon emissions as a result of human activities (deforestation, combustion of fossil fuels) have led to the enhanced greenhouse effect, which threatens to have a profound impact on the world's climate.<br><br>• Carbon stored by vegetation has a significant effect on the atmosphere, whether deforestation (carbon source) or afforestation (carbon sink). |

⬆ **Figure 3** *Impacts of the carbon cycle on land, ocean and atmosphere*

The carbon cycle can have significant regional impacts on climate.

◆ Vegetation plays a pivotal role in the carbon cycle. It also impacts on global climates, by removing carbon dioxide and releasing water and oxygen. Regions with dense vegetation (tropical rainforests) experience high rates of photosynthesis and respiration. This increases levels of humidity and the amount of cloud cover, which in turn may affect regional temperatures and rainfall.

◆ Regions experiencing widespread deforestation may become drier and less humid – fewer trees mean less photosynthesis.

◆ The proliferation of plankton in the oceans may promote the formation of clouds, through the creation of a chemical substance called dimethylsulphide (DMS).

◆ Volcanic eruptions release carbon dioxide into the atmosphere along with ash and other gases. This absorbs more incoming radiation from the sun and can lead to a cooling effect on Earth – sometimes called a *volcanic winter*.

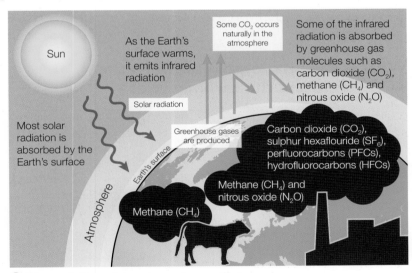

⬆ **Figure 4** *The enhanced greenhouse effect*

## ACTIVITIES

1  a  Look at Figure **1**. What is the major store of carbon?

   b  What process is responsible for the greatest transfer of carbon?

   c  Use Figure **1** to explain why trees are an important part of the carbon cycle despite the fact that they are the smallest store.

   d  For each store calculate whether there is a net gain or net loss of carbon. List the carbon stores and the carbon sinks.

2  a  Look at Figure **2**. What is the evidence that a tree is a carbon sink?

   b  Suggest reasons why the 'Net increase in soil carbon' is variable.

   c  If this was a deciduous tree, how might the carbon budget vary with the seasons?

3  In what ways does the carbon cycle affect the atmosphere and the Earth's climate?

## STRETCH YOURSELF

Consider how human activities in a local drainage basin might affect its natural stores and flows. Construct a diagram to show the impacts of urbanisation, deforestation, and water abstraction.

In this section you will learn about:
- the relationship between the water cycle and the carbon cycle in the atmosphere
- how this relationship links to climate change

## The role of water and carbon in supporting life

Both water and carbon are critical in supporting life on Earth. Carbon is one of six crucially important elements in humans. Stored in the form of glucose, carbon assists cellular respiration; it makes up 18 per cent of the human body.

In trees, the carbon content of leaves and woody matter (stem, branches and roots) is approximately 50 per cent of their biomass. Through food chains, the carbon stored in plants is passed on to animals, where it provides much needed energy for, among other things, breathing, growing and reproducing. Through respiration and decomposition, carbon is returned to the atmosphere in the form of carbon dioxide.

The atmosphere is an important store of both water and carbon. All living organisms need water to survive, albeit in differing amounts – it is needed for drinking and irrigation. It is also a source of power and energy. Carbon in the atmosphere is essential in photosynthesis to create the carbohydrates needed for plant growth. It is also one of the important greenhouse gases that absorb long-wave radiation emitted from the Earth, providing sufficient atmospheric warmth for life to survive.

The storage and cycling of both water and carbon enable life to flourish on land and in water. Changes in the magnitude of the stores (such as the amount of carbon stored as biomass) can have massive local and global implications for flora and fauna.

## What is the relationship between the water cycle and the carbon cycle?

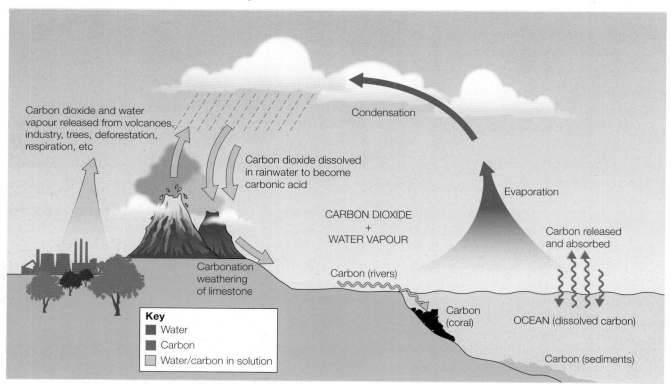

Carbon dioxide and water vapour released from volcanoes, industry, trees, deforestation, respiration, etc

Condensation

Carbon dioxide dissolved in rainwater to become carbonic acid

Evaporation

CARBON DIOXIDE
+
WATER VAPOUR

Carbon released and absorbed

Carbonation weathering of limestone

Carbon (rivers)

Carbon (coral)

OCEAN (dissolved carbon)

Carbon (sediments)

**Key**
- Water
- Carbon
- Water/carbon in solution

An important link between the water cycle and the carbon cycle is the ability of water to absorb and transfer carbon dioxide. Figure **1** shows the many links and connections between carbon and water, particularly in the way that carbon dioxide is soluble in water.

**Figure 1** *The relationship between the water cycle and the carbon cycle in the atmosphere*

The water and carbon cycles do not act completely independently within the atmosphere. One of the key connections involves the absorption of carbon in rainwater, which facilitates key processes and affects the magnitude of both stores and transfers.

The acidity of a solution is measured by its pH value. Pure water has a pH of 7.0 (neutral). However, natural unpolluted rainwater is mildly acidic, with a pH of about 5.6. This acidity comes from the natural presence of three substances, carbon dioxide, nitric acid and sulphur dioxide, found in the troposphere (lowest level of the atmosphere). Carbon dioxide is present in the greatest concentration and is therefore the primary source of acidity in unpolluted rainwater. If air is polluted, say with high concentrations of sulphur dioxide from the combustion of fossil fuels, rainwater can become very acidic with a pH of 4 or even lower. In 1982, the pH of a fog on the west coast of the United States was measured at 1.8!

Acidic rainwater affects weathering. On contact with carbonate rocks, such as limestone and chalk, the acidic rainwater (a mild carbonic acid) converts calcium carbonate into calcium bicarbonate, which is soluble. This process is called *carbonation*. The dissolved carbon is then carried away by rivers to the oceans where it is used for shell growth and ultimately buried to form new limestone deposits. Some carbon is transferred directly back to the atmosphere.

## Feedback links to climate change

A **feedback** is a return or 'knock-on' effect that usually leads to a change in the effectiveness of one or more processes. We have seen that positive feedback enhances the outcome driving a system in one direction and leads to instability, whereas a **negative feedback** works against the outcome, leading to stability and a state of equilibrium. It is possible to identify a number of ways in which water and carbon feedbacks are linked to climate change.

### Water cycle feedback loop

Ice reflects radiation from the Sun so less heat is absorbed by the surface. The extent of the Arctic ice has been shrinking more quickly in recent years exposing more water (Figure **2** and 4.13), with subsequently less reflection and more absorption of heat from the Sun. This warms the water and further melts and reduces the ice coverage. This affects the type and magnitude of water transfers between land, ocean and atmosphere. The local and regional impacts of these changes could have profound effects on life on Earth by affecting patterns of precipitation and the availability of freshwater. Political and economic implications are also considerable – for example, no Arctic sea ice will affect trading routes, the exploitation of resources and the development of settlements.

**Figure 2** *Melting Arctic sea ice, Greenland*

## Carbon cycle feedback loop

Warmer temperatures in the Arctic are having two opposite effects on the carbon cycle.

◆ Higher temperatures have increased the growing season for plants and this has increased the absorption of carbon from the atmosphere.

◆ Higher temperatures have, however, started to melt the permafrost, particularly in parts of Siberia. Organic matter (plant roots and animals) trapped in the frozen ground act as an important carbon store. It is estimated that there is more carbon currently stored in the permafrost than exists in the atmosphere. On melting, the organic matter in the permafrost starts to decompose as oxygen is introduced. The bacteria involved in decomposition produce carbon dioxide and methane as a waste product. These gases bubble to the surface and escape to the atmosphere (Figure **3**).

Currently, the Arctic acts as a net carbon store. However, scientists are concerned that if the scale of permafrost melting increases, the balance might tip such that the Arctic becomes a net carbon source. In this scenario, the higher temperatures and melting permafrost will become part of a highly destructive positive feedback.

## Water cycle/carbon cycle feedback loop

Phytoplankton are microscopic plant-like organisms that live in water. In common with terrestrial plants they use the energy of the Sun, together with carbon dioxide (dissolved in the water), to photosynthesise, live and grow. They are the primary producers in aquatic ecosystems sustaining the food web. They are also important stores of carbon.

Marine phytoplankton releases a chemical substance called dimethylsulphide (DMS) that may promote the formation of clouds (condensation) over the oceans. Increases in phytoplankton populations associated with warmer temperatures and more sunshine could therefore lead to an increase in cloudiness and global cooling. This is because clouds reduce the amount of solar radiation reaching the Earth's surface. Of course, less sunshine might lead to a reduction in the amount of phytoplankton, thereby reducing this cooling effect. This complex feedback loop (though not all phytoplankton species react in the same way) is an example of a negative feedback (Figure **4**).

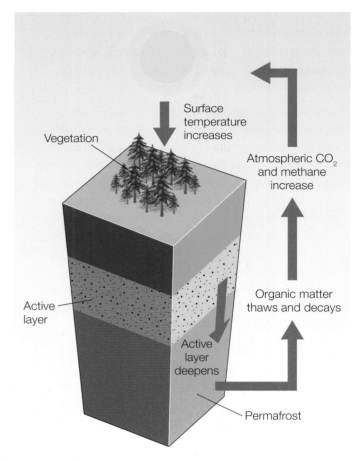

▲ **Figure 3** *Melting permafrost: a carbon cycle positive feedback*

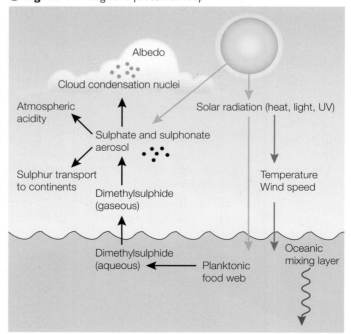

▼ **Figure 4** *A negative feedback loop*

## EXTENSION

### Mountain thermostats – how mountain vegetation controls global temperatures

Research carried out in the Peruvian Andes by scientists from the Universities of Oxford and Sheffield has found that, through an important negative feedback loop, the rates of weathering in mountains caused by vegetation growth play a major part in controlling global temperatures; in effect, acting as a thermostat.

In warmer climates, tree roots grow faster and deeper aided by the decomposition of leaf litter (decomposers are more effective in warm, moist conditions). This enables acidic water to react with carbonates, increasing the rate of weathering.

By sequestrating carbon dioxide from the atmosphere to facilitate weathering, there is a subsequent lowering of global temperatures, decreasing the rate of vegetation growth. This self-regulating process maintains the global temperature balance, keeping the climate relatively stable and maintaining life on Earth despite repeated global climatic events.

To what extent this regulatory process will be able to mitigate recent man-made global warming is an interesting question.

▼ **Figure 5** *The valley in the Southern Peruvian Andes where fine root growth and organic layer thickness were measured over several years.*

## ACTIVITIES

1   Work in pairs to produce a summary table describing, for each of water and carbon, how stores and cycles are vital in supporting life on Earth.

Ⓢ 2   Draw a flow diagram in the form of a cycle to show the links between the water cycle and the carbon cycle in the context of carbonation weathering.

3   Study Figure **3**.
   a   How does this feedback loop operate and why is it a positive feedback?
   b   Assess the regional and global impacts of the Arctic becoming a net carbon source in the future.

4   Study Figure **4**. Consider how global warming may impact on the negative feedback loop. Use a simple diagram to illustrate some of these possible effects.

Ⓢ 5   Attempt to draw a diagram similar to Figure **4**, to show the mountain thermostat negative feedback loop.

## STRETCH YOURSELF

Find out more about why permafrost is melting and how it results in the release of carbon dioxide and methane.

- Are the causes entirely natural or are human activities to blame?
- What evidence is there that this is affecting the climate?
- Can anything be done to reverse the process?

In this section you will learn how human intervention in carbon cycle transfers can mitigate the impacts of climate change

## Which carbon cycle transfers can be modified by human intervention?

### Modifying industrial combustion

Carbon capture and storage (CCS) uses technology to capture carbon dioxide emissions from coal-fired power stations and industry. The gas is then transported it to a site where it can be stored and prevented from entering the atmosphere. Scientists estimate that this could cut global carbon emissions by up to 19 per cent.

Figure **1** shows how carbon capture works. Once captured, the carbon gas is compressed and transported by pipeline to an injection well. It is then injected as a liquid into suitable geological reservoirs, such as underground aquifers and deposits of fossil fuels.

In 2014 Boundary Dam in Canada's Saskatchewan province became the world's first commercial carbon capture coal-fired power plant (Figure **2**). It aims to cut carbon dioxide emissions by 90 per cent by trapping $CO_2$ underground before it reaches the atmosphere. Saskatchewan's state-owned electricity provider expects to reduce greenhouse gas emissions by about 1 million tons a year, the equivalent of 250 000 cars.

CCS is viewed with suspicion by environmental campaigners because its economic viability, so far, depends on using the $CO_2$ to increase oil production (it can be used to force oil out of the ground). Several projects have been initiated, but face long delays and cost overruns.

### Modifying photosynthesis

Trees act as carbon sinks, removing carbon dioxide from the atmosphere through photosynthesis and storing it within their biomass or the soil. Trees also release moisture into the atmosphere and help to moderate the Earth's climates. Plantation forests, which comprise an estimated 7 per cent of the global forest area, are particularly effective in absorbing carbon dioxide compared to natural forests. For some time, this has been recognised by the IPCC as a legitimate option for countries wishing to reduce their carbon emissions.

### Modifying land use change

Farming practices are the most common cause of land-use change, apart from deforestation. *Carbon farming* is where one type of crop is replaced by another that has greater productivity and can absorb more carbon dioxide from the atmosphere.

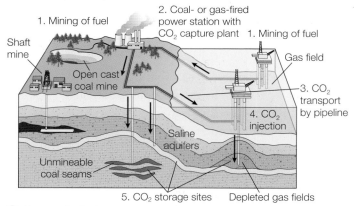

▲ **Figure 1** *Carbon capture and storage*

▲ **Figure 2** *The Boundary Dam CCS power station, Canada*

### Protecting mangroves in Sri Lanka

In 2015 Sri Lanka became the first nation to protect all of its mangrove forests. Having lost an estimated 76 per cent during the last century, the government decided to conserve the remaining forests and seek to expand their coverage.

Mangroves absorb more carbon dioxide than other forests and fix it into the soil where it is stored for hundreds of years. Furthermore, mangrove forests do not burn due to the swampy environment and the lack of fuel.

The project will cost £2.2 million over five years and will protect over 21 000 acres of mangrove forest. A further 9 600 acres will be replanted.

## Modifying deforestation

There are several strategies aimed at reducing the rate of deforestation, which is a major cause of carbon emissions.

◆ Consumers are encouraged to only buy wood certified by the Forestry Stewardship Council (FSC) – timber products that have been grown sustainably.

◆ Countries, organisations and individuals make carbon offset payments to offset their carbon emissions. This might involve paying for existing forests to be protected, developing renewable energy alternatives or planting trees.

◆ In Malaysia, the Selective Management System is a sustainable approach to logging by felling selected trees and planting replacements.

### Government policies in Brazil

Brazil once had the highest rates of deforestation in the world. In 2005, the government embarked on an ambitious plan to drastically reduce rates of deforestation (Figure **3**). This involved requiring landowners to preserve 80 per cent of virgin forest, with infringements punishable by large fines or imprisonment. The government also created protected reserves. By 2009, the rate of deforestation fell by 75% compared with 2004.

In 2018, following the election of Jair Bolsonaro, the rate of deforestation rose to the highest level for a decade. In 2019, over 30,000 forest fires raged across the Amazon due in part to Bolsonaro's stripping back of environmental protection.

**⚠ Figure 3** Cattle grazing on cleared land at the edge of Brazilian rainforest

## Political initiatives: the Paris Agreement

At the Paris climate conference (COP21) in December 2015, 195 countries adopted the first universal legally binding global climate deal, due to be enforced by 2020. The agreement sets out an action plan.

◆ Aim to limit the average global temperature increase to 1.5 °C above pre-industrial levels.

◆ Meet every five years to set more ambitious targets.

◆ Report to each other and the public on the implementation of their individual plans to reduce emissions.

◆ Strengthen the ability to adapt to and be resilient in dealing with the impacts of climate change.

◆ Provide adaptation support for developing countries.

◆ Developed nations will continue to support initiatives in developing countries aimed at reducing emissions and building in resilience to the impacts of climate change.

### ACTIVITIES

1 Study Figure **1**.

**(S)**

    **a** Draw a simplified version of the diagram and add detailed annotations to describe the process of carbon capture.

    **b** Consider the advantages and disadvantages of carbon capture in reducing carbon emissions.

2 Describe the reasons why the Sri Lankan government has decided to protect and extend the coverage of mangrove forests.

3 What is carbon farming and how does it modify carbon transfers?

4 In 2018, the UN Climate Change Conference was held in Katowice, Poland, to ratify the Paris Agreement of 2015. Find out about the key outcomes of the conference. Consider the increasingly significant role of young activists such as Greta Thunberg.

### STRETCH YOURSELF

Find out more about carbon capture and storage. The technology has been around for some time, so why has there been little commercial development? What hurdles need to be overcome and why isn't it popular with environmentalists? Do you think it will ever become a widely accepted alternative?

In this section you will learn about the water cycle in tropical rainforests and how it is affected by human activity and environmental change

## What are the main characteristics of tropical rainforests?

Tropical rainforests are found in a broad belt from Central and South America, through central parts of Africa, south-east Asia and into the northern part of Australia (Figure **1**).

- Annual rainfall 2000+ mm; average 27 °C temperatures throughout the year – ideal for plant growth.
- Home to 200 million people and to about half of the world species of plants and animals.
- A rich biodiversity of plants and animals.
- Plants form distinct layers, with some trees reaching heights of 45 m.
- They absorb huge quantities of $CO_2$ and emit 28 per cent of the world's oxygen.

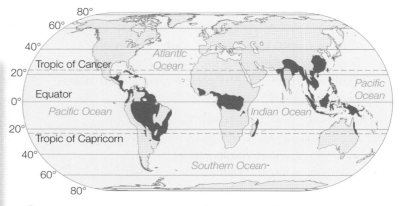

**Figure 1** *Global distribution and characteristics of tropical rainforest*

## The tropical rainforest water cycle

The water cycle operating in a tropical rainforest setting is shown in Figure **2**. Precipitation is very high, mostly exceeding 2000 mm per year. This is due to the high humidity and unstable weather conditions associated with the tropics. Rainfall occurs on most days and often takes the form of torrential downpours.

The dense forest canopy intercepts up to 75 per cent of this rainfall. Some drips to the ground from leaves or flows down tree trunks as stemflow. About 25 per cent of the rainfall is evaporates. Of the remaining 75 per cent, approximately half is used by the plants and eventually returned to the atmosphere by evapotranspiration. The other half infiltrates into the soil – stored temporarily on the ground surface or flowing overland to river channels. In effect this is runoff.

## What is the impact of human activity and environmental change?

Half of the world's rainforests have already been wiped out to make way for commercial farming (plantations and ranching), mining, logging and settlements. Although deforestation appears to be slowing down in some regions, it is still a widespread and serious issue in countries such as Indonesia.

Look at Figure **3**. Notice that the removal of the trees has a massive impact on the rainforest water cycle, almost wiping out some of the components and significantly affecting others.

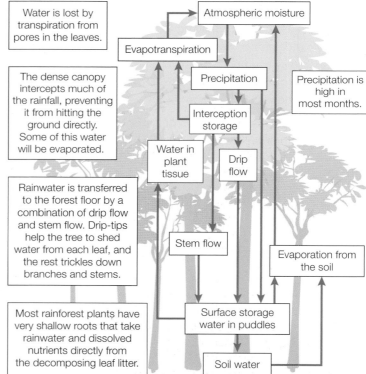

**Figure 2** *The water cycle in a tropical rainforest setting*

Atmosphere becomes less humid as evapotranspiration is reduced.

With few trees, most rainfall reaches the ground immediately, compacting the soil and encouraging overland flow.

Exposed to the sun, the soil will become very dry and vulnerable to erosion.

**Figure 3** *Deforestation in the Amazon Basin, Brazil*

Few trees remain, so very little interception of rainfall or evaporation off leaves. Transpiration will be virtually zero.

Rates of runoff will increase, with an increased risk of flooding.

## Can deforestation affect climate and rainfall patterns?

For several decades scientists have tried to understand the impacts of deforestation on climate and rainfall. Rainforests allow a considerable amount of water to be returned to the atmosphere through evapotranspiration. When forests are replaced by pasture or crops evapotranspiration is typically reduced, leading to reduced atmospheric humidity and suppressing precipitation (Figure **4**).

### ACTIVITIES

**S**

1 Study Figure **2**. Make a large copy of the tropical rainforest water cycle. Use colours and a key to identify the inputs, stores, flows and outputs. Write the percentages given in the text to indicate the relative importance of the flows.

2 Study Figure **3**. Draw a simple diagram with detailed annotations to show the impacts of deforestation on the tropical rainforest water cycle. Use Figure **2** to provide you with a checklist of the stores and flows to consider.

3 Study Figure **4**. Notice that the right side of the diagram shows the impacts of deforestation on the rainforest water cycle.

  a Describe and account for the changes to:
  • evapotranspiration  • runoff
  • soil water transfers.

  b Suggest why the amount of precipitation may be expected to decrease downwind of an area that has been deforested.

  c Should climate scientists be concerned about the impacts of deforestation on global climates?

  d Suggest how sustainable forest management could minimise the impact on the rainforest water cycle.

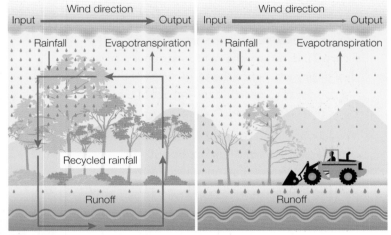

**Figure 4** *Impact of deforestation on the tropical rainforest water cycle*

## Deforestation and rainfall

Some previous studies have shown an *increase* in local rainfall downwind of a deforested area. However, a recent study by the University of Leeds (Spracklen et al., 2012) provided the first observational evidence of a significant effect on remote rainfall (i.e. hundreds to thousands of kilometres downwind).

'We discovered that for more than 60 per cent of the tropical land surface, air that had passed over extensive forest in the preceding few days produced at least twice as much rain as air that had passed over little forest. We estimated that future deforestation of the Amazon rainforest could lead to 20 per cent declines in regional rainfall'
(Spracklen et al., 2012).

www.see.leeds.ac.uk/admissions-and-study/research-degrees/icas/spracklenarnold/

### STRETCH YOURSELF

On an outline map of South America, draw the extent of the Amazon rainforest and the prevailing wind directions. Refer to the internet or an atlas to assist you. Explain the hypothetical impact of widespread deforestation in South America (refer to the University of Leeds research). Consider the social, economic, political and environmental implications. A similar exercise could be undertaken for other rainforests.

In this section you will learn about the carbon cycle in tropical rainforests and how it is affected by human activity and environmental change

### The tropical rainforest carbon cycle

Look at Figure **1**. It shows the carbon cycle operating in a tropical rainforest setting. Notice the following characteristics:

◆ The warm and wet tropical climate is ideal for plant growth. This promotes the process of photosynthesis, which absorbs huge quantities of carbon dioxide from the atmosphere (see Figure **2**). In return, the rainforests emit a great deal of oxygen, hence the term 'lungs of the Earth' – a good way to describe the impact of tropical rainforests on the world's climate.

◆ Wood is about 50 per cent carbon, so rainforest plants and trees are a huge carbon store. They are important 'carbon sinks' in mitigating the effects of global warming.

◆ Respiration by plants, trees and the many animals in the rainforest returns carbon dioxide to the atmosphere.

◆ Decomposition is an active process in tropical climates. Decomposers such as bacteria and fungi thrive in the warm and wet conditions (Figure **3**). This process releases carbon dioxide back to the atmosphere.

◆ Some carbon may also be stored within the soil or dissolved and then removed by streams as an output from the rainforest system.

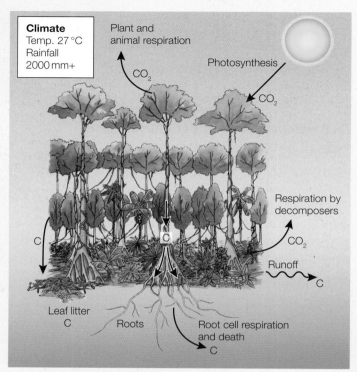

**Climate**
Temp. 27 °C
Rainfall 2000 mm+

Plant and animal respiration

$CO_2$

Photosynthesis

$CO_2$

Respiration by decomposers

$CO_2$

Runoff

C

C

Leaf litter C

Roots

Root cell respiration and death

C

**Figure 1** *The tropical rainforest carbon cycle*

**Figure 2** *Carbon absorption (fixing) resulting from photosynthesis in selected ecosystems*

**Figure 3** *Fungi decomposing leaf litter on the rainforest floor*

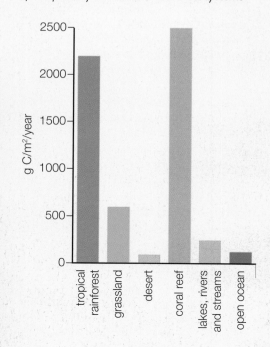

g C/m²/year

2500

2000

1500

1000

500

0

tropical rainforest | grassland | desert | coral reef | lakes, rivers and streams | open ocean

## What is the impact of human activity and environmental change?

Tropical rainforests are cleared to provide land for farming, mining, roads and settlements. One of the most common methods used for clearing is burning. It is cheap and is very effective – often too effective, as fires can burn out of control for weeks and spread, literally, like wild fire!

Look at Figure **4**, which shows an area of burnt rainforest in the Amazon. The blackened stumps and ground surface is effectively carbonised wood. Apart from this burnt debris, the rest of the carbon stored in the trees has been emitted into the atmosphere as carbon dioxide. Deforestation is one of the main contributors to the greenhouse gases that are responsible for climate change.

Consider other effects of deforestation on the carbon cycle stores and flows.

◆ Photosynthesis pretty much ceases, at least until new plants start to colonise the area. This is by far the most significant impact of deforestation on the carbon cycle.

◆ Plant and animal respiration also drops to almost zero.

◆ Rain washes ash into the ground, increasing the carbon content of the soil. Carbon in runoff may increase too.

◆ Decomposers will be largely absent from this environment.

Replacing rainforest with alternative land uses, such as crops and pasture, reintroduces stores and flows, although operating at much less effective levels – compare grassland absorption with tropical rainforests in Figure **2**.

Human activity in the rainforest can take place with minimal impact on the carbon cycle. In parts of Malaysia forests are managed by strict regulation of selective logging followed by replanting, which is both sustainable and has little impact on the carbon cycle.

**Figure 4** *Burnt tropical rainforest in the Amazon Basin, Brazil*

## Impact of deforestation in Indonesia on the carbon cycle

Indonesia's rainforests are one of the most biologically and culturally diverse landscapes in the world. As recently as the 1960s, 80 per cent of Indonesia was rainforest, but since then rapid development has decimated large areas of the natural rainforest (Figure **5**). Demand for paper, pulp, plywood and palm oil, together with allegations of corruption and disputes over land rights, has led to widespread deforestation, endangering species and disrupting the lives of indigenous people. The destruction of the rainforest has had significant impacts on water and carbon cycles, altering the magnitude of important stores and flows. (See 1.11 and 1.15 to find out about the generic impact of deforestation on the water cycle and the carbon cycle.)

Indonesia has one of the highest deforestation rates in the world – just under half of the original forest cover now remains. Estimates vary, but conservative studies suggest that more than a million hectares (2.4 million acres) of Indonesian rainforest is cleared and lost each year, with about 70 per cent occurring in forests on mineral soils and 30 per cent in carbon-rich peatland forests. Once exposed, the peatlands are easily eroded by wind and rain, and increased rates of decomposition of organic matter cause the release of carbon into the atmosphere in the form of carbon dioxide. The rainforests and their soils are no longer a carbon sink, they are now a carbon source, increasing emissions into the atmosphere.

🔺 **Figure 5** *Workers on a palm plantation, Sumatra, Indonesia*

🔻 **Figure 6** *An Indonesian soldier tries to extinguish a forest fire in a peat land forest in Ogan Komering Ilir in South Sumatra, Indonesia.*

One of the major environmental characteristics (and catastrophies) of deforestation has been the prevalence of fires, often very widespread and left to burn out of control. Burning is a very quick way to clear land prior to developing for commercial purposes, such as growing oil palm. Over the last couple of decades, fires have raged in Indonesia, sending black smoke into the atmosphere and releasing huge quantities of carbon that had been stored for hundreds of years. Indonesia is now the world's third largest emitter of greenhouse gases after the USA and China. It is estimated that 85 per cent of its emissions are derived from rainforest and peatland degradation and loss.

In 1997–98, the growth rate of carbon dioxide in the atmosphere suddenly reached the highest on record. Much of this increase was attributed to forest fires in Indonesia that began in 1997 and lasted into 1998. These fires also spewed carbon-containing smoke particles (called black carbon), causing haze and spreading air pollution to nearby countries (Figure **6**).

By the time the forest fires were finally over, more than 8 million hectares of land had been burnt. Scientists believe that the extra carbon these fires pumped into the atmosphere was more than all the carbon that living things on Earth remove from the atmosphere in one year. That's a lot of carbon!

In 2015 over 10,000 fires raged across the Indonesian islands of Sumatra and Kalimantan, caused mostly by illegal slash and burn practices. A strong El Niño event created hot and dry conditions, exacerbating the fire risk as normally wet peatlands dried out. Thick smoke affected the health of thousands of people. Scientists estimate that the fires pumped carbon into the atmosphere equivalent to the UK's total annual emissions.

**Figure 7** *NASA data from Global Forest Watch shows fire activity 10–17 August 2015. Many fires are occurring in primary forest.*

## ACTIVITIES

1 Study Figure **1**. What are the main characteristics of the tropical rainforest carbon cycle?

2 Study Figure **2**.
  a Explain why tropical rainforests play a vital part in absorbing carbon dioxide from the atmosphere.
  b Assess the impact on the carbon cycle of replacing tropical rainforest with grassland for grazing cattle.

3 Draw a diagram based on Figure **1** to show the impacts of deforestation by burning on the stores and flows in the carbon cycle. Add detailed annotations to describe the impact.

4 How can selective logging and replanting minimise the effects of deforestation on the carbon cycle?

5 Study the information about deforestation in Indonesia.
  a Describe the concentration of forest fires in Figure **7**.
  b To what extent do human and physical factors affect the carbon cycle stores and transfers? Consider using a simple sketch to show these effects.
  c Suggest how widespread burning and deforestation in Indonesia is affecting the regional water cycle.

## STRETCH YOURSELF

Consider the impact of forest burning in Indonesia on atmospheric and terrestrial carbon and water stores and transfers. With the aid of simple diagrams, suggest how the carbon and water cycles would be different pre- and post-burning. Compare, for example, the 1960s with the present day.

In this section you will apply your understanding of the water cycle to a river catchment at a local scale

Case study

## Where is the River Exe?

The River Exe flows for 82.7 km from its source in the hills of Exmoor, through Tiverton and Exeter, to the sea at Exmouth on the south coast of Devon. It has an extensive network of tributaries and a high drainage density.

Look at Figure **1**. It shows the upper catchment of the River Exe above a long-term gauging station at Thorverton, eight miles north of Exeter. It is this upper catchment of the River Exe that will form our study area.

## Characteristics of the Exe upper catchment?

◆ Physical – the area of the upper catchment is 601 km². Its maximum elevation of 514 m is in the north. The land is much flatter in the south – the lowest elevation is 26 m.

◆ Geology – an estimated 84.4 per cent of the catchment is underlain by impermeable rocks, predominantly Devonian sandstones, which accounts for the extensive drainage network.

◆ Land use – most of the land is agricultural grassland (67 per cent), with some woodland (15 per cent) and arable farmland. On the high ground of Exmoor, there are moors and peat bogs (3 per cent).

### Water balance

The water balance for the Exe catchment can be expressed as:

precipitation (1295 mm) = evaporation +/− soil water storage (451 mm) + runoff (844 mm)

Rainfall is high, particularly over Exmoor. Much of it is absorbed by the peaty moorland soils. However, if saturated or where drainage ditches have been dug, water can flow off the hills rapidly.

Runoff accounts for some 65 per cent of the water balance, which is quite high compared with other UK rivers. There are two main reasons for this:

◆ The impermeable nature of the most of the bedrock reduces percolation and baseflow.

◆ Drainage ditches on Exmoor reduce the amount of soil water storage.

Figure **3** shows the annual hydrograph for the River Exe using data from the gauging station at Thorverton. It shows that the Exe responds relatively slowly to rainfall events. After flooding, discharge recedes slowly reflecting the rural nature of the lower catchment – 82 per cent of the land use is either woodland or grass.

The pattern of discharge over the course of a year is called the *river regime*.

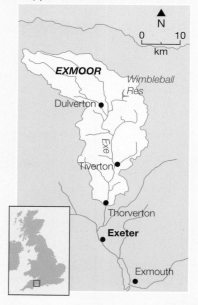

**Figure 1** *The River Exe upper catchment, Devon*

**Figure 2** *River Exe at Dulverton, Exmoor*

**Figure 3** *River Exe: Annual hydrograph 2012–13 (black line); the data also shows the lowest (pink) and highest (purple) flows on each day over the period recorded (1956–2013)*

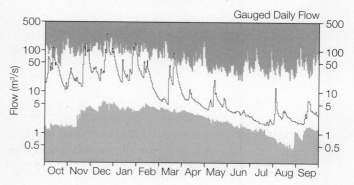

# How have recent developments affected the water cycle?

The water cycle and water balance of the River Exe catchment have been significantly affected by two recent developments.

1 The construction of Wimbleball Reservoir.

2 The restoration of peatland on Exmoor.

## Wimbleball Reservoir

In 1979 the River Haddeo, an upland tributary of the River Exe, was dammed to create Wimbleball Reservoir (Figure **4**). The reservoir has a surface area of 150 hectares (one hectare (ha) is approximately the size of a football pitch).

The reservoir supplies water to Exeter and parts of East Devon. It regulates water flow, ensuring a steady flow regime during the year. This prevents the peaks and troughs of water discharge that make flooding or drought more likely.

Figure **4** shows how Wimbleball has an important role to play in the supply of water to the region, hence the term 'strategic supply area'.

## Peatland restoration on Exmoor

For decades, drainage ditches have been dug in the peat bogs of Exmoor to make it suitable for farming. This has increased the speed of water flow to the Exe, which reduces water quality as more silt is carried downstream. Peat has also been dug as a fuel, leaving behind ugly scars in the landscape. As the peat surface has dried out, decomposition has occurred, releasing carbon from this important carbon store in the form of carbon dioxide and methane.

The Exmoor Mires Project works to restore the peat bogs (mires) by blocking the drainage ditches with peat blocks or moorland bales. This increases water content and returns the ground to the saturated, boggy conditions that would naturally occur in this moorland environment (Figure **5**). These saturated conditions help to retain carbon stored within the peat. (Also see 6.12.)

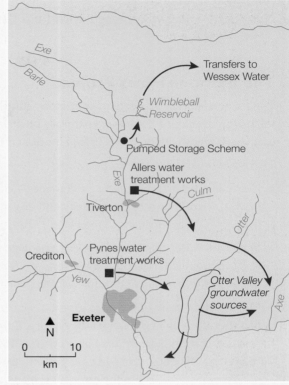

**Figure 4** *Wimbleball Strategic Supply Area (SSA)*

**Figure 5** *Blocking drainage ditches on Exmoor*

## ACTIVITIES

**S** 1 Study Figure **3**. Describe and suggest reasons for the pattern of the annual hydrograph (river regime) of the River Exe.

2 Suggest reasons why runoff accounts for a high proportion of the water budget (65 per cent) for the River Exe catchment.

3 To what extent have the recent developments in the Exe catchment helped to ensure a sustainable water supply?

4 Suggest how each of the following factors affects the water cycle in the River Exe catchment:

a 81.8 per cent of the land use is woodland or grass.

b Only 0.6 per cent of the catchment is urbanised.

c Most of the catchment is underlain by impermeable rocks.

d An extensive network of ditches drains Exmoor's peatlands.

e Wimbleball Reservoir has been built on an upland tributary of the River Exe.

In this section you will engage with field data of a river catchment

**Case study**

## The Exmoor Mires Project

The Exmoor Mires Project aims to restore 2000 ha of Exmoor to the boggy conditions that would naturally be present by blocking drainage ditches with peat blocks and moorland bales.

There are several benefits to this work:

◆ More water storage in upper catchments – water transfer is slowed, increasing storage capacity and ensuring a steady supply of water throughout the year.

◆ Improved water quality – slower throughflow means that less sediment is carried into the rivers. Water is cleaner, less expensive to treat and good for wildlife such as salmon.

◆ More carbon storage – peat is essentially carbon and water, and therefore an important carbon store. Dry peat, however, releases carbon dioxide through oxidation. By encouraging the re-wetting of peat and active peat growth, carbon dioxide is naturally absorbed from the atmosphere and stored.

◆ Improved opportunities for education, leisure and recreation. Peatland habitats are very biodiverse, with many species of plants, birds, butterflies and insects.

◆ Improved grazing and water supply for animals – animals benefit from having year-round drinking water, as well as improved grazing during the drier parts of the year.

⬤ **Figure 1** *Blocked ditches and monitoring sites on Exmoor*

By 2015, over 1000 ha of peat moorland had been restored and nearly 100 km of ditches blocked, raising the water table by 2.65 cm. This has reduced the amount of water that drains from the monitoring area by two-thirds.

A great deal of scientific research has been conducted, including a detailed study of the impact of peat restoration on water tables.

## Data collection

To measure the effects of the project on water tables and water transfers, scientists have established three experimental stations ('pools'), involving the installation of dipwell *transects* (surveys along a line) across newly blocked ditches (Figure **2**).

An electric contact dipmeter records the depth of the water table. This is inserted into the dipwell and when the electrodes make contact with water, a buzzer or light is activated and a depth measurement can be taken.

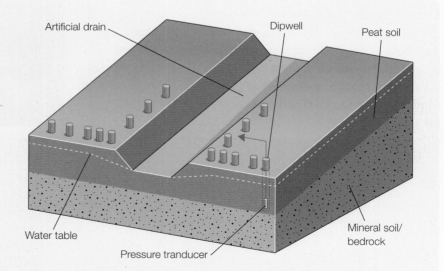

⬤ **Figure 2** *Experimental station with dipwells to measure changes in the water table*

Look at Figure **3**. Notice that the greatest concentration of dipwells is closest to the drainage ditches, where the greatest changes in the water table are likely to take place. This is an example of a stratified sampling strategy, where the sites are deliberately chosen to reflect external factors.

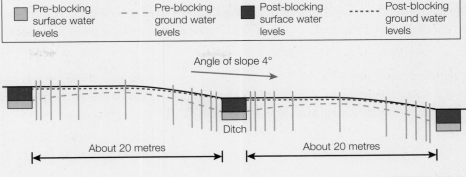

▲ **Figure 3** *Stratified sampling to measure changes in the water table using dipwells*

## Results

The results of monitoring so far indicate that:

◆ Water tables have started to rise, meaning that more moisture is being retained within the soil.

◆ Storm flow and flood peaks have been reduced and baseflow has increased.

Look at Figure **4**, which shows the recorded changes in the water table for one of the experimental sites at Exe Head. Notice the significance of the dates listed.

## Conclusion (South West Water report 2012)

*'The increases in water tables seen reflect an increased storage of water in the peat mass following restoration and it is consistent with similar changes recorded in other re-wetted peatlands in the UK and Ireland.'*

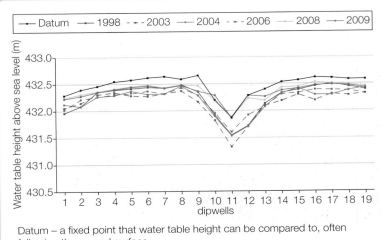

Datum – a fixed point that water table height can be compared to, often following the ground surface
Aug 1998 – dipwell monitoring site established prior to restoration work
Dec 2003 – original ditch blocking
Dec 2005 – ditch infilling and repairs
Mar 2007 – new blocks created and repair of old blocks
2008 – volunteers refurbish dipwells

▲ **Figure 4** *Mean annual water table heights from Exe Head*

## ACTIVITIES

1 With reference to Figures **2** and **3**, describe and explain the sampling strategy used to select dipwell sites.

2 Draw a diagram to show how dipwells enable groundwater levels (water table) to be measured.

3 Study Figure **4**. The 1998 line is the pre-restoration situation.

   a Describe in detail the overall pattern of the water table along the transect.

   b What is the main factor affecting the pattern?

   c How did the hot, dry summer of 2003 affect the water table at Exe Head? Support your answer by using comparative statistics, for example percentage differences between different years.

   d Comment on the success of each phase of restoration in raising the water table. Use figures from the vertical axis to support your comments.

4 Do you think the conclusion is justified?

5 Assess the implications of the conclusion for the water cycle and water balance in the Exe catchment.

6 What have been the economic, social and environmental benefits of the restoration of Exmoor's peatlands?

## STRETCH YOURSELF

Use nrfa.ceh.ac.uk to investigate the water balance and characteristics of a river in your area. Use 'Search data' to find your river. Make a comparison with the River Exe.

# Practice questions

The following are sample practice questions for the Water and carbon cycles chapter. They have been written to reflect the assessment objectives of the Component 1: Physical geography Section A of your A Level.

These questions will test your ability to:

◆ demonstrate knowledge and understanding of places, environments, concepts, processes, interactions and change, at a variety of scales [AO1]

◆ apply knowledge and understanding in different contexts to interpret, analyse and evaluate geographical information and issues [AO2]

◆ use a variety of quantitative, qualitative and fieldwork skills [AO3].

---

**1** Explain how the relative magnitude of global water stores varies over time. (4)

**2** Study Figure **1**, a flood hydrograph presenting the responses of the River Leen and the nearby Dover Beck to rainfall received in the local area on the 28th March 2016. Analyse the data shown. (6)

**Tip**

Flood hydrographs have two y-axes. Ensure you read from the correct axis when quoting data.

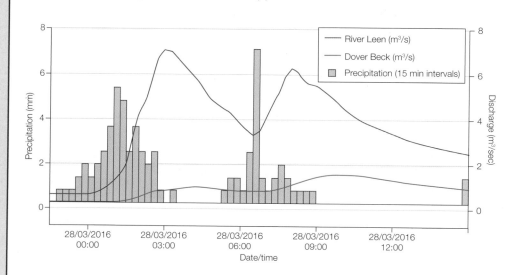

◀ **Figure 1** Flood hydrograph for River Leen and Dover Beck, Nottinghamshire, 28 March 2016

**3** Study Figures **1** (above), **2** and **3**. Using Figures **1–3** and your own knowledge, assess the factors that influence the varying response of rivers to rainfall events. (6)

▲ **Figure 2** River Leen gauging station, Triumph Road, City of Nottingham

▲ **Figure 3** Dover Beck gauging station, Lowdham, Nottinghamshire

**4** With reference to a river catchment at a local scale, assess the influence of human activities on the water balance. (20)

5  Explain the concept of dynamic equilibrium in relation to the carbon cycle at the scale of a lithosere. (4)

6  Study Figures **1** and **2** on page 38. Analyse the data shown. (6)

7  Study Figure **4** on page 36. Using Figure **4** and your own knowledge, evaluate human interventions in the carbon cycle designed to mitigate the impacts of climate change. (6)

8  Critically assess the role of natural and human factors in driving change in major stores of carbon over time. (20)

9  Outline the processes by which natural variation in the carbon cycle occurs within a single year. (4)

10  Study Figure **5** on page 37 and Figure **2** on page 48. Analyse the data shown. (6)

11  Study Figure **6**. Using Figure **6** and your own knowledge evaluate the impact of feedbacks between the water cycle and carbon cycles on the pace of climate change. (6)

Scientists have made a surprising discovery about the glaciers in Canada's Arctic. For decades, the rivers have been absorbing carbon dioxide from the atmosphere faster than the Amazon rainforest does.

The findings challenge the accepted understanding of rivers as sources of carbon emissions. It is true that temperate rivers transport a huge amount of organic material; when these organisms decompose they release a large amount of carbon dioxide into the atmosphere – much more than is absorbed. However, glacial rivers cannot host as much life, in part because of the amount of silt they contain; this means far less decomposition and minimal carbon output.

In addition, the silicate and carbonate sediments from the glaciers get thrown around in the meltwaters, leading to chemical weathering, which also removes carbon dioxide from the atmosphere. The scientists discovered this happening as far as 26 miles (42km) from the source of the river. This means that when the glaciers are melting most, glacial river water could absorb around 40 times as much carbon as the Amazon rainforest itself.

However, scientists warn that we must not rely too heavily on glaciers to absorb excess carbon dioxide, as they are a swiftly dwindling natural resource.

⬆ *Figure 6  Meltwater rivers absorb more carbon dioxide than the Amazon*

12  Assess the value of a taking a systems approach to understanding the carbon cycle on Earth, given the international scientific consensus about the disruption caused by human activities. (20)

*The Atacama Desert bright with colour - what do you think caused one of the driest places on Earth to bloom with flowers (which happens once every five to seven years)?*

## Specification key ideas

## Your exam

You must answer one question in Section B of Paper 1: Physical geography written exam, from a choice of three: Hot desert systems and landscapes or Coastal systems and landscapes or Glacial systems and landscapes. Paper 1 makes up 40% of your A Level.

## Your key skills in this chapter

In order to become a good geographer you need to develop key geographical skills. In this chapter, whenever you see the skills icon you will practise a range of quantitative and relevant qualitative skills, within the theme of 'Hot desert systems and landscapes'. Examples of these skills are:

◆ Interpreting photographs and remotely-sensed images  2.9, 2.10, 2.11

◆ Using atlases and other map sources  2.13

◆ Drawing and annotating sketch maps  2.16

◆ Measurement and geospatial mapping  2.4

◆ Presenting field data and interpreting graphs  2.2, 2.4, 2.12

◆ Analysing data, including applying statistical skills to field measurements  2.16

◆ Drawing and annotating diagrams of physical systems  2.1, 2.5, 2.6

## Fieldwork opportunities

Should you be fortunate enough to visit a hot desert, for example in Morocco, there are many opportunities for fieldwork.

◆ Sand dunes – you could make a study of sand dunes, considering variations in sand dune profiles and investigating causal links with the prevailing wind direction. You might investigate sand dune asymmetry as suggested by the research described in section 2.16.

◆ Role of wind and water – you could investigate the contrasting roles of wind and water in an area of desert, looking for evidence of wind and water action. This might involve studying the desert floor to look for signs of water transport or wind action (such as ventifacts). Sediment analysis techniques could be employed to investigate whether sediment is water carried or wind-blown. Wind-blown sediment (loess) will be better sorted and more spherical than water carried sediment. The presence of desert landforms will also help to support your investigation.

◆ Water cycle – aspects of the water cycle can be investigated, for example rates of infiltration within a drainage basin. The study of sediment in a river channel or on an alluvial fan can give an indication of water velocities using the Hjulstrom Curve.

◆ Desertification – evidence for desertification can be investigated, making use of secondary data (maps etc.) and interviews with local people. It may be possible to study an area that has been restored enabling comparisons to be made with an area currently suffering from desertification.

◆ Sustainable management – you may be able to study a sustainable management project, making use of primary and secondary data (including interviews with local people) to investigate the impacts of sustainable management on an area of desert.

◆ The specification does allow you to study aspects of aeolian processes in the UK in coastal sand dune environments, so you may consider studying sand dune profiles in exposed situations. It would be interesting to investigate the role of the winds (prevailing and onshore) in shaping sand dunes.

In this section you will learn about hot deserts, their distribution and how they function as a physical system

## What is a hot desert?

A *hot desert* is commonly defined as a large, dry, barren region with little vegetation cover and very high temperatures (Figure **1**). Dry polar regions are sometimes referred to as *cold deserts*. The key characteristic of any desert is its lack of precipitation or its aridity. It is commonly accepted that deserts have less than 250 mm of precipitation a year, most of it in short, intense storms. A semi-arid region (semi-desert) experiences annual precipitation of 250–500 mm.

▲ **Figure 1** *Desert landscape, Jordan*

## Where are hot deserts located?

Figure **2** shows the location of hot deserts. Aridity gradually becomes less extreme further from the centre of the desert. Around the edges of hot deserts are less extreme semi-arid regions.

Notice that many of the world's hot deserts are found close to the Tropics of Cancer and Capricorn. Some are also found in continental interiors (e.g. Africa and Australia) or adjacent to mountain ranges (e.g. the Andes in South America). Hot deserts cover 25–30 per cent of the Earth's land surface.

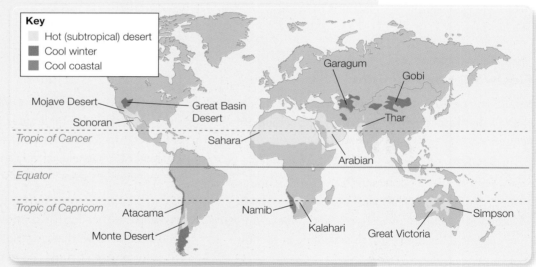

**Key**
- ▢ Hot (subtropical) desert
- ▢ Cool winter
- ▢ Cool coastal

Garagum
Gobi
Mojave Desert
Great Basin Desert
Sonoran
Thar
*Tropic of Cancer*
Sahara
Arabian
*Equator*
*Tropic of Capricorn*
Atacama
Namib
Simpson
Monte Desert
Kalahari
Great Victoria

▶ **Figure 2**
*Distribution of the world's deserts*

## Desert landforms and landscapes

What is the difference between a *landform* and a *landscape*? It's really a matter of scale. An individual landform may extend for several hundred metres, a desert landscape can cover thousands of square kilometres.

Look at Figures **1** and **4**. Notice that there are several different landforms that collectively form the overall landscape. Many of these are linked, both physically and also by the processes that formed them. So it is important to look at the context of a landform when studying its characteristics and formation.

There are three main types of desert landscape:

- ◆ Hamada – bare rocky surfaces such as plateaux
- ◆ Reg – a stony desert where rock fragments are scattered over a large plain
- ◆ Erg – predominantly sand (sometimes called a 'sand sea').

Desert landscapes result from complex interactions between geological structures and the processes of weathering, mass movement, wind and water. Landforms have developed over long periods of time, during which the landscape-forming processes have changed.

# What is the desert landscape system?

Look back to 1.1 to remind you of the systems concept and the key terms. The systems concept helps you to understand the interrelationships and connections between processes and landforms.

A desert system is an open system. There are inputs, such as rainfall and temperature, and outputs, such as water and sediment (Figures **3** and **4**).

**Figure 3** *Elements of a hot desert landscape system.*

| Systems term | Hot desert |
|---|---|
| Input | Precipitation, solar radiation, descending air at the ITCZ |
| Output | Runoff, reradiation of longwave radiation from the Earth's surface into the atmosphere, evaporation |
| Energy | Latent heat associated with changes in the state of water<br>Energy associated with flowing water and moving air |
| Stores/components | Playas (salt lakes), sand dunes |
| Flows/transfers | Wind-blown sand, surface runoff, salinisation, sediment transfer |
| Positive feedback loop | The presence or absence of vegetation can affect the regional climate. If vegetation is removed (overgrazing or deforestation) this will reduce the moisture emitted into the atmosphere. Reduced humidity may lead to less rainfall, which then reduces the extent of vegetation further. |
| Negative feedback | Intense weathering of a slope leads to a build-up of an apron of scree against the mountainside. Without removal (erosion) this apron extends up the mountainside, protecting the lower slopes from weathering. |
| Dynamic equilibrium | Seasonal cycles of winds in some desert regions can lead to small-scale and short-term adjustments in sand dune profiles, but from year to year their shape remains broadly the same. |

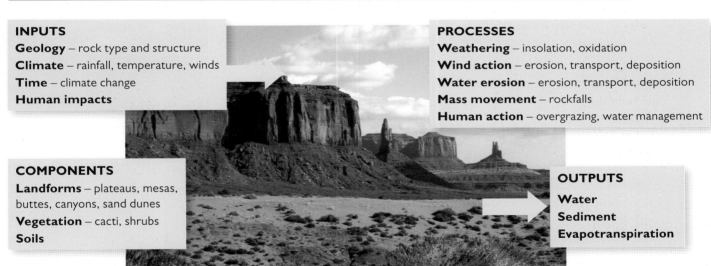

**INPUTS**
**Geology** – rock type and structure
**Climate** – rainfall, temperature, winds
**Time** – climate change
**Human impacts**

**PROCESSES**
**Weathering** – insolation, oxidation
**Wind action** – erosion, transport, deposition
**Water erosion** – erosion, transport, deposition
**Mass movement** – rockfalls
**Human action** – overgrazing, water management

**COMPONENTS**
**Landforms** – plateaus, mesas, buttes, canyons, sand dunes
**Vegetation** – cacti, shrubs
**Soils**

**OUTPUTS**
**Water**
**Sediment**
**Evapotranspiration**

**Figure 4** *The hot desert system, Monument Valley, Utah*

## ACTIVITIES

1 Study Figure **1**. Describe the main characteristics of this hot desert area. What type of desert landscape is it?

2 Study Figure **2**. Describe the distribution of hot deserts.

3 What is the difference between a desert landform and a desert landscape?

4 Work in pairs or small groups to draw a systems diagram (see Figure **3**, 1.1) to represent the hot desert system. Use boxes to represent the components (including inputs and outputs) and arrows to represent the links/transfers (processes). Use the internet to find photos to illustrate your diagram. (This activity could be repeated as a revision exercise at the end of this chapter.)

## STRETCH YOURSELF

Find out about the *badlands* in the USA. What are their characteristics? Suggest how they form a positive feedback loop. Is anything being done to arrest this loop and establish a negative feedback loop? What are the chances of the area reaching a dynamic equilibrium? You may find www.britannica.com/science/badland useful. You could also search for 'Badlands National Park' on YouTube.

In this section you will learn about characteristics such as climate, soils and vegetation of hot desert environments and their margins

## Desert climates

Figure **1** shows climatic conditions that are representative of the vast Saharan desert that encompasses most of Algeria and much of the territory of neighbouring countries. Notice the almost complete lack of rainfall and the extremely high average temperatures in the summer.

Deserts exhibit a number of other climatic characteristics that are not evident on a climate graph:

◆ Many hours of sunshine because of the absence of clouds. Deserts are located in zones of high atmospheric pressure, where the air is sinking and becoming warm and dry.

◆ The lack of cloud cover results in large diurnal (daily) temperature variations of up to 30 °C. In the winter, frosts occur at night and it can even snow!

◆ Strong winds and sandstorms, especially in areas where there are significant variations in temperature.

◆ Thunderstorms can be triggered by intense convective activity.

◆ Some coastal deserts, for example the Atacama in South America, can have fog rolling in off the sea. An almost constant drizzle during the winter affects the Andean hillsides in Peru, yet this only produces less than 50 mm of precipitation on the ground.

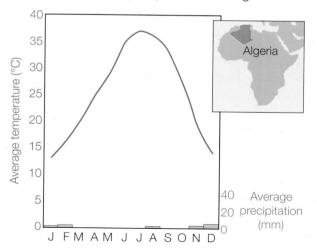

▲ **Figure 1** Climate graph for In Salah, Algeria

## Desert soils

The climate is too dry to support extensive vegetation so soils tend to be poorly developed. When the scarce vegetation dies, the lack of water prevents it rotting to produce humus to enrich a developing soil.

Soils therefore tend to be very dry and sandy. They are commonly referred to as *red desert soils*. The red colour, typical of many hot deserts, is the result of the weathering process of oxidation, where oxygen reacts chemically with iron-rich rocks. They are highly porous and permeable with poor moisture retention. However, the lack of rainfall means that *leaching* (removal of nutrients by water) will be minimal. So, with appropriate irrigation and soil management, crops can be successfully grown in the desert (Figure **2**).

▶ **Figure 2** Pivot irrigation in the Saudi Arabian desert

## Desert vegetation

Deserts are hostile environments for plants. The low rainfall and high temperatures and rates of evaporation create extremely challenging environmental conditions requiring special adaptations. There are three main adaptations:

◆ Drought avoidance – some plants avoid drought by only living for one season. They dry up and die during drought but store moisture, oil, fat, sugar and protein in seeds. Other plants use a long taproot to seek secure water supplies deep underground. Mesquite trees have a taproot that can be up to 20 m in length!

◆ Drought resistance – to resist drought, some plants are able to adjust their metabolism, effectively becoming dormant. Others drop their leaves to conserve moisture.

◆ Water storage – succulent plants , such as cacti and agave, have shallow roots so that they can quickly absorb water during a storm. Water is then stored in fleshy leaves, stems and roots.

Plants that have adapted to living in desert environments are called *xerophytes* (Figure **3**). In some deserts, water may exist close to the surface, resulting in a dense pocket of vegetation called an *oasis*. Oases may form as a result of water being trapped near the surface by an impermeable layer or as a result of shallow underground drainage. Lush vegetation growth may occur, including trees such as date palms, forming important habitats for birds and animals.

It is possible to identify interactions between climate, soil and vegetation within a desert. Climate clearly affects the species of vegetation that can survive, as well as the development of the soils. The soil affects drainage, which, in turn, affects plants. Plants can also have an impact on soil development and can create microclimates (Figure **4**).

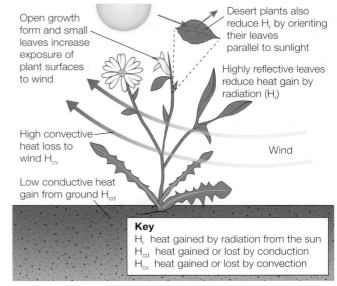

Open growth form and small leaves increase exposure of plant surfaces to wind

Desert plants also reduce $H_r$ by orienting their leaves parallel to sunlight

Highly reflective leaves reduce heat gain by radiation ($H_r$)

High convective heat loss to wind $H_{cv}$

Wind

Low conductive heat gain from ground $H_{cd}$

**Key**
$H_r$ heat gained by radiation from the sun
$H_{cd}$ heat gained or lost by conduction
$H_{cv}$ heat gained or lost by convection

**Figure 3** *Plant adaptations to living in hot deserts*

**Figure 4** *Plant interactions between climate, soil and vegetation in a hot desert*

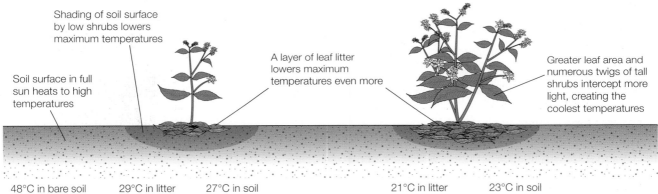

Shading of soil surface by low shrubs lowers maximum temperatures

A layer of leaf litter lowers maximum temperatures even more

Soil surface in full sun heats to high temperatures

Greater leaf area and numerous twigs of tall shrubs intercept more light, creating the coolest temperatures

48°C in bare soil away from shrubs

29°C in litter under low shrub

27°C in soil under low shrub

21°C in litter under tall shrub

23°C in soil under tall shrub

## ACTIVITIES

1 Make a copy of Figure **4**. Add labels from Figure **3** to show how the plant copes with extreme heat. Add arrows and labels to show the interactions between climate, soils and vegetation.

**S** 2 Study the climate data below for Hall's Creek – a semi-arid desert environment.
   a Draw a climate graph from the data.
   b How does the total annual rainfall compare to the definition of a semiarid region?
   c Which months experience the highest temperatures?
   d Why do deserts experience extremes of temperature?
   e Describe and suggest reasons for the pattern of rainfall.

|  | J | F | M | A | M | J | J | A | S | O | N | D |
|---|---|---|---|---|---|---|---|---|---|---|---|---|
| Temperature (°C) | 30 | 29 | 28 | 26 | 21 | 19 | 18 | 21 | 24 | 28 | 31 | 31 |
| Precipitation (mm) | 137 | 107 | 71 | 13 | 5 | 5 | 5 | 2 | 2 | 13 | 35 | 79 |

(Source: DC Money 'Climate, soils and vegetation' UTP)

## STRETCH YOURSELF

Study two or three different plants from around the world that have adapted in different ways to desert environments. Find a photo for each and add annotations to show how they cope with the hostile conditions. Try to indicate some interactions between climate, soil and vegetation.

In this section you will learn about the causes of aridity by studying the roles of atmospheric circulation, continentality, relief and ocean currents

## What causes aridity and the formation of deserts?

Most of the world's hot deserts are found in a broad, but discontinuous belt between 20 and 25 degrees latitude, but some do extend a few degrees north and south of this (Figure **2**, 2.1). These deserts include the Sahara in North Africa, the Arabian Desert in the Middle East, the Kalahari Desert in southern Africa and the Great Victoria Desert in Australia.

There are several factors that account for the distribution of hot deserts, including atmospheric processes, continentality, relief and ocean currents.

### Atmospheric processes

The main factor affecting the distribution of hot deserts is the **global atmospheric circulation system** (Figure **1**). Air rising in the Intertropical Convergence Zone (ITCZ) diverges and spreads both north and south to two distinct tropical circulation cells – the Hadley cells. The air moving poleward cools and converges with the equatorward limb of the Ferrel cells, causing it to sink towards the ground. As the air sinks, it warms and becomes drier, establishing a broad belt of high pressure (anticyclone) roughly 30 degrees north and south of the Equator – the subtropical high. Sinking air associated with anticyclones results in cloudless skies and the subsequent lack of rainfall associated with deserts. The lack of cloud accounts for the high sunshine totals and the extremes of temperature. High rates of evaporation are common under these conditions, which further increases the levels of aridity on the ground.

### Continentality

**Continentality** is the way that a large land mass affects weather and climate. Aridity tends to be higher in continental interiors, such as in central Africa, Asia and Australia. This is because the influence of moist airstreams from the oceans is reduced. Look at Figure **2** in section 2.1 and consider, for example, the location of the Great Victoria Desert in Australia. Here, where distance from the subtropical high pressure area means it has less of an influence, continentality plays a significant role. This type of desert tends to be semi-arid rather than arid, as occasional storms bring welcome rain.

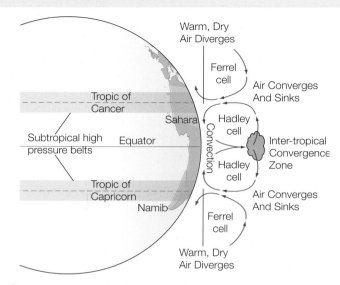

▲ **Figure 1** The role of atmospheric processes in forming deserts

### The importance of latent heat

When water is evaporated it changes its state from a liquid to a gas. This needs energy called *latent heat* to be extracted from the environment, causing a drop in temperature. This is why it feels cool after a rain shower in the summer. The opposite occurs during condensation, when latent heat is released into the air. This heating promotes further rising of the air and cloud formation.

Latent heat is partly responsible for hot deserts experiencing a larger diurnal (daily) range in temperature than moist climates. In a dry climate, evaporational cooling and the release of latent heat is at a minimum during the daytime, resulting in very high temperatures. During the night there is little water vapour to trap longwave radiation, which results in rapid cooling.

On a more global scale:

◆ As it rises at the ITCZ, air gains heat from the release of latent heat as water vapour condenses.

◆ It then rains torrentially.

◆ As the air sinks at the tropics it does not lose heat because the water that condensed has fallen as rain.

◆ Air is warmed by compression and ends up hotter than it was when it started to rise at the ITCZ.

## Relief

The physical landscape or *relief* of an area can have an impact on the formation of deserts. Look at Figure **2**. Notice that the prevailing or dominant wind direction is from the south-east. Moist air from the Atlantic Ocean passes over Brazil towards the Andes Mountains. Here it is forced to rise causing it to cool and condense to form clouds and rain. Notice that most of the rain falls on the windward slopes. When the air descends on the leeward (sheltered) side, it becomes warmer and drier. This area is said to be in a *rainshadow*. You can see how this contributes to the aridity of the Atacama Desert in South America.

## Cold ocean currents

Some deserts occur on coastlines adjacent to cold ocean currents, for example, the Atacama Desert in South America. The air that is in contact with these cold ocean currents is cooled. Cold air, being denser than warm air, tends to sink and remain close to the ground. Aridity is increased as rain formation is suppressed by the air's tendency to sink. Despite the fog and drizzle often associated with the cold and moist sea air (Figure **3**), the amount of overall precipitation is small.

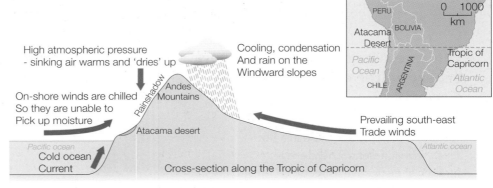

▲ **Figure 2** *The causes of aridity in the Atacama Desert, South America*

### The formation of the Atacama Desert

Several factors often combine to account for the aridity experienced in deserts. The Atacama Desert in South America results from a combination of the global atmospheric circulation system (lying close to the Tropic of Capricorn), the effect of a cold ocean current (the Peruvian current) and being on the leeward side of the prevailing south-east trade winds (Figure **2**).

▲ **Figure 3** *A fogbound Atacama Desert, South America*

## ACTIVITIES

1  Study Figure **1**.

   **a** With the aid of a simple diagram explain why air subsides at the Tropics to form the subtropical high-pressure zone.

   **b** Give a reasoned description of the typical weather conditions experienced under anticyclonic conditions in this zone.

   **c** Why does the global atmospheric circulation system lead to aridity in the Tropics?

   **d** Suggest reasons why some parts of this region are not deserts.

2  Study Figure **2**.

   **a** With the aid of a simple diagram, describe how the rainshadow effect contributes towards aridity in the Atacama Desert.

   **b** What other factors account for the formation of the Atacama Desert?

## STRETCH YOURSELF

Make a case study of one of the following deserts to discover more about its formation: Namib Desert, Kalahari Desert, or the Great Basin, California. For your chosen desert find out why it is an arid region. Use diagrams to support your answer and explain the processes involved.

In this section you will learn about the water balance in hot deserts and the aridity index

## What makes the hot desert water cycle distinctive?

Look at Figure **1**. It shows the operation of the water cycle in a hot desert environment. There are a number of distinctive aspects:

◆ Rainfall is low and sporadic. It mostly occurs as heavy downpours that trigger overland flow, resulting in high levels of evaporation.

◆ While the actual evapotranspiration (ET) is low, reflecting the low annual precipitation, there are very high rates of *potential evapotranspiration* (PET). This is the amount of water that would be evaporated and transpired if there was sufficient water available. Rates of PET can be as high as ten times the annual precipitation.

◆ Water from aquifers can reach the ground naturally through springs or be abstracted using wells. Fossil aquifers may exist as a legacy of a previously much wetter climate.

◆ Rivers in deserts have distinctive sources: exogenous, endoreic and ephemeral (see 2.9).

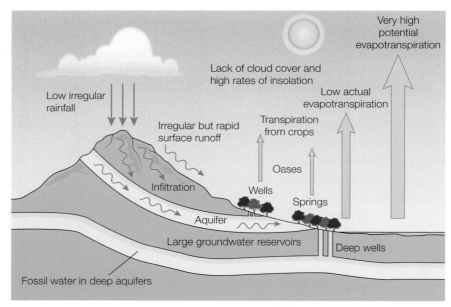

▲ *Figure 1*  The desert water cycle

## What is the aridity index?

The UNESCO **aridity index** is one of the most common measures of aridity. It is the ratio between mean annual precipitation (P) and mean annual potential evapotranspiration (PET).

Arid regions have a P/PET ratio of less than 0.20. This means that rainfall supplies less than 20 per cent of the amount of water needed to support optimum plant growth. Semi-arid regions have an aridity index of 0.2–0.5 (20–50 per cent).

Look at Figure **2**. Most of Africa is classified as dry subhumid, semiarid or arid. It is in these areas that 40 per cent of Africa's population live, which makes them susceptible to climate change and desertification.

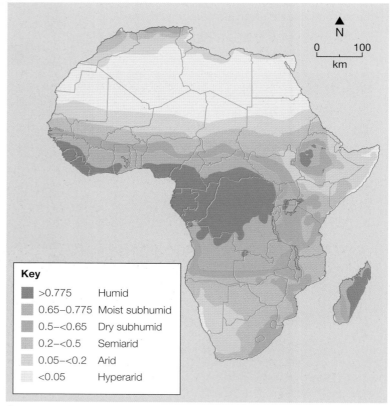

**Key**

| | |
|---|---|
| >0.775 | Humid |
| 0.65–0.775 | Moist subhumid |
| 0.5–<0.65 | Dry subhumid |
| 0.2–<0.5 | Semiarid |
| 0.05–<0.2 | Arid |
| <0.05 | Hyperarid |

▲ *Figure 2*  Patterns of aridity in Africa

# What is the hot desert water balance?

The water balance shows the relationship between precipitation, soil moisture storage, evapotranspiration and runoff. It can be expressed as:

> precipitation = evaporation +/− soil water storage + runoff

In a hot desert, the water balance has important implications for food supply and the need for irrigation. This is primarily due to the very high rates of PET and the strain this puts on water supply and plant growth.

When PET exceeds precipitation, as it does in many hot deserts, there is no surplus water (runoff) for agriculture, industry or domestic consumption. The knock-on effect of this *water deficit* is that water has to be transferred from elsewhere or abstracted from fossil aquifers, which may be unsustainable.

Look at Figure **3**. It shows the water balance graph for Baghdad, Iraq. Baghdad is inland and experiences an average annual rainfall of just 125 mm. Baghdad is situated in a typical hot desert environment (Figure **4**).

## ACTIVITIES

1  Study Figure **1**. What are the distinctive characteristics of the hot desert water cycle?

2  What is potential evapotranspiration and why is it such an important aspect of the hot desert water balance?

**S**

3  Study Figure **2** and refer to an atlas map of Africa.

   a  Attempt to draw an aridity profile from the north coast of Algeria through West Africa to the Nigerian coast. Assume an equal vertical scale through the different desert catagories. Your horizontal scale is the distance across West Africa. Label the Sahara Desert and identify the different countries.

   b  Use your profile and Figure **2** to describe the pattern of aridity in West Africa.

4  Using evidence from Figures **3** and **4** outline the water supply challenges facing Baghdad.

## STRETCH YOURSELF

Use the internet to research aridity values for a selection of deserts. The Atacama Desert has a value of 0.05, making it a 'hyperarid' desert. See if you can categorise other world deserts.

### Baghdad

Low rainfall and high summer temperatures result in a high rate of potential evapotranspiration and a significant water deficit. Irrigation is needed for most of the year to supply the farms around the city. Guaranteeing a reliable water supply is a major issue.

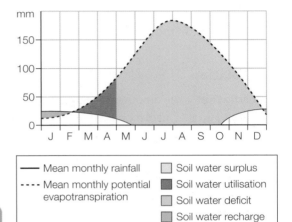

— Mean monthly rainfall
- - - - Mean monthly potential evapotranspiration

☐ Soil water surplus
☐ Soil water utilisation
☐ Soil water deficit
☐ Soil water recharge

⬆ **Figure 3**  *The water balance graph for Baghdad, Iraq*

⬆ **Figure 4**  *Baghdad – Iraq's desert city*

In this section you will learn about the sources of energy and sediment in hot desert environments

## What are the sources of energy in hot deserts?

Energy is required to develop landscapes and enable life to exist. The fundamental source of energy in desert regions is insolation from the sun. Insolation drives the atmospheric circulation which, in turn, determines surface pressure belts. Air moving between high and low pressure belts on the ground surface create the winds that blow across the deserts, shaping sand dunes and abrading rock outcrops. Precipitation, triggered by the atmospheric circulation as well as localised convective activity (caused by insolation), feeds the rivers that flow across the land, creating narrow corridors of life in the vast expanses of desert.

### Insolation

**Insolation** is the amount of heat (shortwave radiation) that reaches the Earth's surface. Insolation is generally high in desert regions due to the lack of cloud cover that would otherwise absorb, scatter or reflect the sun's radiation. As a direct source of energy, it is important in a variety of ways:

◆ The nature and extent of weathering is significantly determined by insolation. For example, extremes of hot and cold, typically experienced in deserts on account of the lack of cloud cover, can lead to intense insolation weathering where the outer 'skin' of rocks can break away (see 2.6).

◆ The hot desert air is responsible for very high rates of potential evapotranspiration, one of the key features of the desert water balance.

◆ Variable warming of the desert surface can lead to localised air movement (winds), which can shape desert landforms.

▲ **Figure 1** Wadi Rum, Jordan

### Water

Although water is scarce in deserts, it is a very powerful force for landscape development. Following an infrequent heavy storm, water will either flow as a sheet across the landscape or in gullies and dry valleys called wadis (Figure **1**). The power of this water is immense and rates of erosion can be very significant in a short period of time.

A few rivers, such as the Nile in Egypt and the Euphrates in Iraq, have their sources outside the deserts – such rivers are referred to as exogenous (see 2.9). Fed by water from areas with wetter climates, they are able to flow continuously across deserts, transporting sediment and creating fluvial landforms. Meltwater-fed rivers are important sources of water and energy in some desert regions. For example, water from the High Atlas in Morocco supports settlements on the fringe of the Sahara Desert.

### Wind

Wind, driven by the atmospheric circulation system, is an important secondary source of energy in hot deserts. About 20–25 per cent of desert surfaces are composed of sand and smaller-sized sediment, which can be moulded or transported by the wind. Most deserts are affected by strong seasonal trade winds or local winds generated by convective activity.

The Harmattan wind is a dry and dusty hot wind that blows from the Sahara Desert over the West African subcontinent during the November–April dry season. It is often associated with powerful dust storms (Figure **2**). The Khamsin is a Saharan wind that blows hot air northwards along the Nile Valley towards Cairo during the spring.

# Sediment sources, sediment cells and sediment budgets in hot deserts

## Sediment sources

Look at Figure **2**. There are several sources of sediment in hot deserts. Some are internal (blue) and others may be external (red).

**Figure 2** *Sediment sources in hot deserts (Monument Valley, Arizona, USA)*

Mass movement (rock falls)

Sediment washed onto the desert plain via gullies and wadis

Sand and finer sediments carried into the desert by winds or washed in by rivers from nearby mountains

Weathering of exposed rocks (exfoliation, chemical weathering)

Mass movement (soil creep, talus creep)

Erosion of rocks exposed on the flat desert plain

## Sediment cells

The movement of sediment can be described as a **sediment cell**, a concept that is commonly applied to coastal environments (see 3.3). A sediment cell has the following components:

◆ *inputs* – the sources of sediment (Figure **2**)

◆ *transfers* – the movement of sediment by wind and water

◆ *sediment sinks* – areas of deposition (e.g. alluvial fans, scree slopes, valley floors)

◆ *outputs* – sediment may be transported away from the desert by wind or water.

## Sediment budgets

A **sediment budget** considers the relative amounts of sediment in each part or component of the hot desert system. Few studies have been carried out as years of observation are required and field conditions are difficult.

In 1966, Leopold, Emmett and Myrick completed a seven-year study of erosion, transportation and deposition in the desert near Santa Fe, New Mexico, USA. They calculated a sediment budget for the desert (Figure **3**).

The results suggested that surface erosion by water (slopewash) was the main process of erosion. Only 22 per cent of the total eroded material was deposited.

| | Estimated average rates (tonnes/km²/yr) | % of total erosion |
|---|---|---|
| **Total erosion** | **5452.9** | **100** |
| Surface erosion (slopewash) | 5335.2 | 97.8 |
| Gully erosion | 78.5 | 1.4 |
| Mass movement | 38.4 | 0.7 |
| | | |
| **Total deposition** | **1205.5** | **22** |
| Deposition in channels | 564.9 | 10 |
| Trapped in reservoir | 640.6 | 12 |

**Figure 3** *Sediment budget for the desert at Santa Fe, New Mexico, USA*

### ACTIVITIES

1 Study Figure **1**. How have insolation, water and wind contributed to shape this landscape?

2 To what extent do low-frequency, high-magnitude events cause significant change in desert environments?

3 Draw a simple diagram to show the operation of a sediment cell in a desert.

4 Study Figure **3**.

 a Suggest reasons why slopewash appears to be the dominant process.

 b Only 22 per cent of the eroded sediment has been accounted for. Where is the remaining 78 per cent?

 c What are the challenges facing scientists wishing to study sediment budgets in hot deserts?

 d Is this study still relevant today, some 50 years on from when it was conducted?

### STRETCH YOURSELF

Find out more about seasonal winds, such as the Harmattan, and local winds caused by convective activity. To what extent is the descending air and the high pressure belt the driving force behind these winds? What is the role of solar insolation in forming the winds? What are the impacts of these winds on people's lives?

In this section you will learn about the processes of weathering in hot desert environments

## What are the processes of weathering in hot deserts?

Weathering is the breakdown or disintegration of rocks in their original place (in situ) at or close to the ground surface. The weathering processes operating in deserts are much the same as those operating in other environments, although their extent and degree of severity varies. In common with all physical environments, weathering processes (together with erosion) are intrinsically interlinked; they combine to shape the landscape and are often interdependent.

### Mechanical (physical) weathering

The gradual disintegration of rocks without any chemical change taking place is called *mechanical* (physical) *weathering*. In a desert environment thermal fracture (insolation weathering) is perhaps the most important type of mechanical weathering.

### Thermal fracture

Intense temperature fluctuations causes rock to expand when heated and to contract when cooled. This is **thermal fracture**. Scientists believe that the presence of moisture promotes this process. This moisture is most likely to be dew – which forms when the temperatures plummet at night and the air becomes saturated (its dew point temperature).

Thermal fracture is very dependent on the colour, lithology (rock type) and geological structure of the rocks. These factors affect the mechanism of rocks disintegration. Figure **1** suggests that these forms of disintegration are discrete. However, in reality they will often be linked and will operate together on the same outcrop of rock.

- **Granular disintegration** – granular rocks (those made of grains) such as granite or sandstone tend to crumble and break down into individual grains of sand. The coloured minerals in granite (black mica and white quartz) are heated at different rates, which facilitates the break-up of the rock.

- **Block separation** – if rocks have a clear pattern of joints and bedding plains – as in sedimentary rocks such as limestone – whole blocks of rock may break apart.

- **Shattering** – rocks that do not have separate grains or joints tend to shatter to form angular fragments. Basalt is a dark igneous rock that is particularly prone to shattering.

- **Exfoliation** – rocks are poor conductors of heat, so the outer surface is particularly prone to repeated heating and cooling. This can lead to the outer 'skin' peeling or flaking away (Figure **2**). Exposed granite and sandstone are particularly prone to exfoliation.

Granular disintergration

Exfoliation

Block separation

Shattering

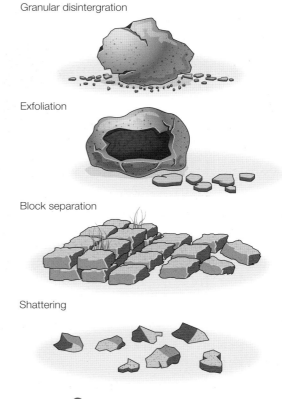

▲ **Figure 1** Thermal fracture processes

▼ **Figure 2** Exfoliation in Wadi Rum, Jordan (the camera lens cap gives an idea of scale)

## Salt crystallisation

When it rains, the water dissolves salts in the soil. The high rate of evaporation draws water to the surface by capillary action and salt crystals are deposited on the ground surface. Over time, these crystals may grow and expand, causing stresses that eventually break up rocks. This is **salt crystallisation.**

## Frost shattering

Temperatures need to fall below 0 °C for **frost shattering** to occur. It is therefore not common in hot deserts but can occur in mountain regions during the winter. Water, often in the form of dew, freezes and expands in confined cracks and pores within rocks. Repeated cycles of freezing and thawing cause these rocks to shatter. Frost shattering is common in upland or mountainous regions, such as in parts of the Middle East and the northern fringes of the Sahara Desert.

## Chemical weathering

Rocks that change as a result of chemical action, usually in the presence of water, is called **chemical weathering**. Although water may be scarce, when it is present the high temperatures often result in an intensification of weathering, as some chemical processes are enhanced by high temperatures. The exception to this is the process of carbonation – dissolving of limestone by acidic rainwater (involving the solubility of carbon), which is most effective in cold environments (see 4.6).

Rocks containing salts are vulnerable to being dissolved. This is *solution*. Once dissolved, the high rates of evaporation will lead to the deposition of salty deposits on the ground. These salts can lead to crystal growth, which is capable of corroding buildings and metal structures.

Flaking or pitting of rock surfaces is evidence of chemical weathering in hot deserts. Underground deposits of rock salt (halite) are sometimes dissolved to create tunnels, as can be found in the Atacama Desert.

⬥ **Figure 3** *Weathering in the Jordanian desert*

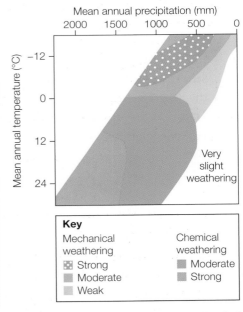

Mean annual precipitation (mm)

**Key**

Mechanical weathering
- ▨ Strong
- ▨ Moderate
- ▨ Weak

Chemical weathering
- ▨ Moderate
- ▨ Strong

## ACTIVITIES

1. Study Figure **1**. Draw sketches with detailed labels to show why each thermal fracture process happens. Comment on lithology (rock type), geological structure and colour.

2. Study Figure **3**. Use a labelled sketch to show the evidence that weathering is both active and varied in this environment.

3. Assess the importance of water in weathering in hot deserts.

4. On Figure **4**, mark 250 mm of rainfall (the accepted maximum rainfall for deserts). Give an account of the variations in type and intensity of weathering processes in desert environments.

5. Why is biological weathering of little significance in hot desert environments?

## STRETCH YOURSELF

Use the internet to find out more about the role of salt in weathering processes in hot desert environments. Consider the factors affecting the process of solution in dissolving salts – why is solution an active process given the lack of water? What is the evidence that solid salt deposits cause weathering?

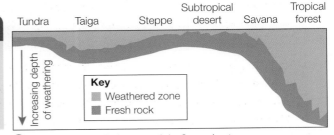

**Key**
- ▨ Weathered zone
- ▨ Fresh rock

⬥ **Figure 4** *The Peltier model of weathering*

In this section you will learn about the role of wind in hot desert environments

## What are the processes of wind action in hot deserts?

Hot deserts are affected by both seasonal and local winds. Wind was once considered to be the main agent of landscape change in deserts, because of the apparent absence of water. Now it is acknowledged that water does have a significant role to play particularly during low frequency, high magnitude storm events when a great deal of erosion, transportation and deposition can take place.

Wind action (erosion, transportation and deposition) is active in sandy (erg) deserts, where loose sand and finer particles can be picked up by the wind. Particles carried by wind become very rounded, even spherical due to the process of *attrition*. This is a distinguishing characteristic of wind-blown sand. Sandy deserts, with their extraordinary range of sculpted landforms and impressive, often evocative sand dunes, form a very distinctive desert landscape, quite different from rocky deserts.

### Wind erosion

There are two main types of erosion by wind – deflation and abrasion.

**Deflation** is the removal of loose material from the desert floor, often resulting in the exposure of the underlying bedrock. Over time the desert surface is lowered and can resemble a cobble pavement as only the larger stones are left behind (Figure 1). Moisture collected in a particular locality may increase the rate of deflation by assisting the break-up of the surface and making it crumbly.

Strong eddies or localised winds can hollow-out the desert surface to produce a *deflation hollow*. These can be very extensive covering thousands of square kilometres (see 2.8).

**Abrasion** is sometimes referred to as the 'sandpaper effect' but in the case of deserts it is more akin to 'sand blasting'! Wind-blown sand is driven against rock surfaces, carving or sculpting them into a variety of shapes. Much of this erosion occurs within a metre of the desert floor which is where the bulk of sand transportation occurs. The concentration of abrasion close to the ground surface can result in the formation of weird mushroom-shaped features (Figure 2).

Several factors affect the intensity of abrasion including the strength, duration and direction of the wind, the nature of the wind-blown sand (rock type and its angularity) and the lithology and vulnerability of the exposed rock outcrops.

**Figure 1** *The impact of deflation on a desert surface*

**Figure 2** *Mushroom rock carved by abrasion, Wadi Rum, Jordan*

## Wind transportation

Look at Figure **3**. There are three main ways that sand will be transported by the wind.

◆ **Surface creep** – larger particles are rolled along the desert floor.

◆ **Saltation** – sand particles move in a series of leaps as they are picked up by a gust of wind before being dropped again a few centimetres downwind.

◆ **Suspension** – fine sands and clays may be picked up and carried by the wind for considerable distances, sometimes way beyond the margins of the desert itself. Suspension is particularly common when high velocity winds create sandstorms.

> **Did you know?**
> Sand from the Sahara has been transported to Northern Europe and even Florida in the USA, over 8000 km away.

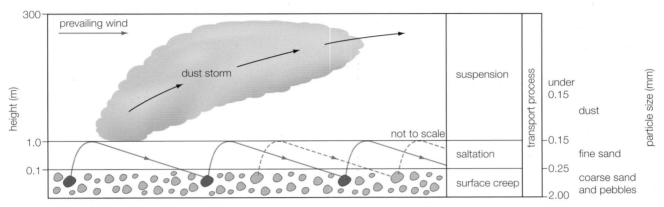

▲ *Figure 3* The processes of wind transportation

## Wind deposition

Deposition of sand will take place when wind velocity falls below a critical value whereby the sand can no longer be carried or moved by the wind. This velocity will vary according to the size (mass) of the sand grains being transported.

In common with deposition in all physical systems, sand will be deposited in sheltered areas protected from the wind, for example on the leeward side of exposed rock formations.

Whilst sand dunes are often considered to be solely depositional features, in many cases, they are extremely mobile and therefore also represent a type of flow, an ongoing surface creep of sand. Only when they are fixed and largely static can they be truly considered to be depositional. In this situation, sand dunes can also be considered to represent a temporary store, in balance with the environmental conditions of the time but vulnerable to change should these conditions alter in the future.

▲ *Figure 4* Saharan dust in London, UK. *Sand from the Sahara has been transported as far as Florida, 8000 km away.*

### ACTIVITIES

1 Study Figure **1**. What is the evidence that wind deflation has been operating on this desert surface? Is there any evidence of abrasion?

2 Study Figures **2** and **3**.
   a Describe the mushroom rock in Figure **2**.
   b What forms of wind transportation are most likely to have been responsible for the erosion of this landform? Justify your answer by using Figure **3**.
   c Explain how the effectiveness of abrasion will be affected by **i)** the characteristics of the wind **ii)** the sediment being transported **iii)** the nature of the rock being sculpted.

### STRETCH YOURSELF

Find out more about where sand is deposited at the local scale in deserts and why? Look at examples of sand being carried great distances away from deserts. See what you can find out about an extensive deposit of wind-blown sand and dust called *loess*.

In this section you will learn about the origins and development of aeolian landforms in hot desert environments

## Landforms resulting from aeolian (wind) processes

Wind is a potent source of energy in hot deserts and is responsible for the formation of a number of distinctive landforms. Some are extremely extensive, covering thousands of square kilometres, whereas others are at a very small scale, involving the shaping of individual rock fragments.

### Deflation hollows

A depression created by strong, gusty winds that erode loose material from the desert surface is called a *deflation hollow*. Sand, temporarily stored on the ground surface, is whisked up by the wind (a source of energy) to be transferred and deposited elsewhere. This is a good example of the desert system in action, affording landscape change by the flow or transfer of energy.

There are numerous deflation hollows in the Sahara Desert some of which are very extensive – one of the largest in the world is the Qattara Depression in Egypt (Figure **1**). Over 50 000 km² in area, this depression has been lowered by the process of deflation to 134 m below sea level. Thousands of cubic kilometres of sand must have been removed to form this depression. Where has it all gone?

Several factors can control the depth attained by a deflation hollow. These include the dynamics of the wind as it becomes affected by the hollow, the presence of consolidated material at depths below the more mobile sand (and possibly the exhumation of bedrock) and the presence of a water table. Indeed, saturated sand explains the presence in many deflation hollows of salt lakes or ephemeral lakes (Figure **2**).

### Desert pavements

A **desert pavement**, or *reg*, is a desert surface covered with rock fragments often resembling a cobbled pavement (Figure **3**). This rocky pavement is formed when the wind blows away the finer sands, leaving behind the larger stones (Figure **4**).

Desert pavements are common and well-developed on alluvial fans. Here, soil-forming processes move finer sediments from beneath the larger rock fragments that form the surface rocky pavement.

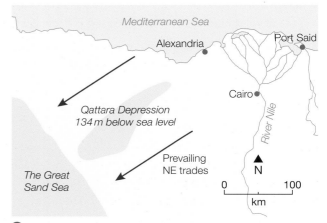

▲ *Figure 1* Location of Qattara Depression, Egypt

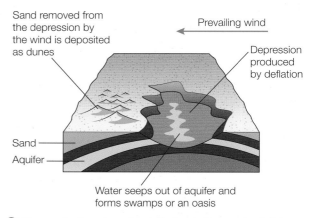

▲ *Figure 2* Creation of a deflation hollow and salt lake

▲ *Figure 3* A desert pavement

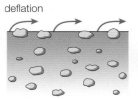
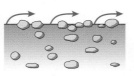

deflation                    deflation                    desert pavement

◀ **Figure 4** *Formation of a desert pavement*

Rocks are initially
dispersed in ground

Rocks start to become
concentrated as fine
sediment is blown away

Pavement forms as a
lag deposit

## Ventifacts

**Ventifacts** are small-scale features found on desert
pavements (Figure **5**). They are individual rocks, usually
the size of pebbles that have a clearly eroded face (*facet*)
that is aligned with the prevailing wind (Figure **6**).
Some ventifacts have a single well-developed facet,
indicating a single wind direction. Those with three
well-developed facets are known as *dreikanter*.

## Yardangs

Resembling the hull of upturned ships, **yardangs** are
elongated ridges separated by deep grooves cut into
the desert surface. They can vary in size from just a few
centimetres in height and length to hundreds of metres
in height and several kilometres in length.

Look at Figure **7**. Notice that the rocks are vertical to
the ground surface. The weaker rocks are eroded by
abrasion to form deep troughs, whereas the tougher
rocks are left upstanding. It is these more resistant
rocks that form the yardangs (Figure **8**). Notice that the
winds are unidirectional – blow in one direction only.
Yardangs do not develop in regions where the winds are
multidirectional.

▼ **Figure 5** *A ventifact on a desert pavement*

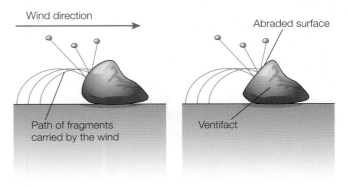

Wind direction

Abraded surface

Path of fragments
carried by the wind

Ventifact

▲ **Figure 6** *Formation of ventifacts*

▲ **Figure 7** *Yardangs developed in white limestone near
Dakhla, Egypt*

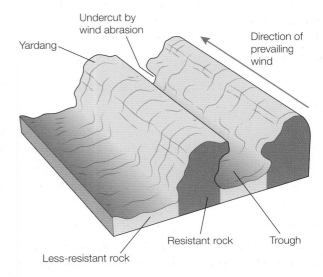

Undercut by
wind abrasion

Yardang

Direction of
prevailing
wind

Resistant rock        Trough

Less-resistant rock

▲ **Figure 8** *Formation of yardangs*

## Zeugen

Similar to yardangs, in that they also form ridges, **zeugen** can, in some cases, be up to 30 m high. The key difference is that zeugen develop in horizontally layered rocks rather than vertical rocks. This gives them a pedestal-like shape, with a flat-topped 'cap rock' protecting the less-resistant underlying layers (Figure **9**). Zeugen are similar (though not the same) as 'mushroom rocks', which are isolated wind-sculpted pillars (see 2.7).

Look at Figure **10**. It shows the main features and processes operating in the formation of zeugen. The primary process is abrasion. With most abrasion being concentrated within a metre or so of the desert surface, zeugen often have a slightly narrower, more eroded lower portion. You can see this in Figure **9**. While wind abrasion may be the dominant process, it seems likely that other processes, such as water erosion and weathering (the concentration of moisture in the form of dew) may be active too.

## Sand dunes

In sandy deserts the wind can form vast and beautifully symmetrical sand dunes. In much the same way as at the coast, desert sand dunes usually start with deposition on the leeward side of an obstacle such as a rock. As more and more sand is deposited, it becomes shaped by the wind. Sand dunes form extensively in the Sahara and in Australia but are relatively rare features of the deserts of the USA.

There are two common forms of sand dune: barchans and sief dunes.

▲ **Figure 9** Zeugen

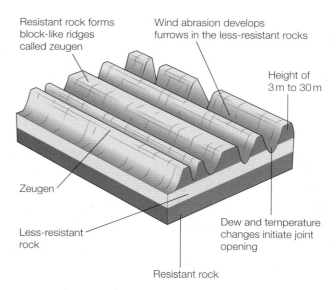

Resistant rock forms block-like ridges called zeugen

Wind abrasion develops furrows in the less-resistant rocks

Height of 3 m to 30 m

Zeugen

Less-resistant rock

Dew and temperature changes initiate joint opening

Resistant rock

▲ **Figure 10** The features and formation of zeugen

▼ **Figure 11** The shape and formation of a barchan sand dune

**Profile**

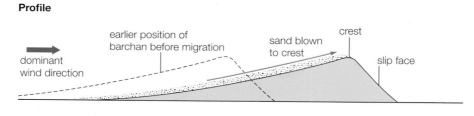

earlier position of barchan before migration

sand blown to crest

crest

slip face

dominant wind direction

**Plan**

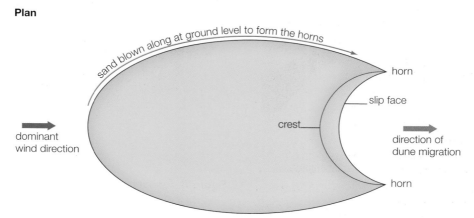

sand blown along at ground level to form the horns

horn

slip face

crest

dominant wind direction

direction of dune migration

horn

**Barchans** are crescent-shaped sand dunes, often found in isolation in deserts where there is a relatively limited supply of sand but a strongly dominant wind direction. They are transverse sand dunes, forming at right angles to the prevailing winds.

Look at Figure **11**. Notice that sand is blown up the gentle windward side of a barchan before sliding down the steeper sheltered side, which has highly mobile extending 'horns' that move forward several metres a year. Barchans are common on the edge of the El Kharga oasis in Egypt and in parts of the Atacama Desert.

**Seif dunes** are elongated linear sand dunes that are commonly found in extensive areas of sand called *sand seas* (Figure **12**). In places they may stretch for several hundred metres and are formed parallel to the prevailing wind direction. It is possible that they may develop from barchans, as Figure **13** explains.

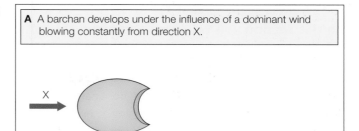

A  A barchan develops under the influence of a dominant wind blowing constantly from direction X.

B  Then the wind changes direction and blows from Y. This causes one of the horns to lengthen.

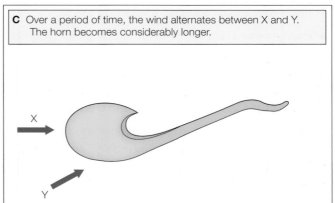

C  Over a period of time, the wind alternates between X and Y. The horn becomes considerably longer.

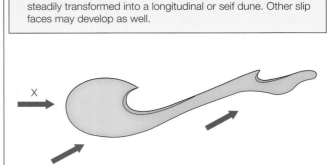

D  As the wind direction continues to alternate, the dune is steadily transformed into a longitudinal or seif dune. Other slip faces may develop as well.

🔺 **Figure 12** *Seif sand dunes in the Sahara Desert*

🔺 **Figure 13** *The development of a seif sand dune from a barchan*

## ACTIVITIES

Ⓢ

1  Study Figure **4**. Draw a fully labelled diagram to describe the formation of a desert pavement.

2  What is the evidence in Figure **5** that the ventifact is affected by multidirectional winds?

3  Compare and contrast the characteristics and formation of yardangs and zeugen. Use diagrams or sketches to support your answer.

4  What are the main differences between barchans and seif dunes? How can seif dunes develop from barchans?

In this section you will learn about the role of water in hot desert environments

## What is the role of water in hot deserts?

Water action is considered to be the dominant process in most of the world's deserts. In mountain ranges, erosion dominates following storm events or periods of snowmelt. On the lowland desert plains, deposition is the main process.

Look at Figure **1**. It shows an **alluvial fan** (see 2.10). This is a depositional features formed by water flowing out of a mountain range onto the flat desert plain. The sediment has been eroded from the mountains and then transported along a river channel before being deposited on the edge of the desert plain.

In the past the climate of the low latitudes was wetter than it is today and many present-day landscapes reflect evidence of this more actively erosive period.

**Figure 1**  *Alluvial fan, Death Valley, USA*

## What are the sources of water in hot deserts?

There are four main sources of water:

◆ **Exogenous rivers** are rivers that have their source in mountains outside desert regions. They have sufficient water to flow continuously despite the high rates of evaporation. Examples of exogenous rivers include the Colorado, the Indus and the Nile.

◆ **Endoreic rivers** are rivers that flow into deserts but terminate usually in a lake or inland sea, such as the River Jordan, which drains into the Dead Sea.

◆ **Ephemeral rivers** are rivers or streams that flow intermittently in desert regions. They might flow after storm events or they might be fed by snowmelt in the spring from adjacent mountains. Flow rates can vary dramatically and in times of flood ephemeral rivers can be powerful forces of erosion.

◆ **Episodic flash floods** are infrequent rainfall events that tend to involve torrential, convectional storms unleashing large amounts of water in a very short period of time. The sunbaked soil results in large amounts of overland flow that is capable of carrying out significant erosion, particularly in mountains where steep gradients increase the rate of flow. Water flows as a sheet over the landscape (**sheet flooding**) or is confined within a channel (**channel flash flooding**). During such high-magnitude events, huge amounts of sediment can be washed out of the mountains to be deposited as vast alluvial plains on the lowlands below.

Figure **2** summarises the main water action processes.

**Erosion**
- Hydraulic action – the sheer force of the water
- Abrasion – sandpaper effect of loose rocks being ground over bedrock
- Corrasion – fragments of rock carried by water gouging or sculpting bedrock
- Solution – dissolving of soluble rocks, such as limestone

**Transportation**
- Traction – rolling sediment along the surface or channel bed
- Saltation – bouncing or leaping motion of particles
- Suspension – sediment carried within the body of water
- Solution – dissolved sediment

**Deposition**
- Takes place where velocity drops, such as on the inside of meander bends or where streams flow out of mountain edges onto flat desert plains (alluvial fan)

**Figure 2a**  *Water action (river) processes*

| | |
|---|---|
| **Splash erosion** | The force of falling rainwater displacing soil particles |
| **Sheet erosion** | Water running as a sheet over impermeable surfaces or compacted soil washing away disturbed particles |
| **Rill erosion** | Sheet wash wears down the soil to form a definite path to form rivulets in the soil, called rills |
| **Gully erosion** | Over time, rills become wider and deeper to form gullies |
| **Bank erosion** | Fast water flow wears away the stream sides, causing the banks to collapse and the channel to widen |

**Figure 2b**  *Water action (sheet wash) processes*

2 Hot desert systems and landscapes

## Episodic flash flood: Morocco

In August 1995 a torrential downpour dumped 70 mm of rain in just a few hours close to the village of Imlil in the semi-arid foothills of the Atlas Mountains in Morocco (Figure **3**).

The flash flood that resulted swept boulders the size of lorries down the valley of the River Reraya and into the village. For much of the year there is little or no water in the river (Figure **4**), yet for just a few hours after the rainstorm it became a raging torrent.

Some 150 people were killed by the flood. Crops of maize, alfalfa and grass were destroyed and irrigation channels blocked with silt. Walnut trees, an important cash crop for the local Berber people, were swept away in the flood – it takes around 15 years for a walnut tree to mature.

The Imlil flood demonstrates how low-frequency but high-magnitude events can cause massive changes to the physical and human geography of a desert environment.

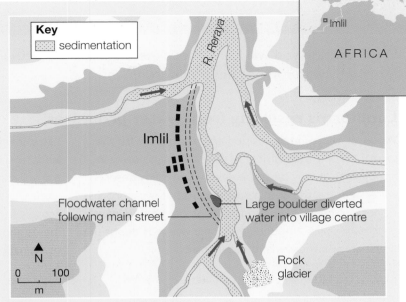

**Figure 3** Location of Imlil, Morocco

**Figure 4** The valley of the River Reraya at Imlil

## ACTIVITIES

1 Study Figure **1**.
   a Draw a sketch of the photo to show the alluvial fan in relation to the mountains in the background and the flat desert plain in the foreground. Add labels to describe the main features in the photo.

   b What is the evidence that the alluvial fan has been formed by the action of water?

   c Why has the alluvial fan formed at this point?

   d Do you think the alluvial fan is being actively formed at the present time?

2 Study Figure **4**.
   a What is the evidence that the River Reraya is an ephemeral river?

   b What is the evidence that the river occasionally experiences high flow?

   c What other factors shown in the photo increase the likelihood of flash flood events in the area?

   d What were the immediate and longer-term impacts of the Imlil flood? Consider social, economic and environmental impacts.

3 With reference to the Imlil flash flood, describe how events of this kind (low frequency, high magnitude) can cause profound landscape change.

In this section you will learn about the origins and development of landforms formed by water in hot desert environments

## Landforms resulting from water processes

Water erosion and deposition are responsible for the formation of a wide variety of distinctive landforms in hot deserts. It may be surprising that, in many deserts, water action is the dominant geomorphological process. Generally speaking, erosion tends to be the dominant process in upland mountain regions, whereas deposition dominates on the lowland desert plains.

Look at Figure **1**. It shows a typical hot desert landscape with many features associated with erosion, transportation and deposition by water. Some of these features, for example wadis and canyons, are still being actively formed by erosion. Others, for example, mesas and buttes, are gradually being diminished by a combination of erosion, weathering and mass movement.

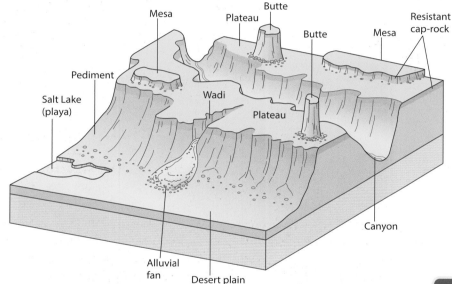

◀ **Figure 1** *Landforms created by water action in hot deserts*

## Wadis

A **wadi** is a dry riverbed that can take the form of a distinct channel in a lowland plain or an incised gully or valley cut into a plateau (Figure **2**). They can range from relatively short features of just a few metres in length, to highly complex, extensive features many hundreds of kilometres in length. Despite the variety in size, form and location, wadis have a number of common features:

◆ Steep edges that are the result of severe erosion during periods of high water flow.

◆ Flat bottoms infilled with sediment – mainly boulders and pebbles as the finer material is washed out or removed by wind action. Notice the extensive gravel deposits on the floor of the wadi in Figure **2**.

◆ They often comprise many channels that split and rejoin to form a complex drainage pattern (*braided*). This is due to the high quantities of sediment being transported and the very 'flashy' nature of the water flow.

▲ **Figure 2** *Gravel road on the bed of Wadi Bani Habib, Oman*

## Alluvial fans and bajada

At the edge of a mountain range, the sediment washed out through a wadi or canyon is deposited to form a delta-like **alluvial fan** (Figure **1**). As the river water spreads from the mountain front, energy is lost and sediment is deposited rapidly. Over time, as water reworks this vast store of sediment, the alluvial fan displays clear stratification (layering) and sorting. Coarser sediment becomes concentrated closest to the mountain range as the finer sediment is washed out onto the desert plain. Alluvial fans can extend for several kilometres away from the mountain edge and can reach thicknesses of up to 300 m.

Extensive alluvial fans emerging from an upland area can coalesce (merge) to form a continuous apron of sediment extending for many kilometres across a mountain front. This feature is called a **bajada** (Figure **3**). Bajadas have similar characteristics to alluvial fans in that they exhibit clear stratification and sorting – coarser sediment is closest to the mountain front and finer material is deposited on the desert plain. **Playas** (salt lakes) may form on lower bajadas where they become extensive flat deposits.

**Figure 3** *An extensive bajada in Death Valley, California, USA*

## Pediments

A **pediment** is a gently-sloping (1–7 degrees), straight or slightly concave erosional rock surface at the foot of a mountain range. A common characteristic is a distinct break of slope (change in angle) between the mountain front and the top of the pediment (Figure **4**). A pediment is often partly or completely blanketed by sediment (such as alluvial fans) washed down by rivers from the nearby mountains.

The formation of pediments is the subject of a great deal of debate. They may well result from more than one process and a single hypothesis may not account for all examples. One theory suggests that, as they do not have any distinct carved channels, they may be formed by intensive sheetwash erosion, combined with extensive surface and subsurface weathering. At the edge of a mountain range, water builds up within the ground resulting in more concentrated weathering. Periodic sheetwash then removes the loose material creating a smooth rock surface.

Both pediments and bajadas form at the foot of a mountain front but make sure you do not to confuse the two! A pediment is bare rock, a bajada is deposited alluvial sediment.

**Figure 4** *The idealised profile of a pediment*

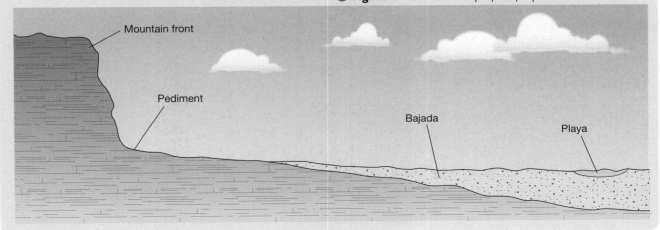

Mountain front

Pediment

Bajada

Playa

## Playas (salt lakes)

Water flows over pediments and their surface deposits (alluvial fans and bajadas) carrying fine material out onto the desert floor. Some of the water present in this area will percolate through the coarse sediment. It will then be transferred as throughflow to collect in a large hollow or depression where it may form a **playa**.

A playa is an enclosed desert lake with no outflow. Playas are extremely flat and any ephemeral water will be very shallow. The large surface area will encourage evaporation and a salty crust will often form around their edges. In some deserts, the accumulation of salt is sufficient for it to be exploited commercially, for example, the Chott el Djerid in southern Tunisia.

**Figure 5** *Mesas and buttes in Monument Valley, Arizona, USA. Mesas are the large, flat-topped features on the far horizon and left-of-centre on the near horizon. Buttes are the isolated rock pinnacles.*

## Canyons

*Canyons* are steep-sided gorges mostly associated with rapid downcutting into plateaus (Figure **6**). Some very deep canyons are formed by exogenous rivers that flow permanently through a desert region. The Grand Canyon in Arizona is a good example. Formed by the Colorado River, the Grand Canyon is 446 km long, has a width of up to 29 km and a depth of up to 1857 m. It has geological significance as it exposes layers of ancient rocks in its walls, which can be clearly seen in Figure **6**. This gives a record of the geological history of North America dating from between 2 billion and 230 million years ago.

**Figure 6** *The Grand Canyon, Arizona, USA*

## Mesas, buttes and inselbergs

Mesas, buttes and inselbergs are relic features of a mountain landscape that have become isolated by river erosion (Figure **5**).

**Mesas** are large plateau-like features often bordered by steep wadis or canyons. The word 'mesa' comes from the Spanish word meaning 'table'.

**Buttes** are smaller pinnacles of rock and represent a more advanced stage of landscape development. They are usually surrounded by flatter desert plains. Both mesas and buttes have extensive scree slopes formed by mass movement (rockfalls) and mechanical weathering.

**Inselbergs** are relic landforms that develop in rocks such as granite or sandstone where there is an absence of layering or variable rock strengths (Figure **7**). They are more rounded than mesas and buttes. Some scientists believe that inselbergs may have been formed during past climates when higher levels of humidity would have resulted in more intense chemical weathering. This is because chemical weathering tends to smooth the edges of exposed rocks. Note that inselberg can be used as an umbrella term to include all relic hills including mesas and buttes. The word inselberg comes from the German word meaning 'island mountain'.

Similar features appear outside desert environments, such as tors in south-west England and bornhardts in the African savannah. It is unlikely that all examples were formed in exactly the same way.

⬤ **Figure 7**  *Inselberg in the Joshua Tree National Park, California, USA*

# The development of characteristic desert landscapes

In this section you will learn about the desert landscapes that develop over time resulting from the interactions between processes and landforms

## The development of desert landscapes

Every desert landscape is unique because of the interaction between the rocks and the processes operating on them, both past and present. These processes are essentially water and wind action, although weathering, mass movement and the role of vegetation are also important drivers in the desert system.

## The importance of time

A desert landscape is like pausing a single frame in a video. What you see in Figure **1** is a landscape that will probably be different in the future and is certainly different from what it was like in the past.

It is unlikely that processes can be seen operating, other than during a sandstorm or following a period of heavy rain. It is therefore difficult to make connections between the landforms and the processes that have been responsible for their formation. Indeed, scientists are increasingly concluding that many desert landforms were formed over very long periods of time. They may even have been sculpted or deposited during different climatic conditions than those of today.

Furthermore, long-term changes, such as uplift resulting from tectonic activity, may well be highly significant, but impossible to witness directly. It is believed that some of the world's great canyons, including the Grand Canyon (see 2.10), owe their considerable depth to tectonic uplift. It is like raising butter upwards through a knife!

Figure **2** shows a typical basin and range landscape associated with tectonic activity. Vertical displacement has taken place along a line called a fault. This landscape exhibits many of the features that you have studied earlier. It demonstrates the complexity of a landscape affected by a range of different processes set in the context of an evolving history.

▲ **Figure 1** *Dawn over Wadi Rum, Jordan*

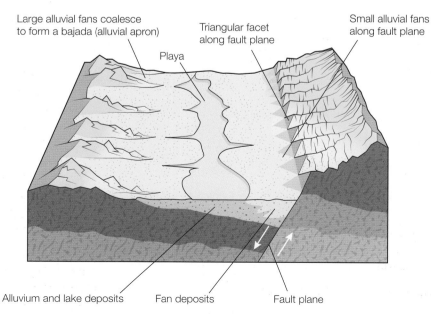

▼ **Figure 2** *Basin and range landscape, typically found in south-west USA*

Large alluvial fans coalesce to form a bajada (alluvial apron)

Playa

Triangular facet along fault plane

Small alluvial fans along fault plane

Alluvium and lake deposits

Fan deposits

Fault plane

## King's arid cycle of erosion

In the 1940s the desert geomorphologist L.C. King developed his *arid cycle of erosion* (also known as the *model of pediplanation*) following his research in Africa. Figure **3** summarises the concept. While this model is clearly highly simplified it does make the point that landforms are dynamic and evolving elements of a changing landscape. Furthermore, it stresses the importance of time in understanding the development of desert landscapes.

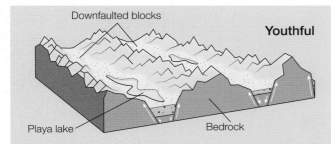

Following tectonic uplift, processes of weathering, mass movement and erosion (primarily water) result in the parallel retreat of mountain fronts, leaving behind a gently angled rock slope – the pediment

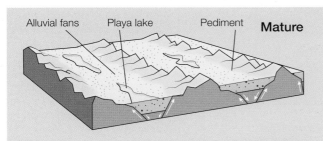

Over time, the uplifted landscape is gradually eroded away, leaving coalescing pediment surfaces and isolated remnants of the mountains – relic hills called inselbergs

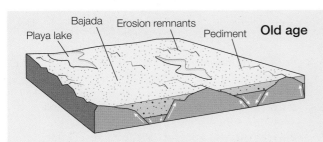

Eventually these isolated hills crumble, leaving behind a largely flat surface known as a *pediplain*. With further uplift, the whole process will start over again.

**Figure 3** *King's arid cycle of erosion*

## Landscape development

Geomorphologists have developed models to describe how landscapes change over time. One of the most influential early models was W.M. Davis' cycle of erosion, developed in the late nineteenth century. Davis suggested that landscapes undergo an evolutionary pattern of change resulting from landscape denudation (primarily river erosion) whereby landscapes pass from youth, through maturity and into old age. While heavily criticised as being over-simplistic and taking no account of changes in the type and magnitude of processes, Davis' model was adapted by other geomorphologists for different environments.

Both King's and Davis' concepts are at the heart of modern-day study in that the systems approach, with its holistic view of the interrelationships between energy flows, stores and transfers, enables a causal understanding to be gained of the links between process and landform. The concepts of positive and negative feedback loops are also a critical part of the systems approach in the way that they control or perpetuate energy flows and transfers. As King also identified, the need to contextualise landforms into the wider desert landscape is extremely important in understanding the development of such a distinctive global environment.

### ACTIVITIES

1 Study Figure **2**.
   a Describe how past and present processes have affected the development of this landscape.
   b How does this landscape fit King's model?

2 a Study Figure **3**. Describe the impact of tectonic uplift on the landscape.
   b What landscape processes act on the landscape as it develops from youth, through maturity to old age?
   c Do you think that, at some stage, deposition becomes more important than erosion? If so, suggest when this happens.
   d What do you think happens after the final stage?

3 Write a critical evaluation of King's arid cycle of erosion. Assess its value and its shortcomings.

4 Assess the importance of time in understanding the development of a desert landscape.

### STRETCH YOURSELF

Find a photo of a desert landscape. Describe the landforms and the present-day processes operating in the landscape. To what extent can King's model help your understanding of the development of this landscape? Contrast King's approach to understanding the development of desert landscapes with the modern-day systems approach. Present your work as a poster using labels to interpret the landscape.

In this section you will learn about the changing extent and distribution of deserts over the last 10 000 years

## How has the extent and distribution of deserts changed?

Scientists have used carbon dating techniques in their study of plant remains, together with pollen analysis, to investigate patterns of the world's vegetation zones thousands of years ago.

Look at Figure **1**. The present-day extent and distribution of deserts has remained almost unchanged for the last 5000 years or so. This reflects a largely stable climate with few major shifts or changes. The atmospheric circulation system, which is the main driving force behind the location of the world's deserts, has also remained stable and there has been no significant change in the other factors controlling desert locations.

However, between the maximum extent of the last glacial period (about 20 000 years ago) and 5000 years ago, there were considerable climatic fluctuations triggering dramatic changes in the extent and distribution of the world's deserts.

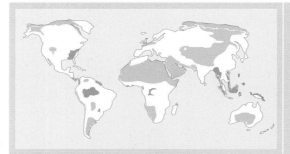

**Last glacial maximum (about 20 000 years ago)**
During the last glacial maximum, aridity was very widespread. Deserts in the far north were cold deserts. Further south, deserts existed in similar locations to the present day but they were far more extensive.

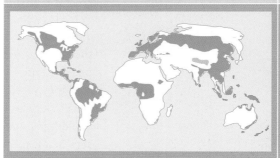

**About 8000 years ago**
During this interglacial period conditions were very much warmer and more humid. Forests were widespread, thriving in warm and wet conditions. Aridity fell dramatically, with many of the present-day deserts becoming grasslands.

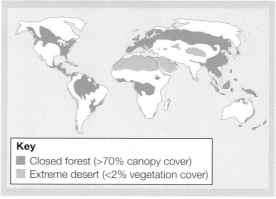

**Key**
- ■ Closed forest (>70% canopy cover)
- ■ Extreme desert (<2% vegetation cover)

**5000 years ago – present**
The monsoon rains over Asia and Africa began to diminish. This was the start of conditions becoming more arid. By 3000 years ago, conditions were much as they are today. This map shows the 'present potential' vegetation cover and does not take account of the fact that large areas of forest have been cleared by people to make way for farming.

 **Figure 1** *Changes in the extent and distribution of deserts (about 20 000 years ago to present day)*

## ACTIVITIES

1 Study Figure **1**.
   a Describe the extent and distribution of deserts during the last glacial maximum, about 20 000 years ago.
   b Which present-day deserts existed 8000 years ago when conditions had become much more moist? Consider which factors affecting the location of deserts were most influential at this time.
   c Describe the changes to the extent and distribution of the world's deserts between 8000 years ago and the present day.

**(S)** 2 Present the information in Figure **2** in the form of a divided bar. Draw a thick vertical or horizontal bar to represent the period from 22 000 years ago to the present day (try a scale of 1 cm = 1000 years). Then divide the bar into the time segments from the table. Devise a colour code to represent the different conditions and shade each section of the bar accordingly. Finally, add labels and annotations to describe the conditions associated with each time period.

## The ever-changing Sahara Desert

The Sahara Desert, with its vast sand seas, is considered to be the most famous and hostile of the world's deserts. It is hard to believe that just a few thousand years ago much of it was covered in grass and parts of it were even swampy!

Figure **2** shows that after a very dry period coinciding with the last ice advance, conditions changed dramatically around 10 500 years ago.

▶ **Figure 2** *Changing conditions in the Sahara (22 000 years ago–present day)*

| Years before present day | Conditions in the Sahara |
|---|---|
| 22 000–10 500 | Very dry – the Sahara extended 250 miles further south than it does today. No human settlement existed in this area. |
| 10 500–9100 | Moist – a dramatic change due to strengthening monsoon rains transforms the desert into grassland. The area becomes settled by migrants from the Nile Valley. |
| 9100–8900 | Slight drying takes place |
| 8900–7900 | Moist conditions return. By now human settlements are well developed and sheep and goats have become domesticated. Forests reach their maximum extent. |
| 7900–6500 | Moderately dry – retreating monsoon rains cause the desert to start drying up. People begin to move back to the Nile Valley. |
| 6500–5000 | Moist conditions return periodically |
| 5000–4100 | Very dry – as dry as present day |
| 4100–3700 | Still dry but slightly moister than present day |
| After 3700 | Remaining about as dry as present day |

Scientists believe that a burst of monsoon rains transformed the desert into grasslands over a period of just a few hundred years – an extremely rapid rate of climate change. Trees sprouted in valley bottoms close to groundwater sources and episodic rainpools attracted animals and people. Elephants, rhinos and hippos roamed over the savannah-like vegetation (Figure **3**). The region became well settled with people moving out of the congested Nile Valley to live on the grasslands of the Sahara.

This moist period lasted until about 7500 years ago when conditions started to become more arid and less hospitable for human settlement. From about 5000 years ago, conditions became more or less as they are today.

**Key**

| | | | |
|---|---|---|---|
| ▥ | Mediterranean forest | ▦ | Mediterranean scrub |
| ■ | Montane forest | ▤ | Extreme desert |
| ▨ | Semi-desert | ▫ | Grasslands |
| ■ | Savanna (a few trees) | ▨ | Scrub |
| ■ | Woodland (open canopy) | ◪ | Tropical rainforest |
| ▨ | Recolonizing forest mosaic | | |

8000 years ago

Present-day potential vegetation

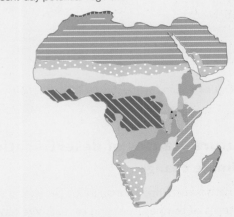

▲ **Figure 3** *Vegetation zones in Africa (a) 8000 years ago (b) present day*

## ◀ ACTIVITIES

**3** Study Figures **2** and **3**. Describe in detail the changing conditions in the Sahara in the last 22 000 years. How have these changes had an impact on human settlement in North Africa.

## STRETCH YOURSELF

Find out more about the conditions that existed in the Sahara or another desert of your choice during the wetter period about 8000 years ago. How did the climate differ from today's climate and what triggered these changes? What changes took place in the types of plants and animals and what effects did the changing climate have on human activities?

In this section you will learn about the distribution and causes of desertification

## What is desertification?

The UN has defined **desertification** as '*the destruction of the biological potential of the land, which can lead ultimately to desert-like conditions*'. It is the destruction of ecosystems and habitats. Land that was once marginal is turned into an unproductive wasteland. Vegetation dies and soil becomes exposed and eroded. Figure **1** shows the areas that are most vulnerable to desertification.

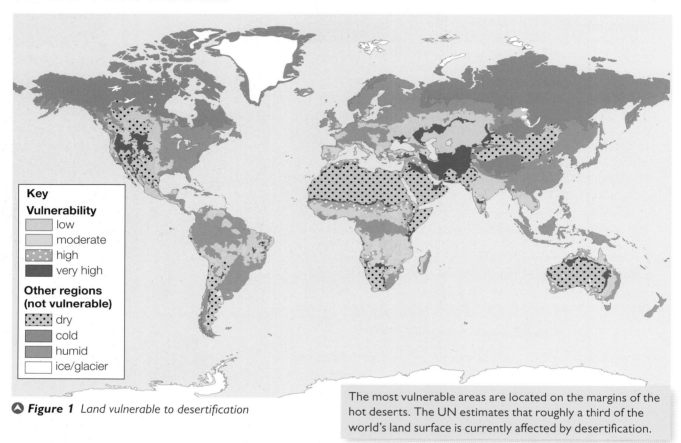

**Key**

**Vulnerability**
- low
- moderate
- high
- very high

**Other regions (not vulnerable)**
- dry
- cold
- humid
- ice/glacier

🔺 *Figure 1* Land vulnerable to desertification

The most vulnerable areas are located on the margins of the hot deserts. The UN estimates that roughly a third of the world's land surface is currently affected by desertification.

## Natural causes of desertification

### Climate change

Climate has changed throughout geological time, altering patterns of temperature and rainfall. We have already seen how climate change during the last 10 000 years has affected the extent of the world's deserts and impacted on ecosystems and human activities (see 2.12).

About 8000 years ago climates in North Africa and the Middle East were much wetter than they are today. Evidence for this includes the substantial fossil aquifers beneath countries such as Egypt and Jordan, as well as fossil plant remains and archaeological evidence (tools, rock art, etc).

Recent climate change is associated with human-induced global warming. An increase in emissions of greenhouse gases since the mid twentieth century has enhanced the natural greenhouse effect – global temperatures have risen by about 1 °C since 1880. Rainfall patterns (Figure **2**) as well as temperatures are changing and droughts have become common in many parts of Africa, for example, over the last few decades (Figure **3**).

# Human causes of desertification

People are not likely to deliberately damage the land on which they depend for their survival. However, they can inadvertently contribute the process of desertification by:

◆ *overcultivation* – intensive farming on marginal land can reduce soil fertility. The lack of organic matter makes it friable and more likely to be washed or blown away.

◆ *overgrazing* – if an area of marginal grassland is overstocked with livestock, the vegetation and soil will suffer – it exceeds the *carrying capacity* (the number of animals that the land can support).

◆ *overirrigation* – if plants are overirrigated, salts can be dissolved in the soil, forming a salty crust on the surface when the water evaporates (salinisation). This creates impermeable, infertile soil.

◆ *population increase* – Africa's Sahel region (Figure **4**) has an estimated population of 260 million and it doubles every 20 years. The rate of increase (3 per cent per year) is greater than the growth rate of food production (2 per cent), putting additional pressure on the land to be productive.

◆ *firewood* – in semi-arid regions most people have to rely on wood for cooking, so land quickly becomes stripped of what few trees there are. This is unsustainable and also exposes the soil to wind and rain, leading to soil erosion.

◆ *tourism* – high concentrations of safari minibuses in countries such as Kenya can cause serious damage to vegetation, leading to soil erosion and ultimately desertification.

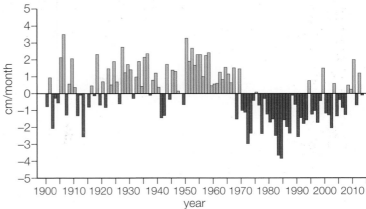

▲ **Figure 2** *Rainfall trends in the Sahel (1900–2013); long-term average is shown by the horizontal line at 0 cm/month*

▲ **Figure 3** *Impact of drought in semi-arid environments*

▲ **Figure 4** *The Sahel region of Africa*

## ACTIVITIES

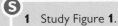

1 Study Figure **1**.

   a On a blank world outline map plot the dry regions (deserts) and the main areas of 'very high' vulnerability to desertification.

   b Use an atlas to label the deserts and name some of the regions/countries at high risk of desertification.

   c Why do most of the high-risk areas appear to be located at the margins of deserts?

   d Are there any exceptions? Suggest possible reasons for them.

2 Study Figure **2**. The anomalies show the variations either side of the average.

   a Describe the rainfall trends in the following three periods:
   1900–1950      1950–1970      1970–2007

   b How do the figures help to explain the recent increase in desertification?

3 To what extent do human causes dominate over natural causes in causing desertification?

## STRETCH YOURSELF

Find a detailed map showing the areas at risk of desertification in the Sahel region of Africa and assess the natural and human factors contributing to the problem.

In this section you will learn about:
- the current impacts of desertification on ecosystems, landscapes and populations
- possible future impacts resulting from climate change

## What are the current impacts on ecosystems?

When considering the impacts of desertification on ecosystems it is important to appreciate the timescales involved. Ecosystems (soils, plants and animals) can take hundreds of years to become established – indeed, soils can take thousands of years to develop fully. In contrast, desertification can be relatively rapid, and subsequent soil erosion by wind and rain can destroy an ecosystem in just a few hours.

Desertification affects ecosystems in several ways:

- Loss of nutrients, due to overexploitation by agriculture, leaves soils impoverished and *biodiversity* is reduced (Figure **1**).

- Loss of topsoil, the most nutritious horizon, due to increasing exposure to wind and rain.

- Increased salinity if irrigation is used.

- Water sources become depleted, the water table falls and plants and animals die.

- As vegetation is destroyed animals have to migrate to find other areas of suitable vegetation (Figure **2**).

## What are the current impacts on people?

Many people living in *drylands* (areas with an aridity index of less than 0.65) are desperately poor. They are highly vulnerable to these impacts of desertification.

- Lack of water: results in hygiene issues and the spread of disease.

- Loss of vegetation: reduces food productivity, forcing people to abandon settlements.

- Dust clouds: affect air quality and cause health problems even for people living in cities far from desert regions (Figure **4**).

- Food insecurity: in India an estimated 25 per cent of the land is turning into desert, compromising the country's ability to feed its people.

- Depleted grasslands: not able to sustain animals. Animals may die or become emaciated. People forced to migrate to already overcrowded cities.

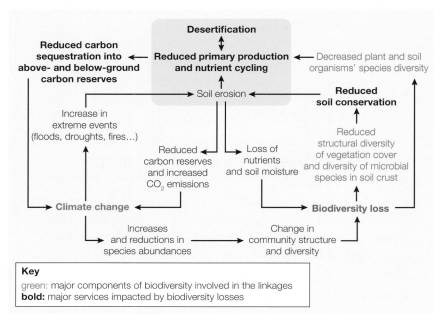

**Figure 1** *Links between desertification, biodiversity loss and climate change*

**Figure 2** *Cattle overgrazing sparse vegetation in the buffer zone surrounding Ranthambore National Park, India*

## What are the current impacts on landscapes?

Desertification will result in the spread of desert conditions as previously vegetated ground becomes dry and barren. Deflation may remove fine particles leaving behind a rocky surface floor. This can affect the landscape in a number of ways:

◆ Formation of sand dunes as loose material is no longer trapped by vegetation.

◆ Soil erosion by water may cause deep gullies to form in the landscape (Figure **3**).

◆ Landslides may occur on steep slopes when vegetation is removed.

◆ Sandstorms may become more common when the protective layer of vegetation is removed (Figure **4**). This may lead to increased rates of abrasion.

## How might future climate change affect desertification?

By the middle of this century, temperatures in the world's drylands could rise by 2 to 5 °C. This will inevitably accelerate the rate of desertification. Look at Figure **1** to see the two-way links between climate change and desertification. Rates of evapotranspiration would increase, soil water would decrease and droughts would become more frequent and severe. Rainfall may decrease as drylands expand.

There are a number of possible future scenarios for local people:

◆ As soils become less productive, settlement and agriculture will become less viable and it is likely that many people will migrate.

◆ Migration to rural areas could lead to further desertification, or migration to cities could put additional pressure on areas which are often already overcrowded.

◆ Some people may choose to stay and adapt, by implementing soil conservation strategies or changing agricultural practices.

⬤ **Figure 3** *Soil erosion and gullying, Knersvlakte Plateau, South Africa*

⬤ **Figure 4** *Sandstorm in Eritrea, near the Sudanese border*

### ACTIVITIES

1 Study Figure **1**.
   a How does desertification affect species diversity?
   b What type of feedback loop is operating between desertification and species diversity? Explain how this works.
   c How does climate change affect species diversity?
   d Describe the two-way link between desertification and climate change. How does desertification affect climate change?
   e How does climate change affect desertification?
   f Attempt to draw a similar diagram replacing 'Biodiversity loss' with 'Local populations'

2 Study Figure **2**.
   a What evidence is there that this ecosystem is being damaged by overgrazing, leading to desertification?
   b If this environment continues to be desertified, what impact will this have on local populations and landscape development?

3 Study Figure **4**. Suggest the possible problems associated with sandstorms and dust clouds in urban areas.

### Did you know?

In Mongolia, more than half of its wheatlands were abandoned between 1985 and 2005, and wheat production plummeted. However, production had recovered by 2016 to two-thirds of its 1985 level.

### STRETCH YOURSELF

How might climate change increase the rate of desertification in certain areas? What will the impact be on local populations and on the natural environment?

In this section you will learn about the causes and impacts of, and responses to desertification in the Badia, Jordan

## Where is the Badia?

The Badia is the name given to the vast desert region in eastern Jordan and extending into Syria (Figure **1**). This area of stony desert is sparsely populated, mostly by traditional Bedouin who herd animals such as sheep, goats and camels (Figure **2**). Annual rainfall is less than 150 mm and much of it falls as torrential storms. Temperatures can soar to over 40 °C in the summer, yet in winter they can fall below freezing.

## What has caused desertification in the Badia?

The Tal Rimah Rangelands is an area of gently rolling hills, not far from the desert town of As Safawi (Figure **1**). It is sparsely vegetated with thorn shrubs and grasses and has been grazed sustainably by local farmers for hundreds of years.

However, following the first Gulf War in 1991, large numbers of sheep were available cheaply from Iraq. Local farmers bought hundreds and grazed them on the Rangelands. This overgrazing tipped the ecological balance and the land quickly became desertified and unproductive. This traditional grazing pasture was abandoned and farmers migrated away from the area, many moving to the capital, Amman.

For some time the Badia has been suffering from rural depopulation, which has had a negative impact on the provision of services – shops, schools, community centres, etc.

## What has been done to address the problem of desertification?

In 2002 the Tal Rimah Rangeland Rehabilitation Project was initiated with financial support from the charity USAID and the US Forestry Service. Following lengthy discussions with local herders, a plan was agreed to turn the wasteland back into sustainable, productive grazing land.

Stone walls were constructed (Figure **3**) to control and retain water – this is called *water harvesting*. Water is diverted along the contours into shallow ditches where drought-tolerant shrubs such as atriplex have been planted (Figure **4**).

The newly introduced plants are extremely well-suited to the hostile conditions:

◆ They grow quickly into shrubs to provide grazing for animals.

◆ They provide fuel for firewood.

◆ Their roots help to hold the soil together.

◆ They encourage greater species diversity by offering a number of different habitats for birds, insects and small animals.

▲ **Figure 1** *Location of Jordan and the Badia*

▲ **Figure 2** *Traditional nomadic herding in the Badia*

## How successful has the project been?

By 2008 some sheep had been reintroduced in a carefully managed pilot study to enable the project's success to be evaluated.

◆ Between 2004 and 2008 the number of plant and animal species increased from 21 to 54.

◆ The flowering plants have attracted butterflies and other insects.

◆ Birds are now nesting in the base of the shrubs.

◆ Animals and reptiles are returning to the area too.

The Tal Rimah Rangeland Rehabilitation Project clearly demonstrates how it is possible to address the issue of desertification to ensure a sustainable outcome. Farmers are once again able to make a living from the land, which means that they are less likely to move away from the area. The local economy benefits, communities stabilise and rural depopulation slows down.

## Evaluating the human responses

Without the cooperation and involvement of local communities, for example in building the low stone walls, the project would not have been a success.

The local people have shown resilience and patience throughout. It was vital not to return to the land too soon and risk damaging it once again. Local people have had to adapt their farming techniques and stock management practices to enable the new plants to become established and avoid the dangers of overgrazing. Gradually, as the new shrubs have become established and a greater variety of plants have colonised the area, so stocking levels have been increased and the future looks promising.

▲ *Figure 3* Stone walls constructed to retain topsoil and harvest water

▲ *Figure 4* Planting atriplex in the Badia rangelands

## ACTIVITIES

1 What caused desertification in the Badia region?

2 What were the impacts of desertification on both the environment and on local communities?

3 Study Figures **3** and **4**. How has the process of desertification been reversed?

4 Why is it important for international aid agencies and local communities to work together at all stages, when trying to solve the problem of desertification?

5 Evaluate the human responses of resilience, mitigation and adaptation to the problem of desertification in the Badia.

## STRETCH YOURSELF

Use the internet to find out about the rehabilitation of the Badia in neighbouring Syria. It was initiated in 1998 and it has achieved a good deal of success. What are the challenges for the future given the political turmoil in the region?

In this section you will learn about the landforms and landscapes of the Mojave Desert

## Case study

### Where is the Mojave Desert?

The Mojave Desert is located in the south-west of the USA. It occupies parts of the states of California, Nevada, Utah and Arizona and covers an area of 124 000 km² (Figure **1**).

'The landscape reflects the cumulative effects of geologic forces or events that have transpired over many millions of years. However, faulting, volcanism and erosion within the past million years, and particularly changing climatic conditions within the last 20 000 years, have had particularly strong effects on the physical appearance of the Mojave Desert landscape today.'

(USGS Western Region Geology and Geophysics Science Center, 2009)

**Figure 1** *Location of the Mojave Desert, California, USA*

**Figure 2** *Characteristics of the Mojave Desert*

| | |
|---|---|
| **Physical geography** | The Mojave Desert is a high desert area, displaying classic basin and range topography. Its highest elevations reach over 3000 m, while its lowest point – the infamous Death Valley – is 86 m below sea level, the lowest elevation in North America. |
| **Climate** | With an annual rainfall of less than 140 mm it is North America's driest desert. Summer temperatures can reach 50 °C in the lowest valleys, whereas winter temperatures can plummet to −7 °C. Rain tends fall as thunderstorms in the summer. The desert is affected by strong winds. |
| **Drainage** | There are a small number of exogenous rivers that flow through the desert, including the Colorado River in the east. |
| **Settlement** | The desert is sparsely populated, although there are a few large cities, primarily Las Vegas, which has a population approaching two million people. |

### Landforms and landscapes of the Mojave Desert

Look at Figure **3**. It shows mountain ranges separated by broad, flat basins – the typical landscape of the Mojave Desert. The granitic mountain range in the distance is subject to weathering, particularly thermal fracture, leading to granular disintegration and exfoliation of the crystalline rocks. There may also be some frost shattering and mass movement in the form of rockfalls.

You can see alluvial fans at the foot of the mountains with deposits spreading out over the desert plain, coalescing in places to form bajadas. Bare rock pediments are exposed in places. Playa lakes are common on the plains – you can see one in the middle right of the photo. In the foreground are extensive sand dunes, most of which are highly mobile.

The presence of water and wind-related landforms shows that both agents are important in landscape development in the Mojave Desert. The deeply eroded canyons of the mountains suggest that water action is particularly important in both sculpting the landscape and providing sand for the wind to form into sand dunes.

**Figure 3** *Basin and range landscape of the Mojave Desert*

# The Mojave National Preserve

The Mojave National Preserve (Figure **4**) is a protected area in the south of the Mojave Desert in southern California (Figure **1**). It is famous for its stunning landscapes and distinctive desert landforms.

There are several examples of classic desert landforms within the Mojave National Preserve.

## Lucy Gray Fan – alluvial fan

Lucy Gray Fan (Figure **5**) is an alluvial fan just to the north of the Mojave National Preserve. It radiates out from a canyon cutting through the Lucy Gray Mountains and drains into the Ivanpah Valley (Figure **4**). It is actually just to the north of the Mojave National Preserve.

Notice that below the mouth of the canyon, the course of the stream divides into several channels. Also notice that the dry lakes appear white on the image. Channels migrate across the fan as they become choked with sediment. In common with all alluvial fans, coarse sediment is found at the top of the fan and finer material is spread out over the desert plain. Some of this reaches the Ivanpah playa.

## Cima Dome – pediment and inselbergs

Figure **6** shows the ancient rocky pediment of Cima Dome (also see Figure **4**). Erosion and weathering have stripped away the mountainous landscape to leave behind an extensive rocky pediment. Notice the isolated rounded granite inselbergs.

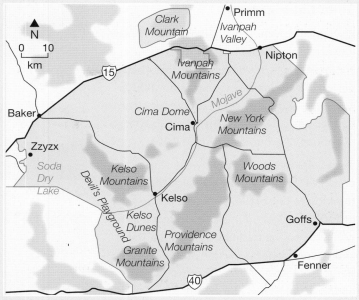

▲ *Figure 4* *The Mojave National Preserve, California*

▲ *Figure 5* *Aerial view of Lucy Gray Fan*

▼ *Figure 6* *Cima Dome pediment and inselbergs*

## Soda Lake – playa

Playas are common landforms in the Mojave Desert. Many exist in places where lakes and marshes formed during the last glacial period. These lakes dried up about 8000 years ago and today only hold water after flash floods or when springs discharge large quantities of groundwater (see 2.10).

Soda Lake is located in the west of the preserve (Figure **4**). It is the largest playa in the Mojave Desert, extending over an area of 150 km² (Figure **7**). Clays and muds are washed into the basin by the Mojave River and springs generate water on the western side of the playa. Winter storms increase discharge into lake. During the summer, salt crusts develop in places on the lake. In late summer and autumn strong winds whisk up the salts and create a dusty haziness in the air that can spread across the region.

## Kelso Dunes – sand dunes

Wind plays an important role in landscape development (see 2.8) in the Mojave Desert. Sources of sand that are shaped by the wind include alluvial fans, weathered rocks and dried lake beds. The Kelso Dunes and neighbouring Devil's Playground (Figures **4** and **8**) form an extensive area of sand deposition in the west of the preserve.

Extending over 120 km², the Kelso Dunes are the largest area of sand dunes in the Mojave Desert. They comprise a mixture of mobile and stabilised (partly vegetated) dunes, the tallest of which rise to over 200 m above the desert floor. Most of the sand originates from the granites of the San Bernardino Mountains. This has been deposited in the Mojave River valley from where it is transported by the wind in an easterly direction to form the Kelso Dunes.

▲ **Figure 7** *Soda Lake*

▲ **Figure 8** *Kelso Dunes*

## ACTIVITIES

1 Study Figure **1**.
  **a** Describe in detail the location of the Mojave Desert.
  **b** The main cause of aridity here is the so-called 'rainshadow effect'. Describe how this operates and why it has led to the formation of the Mojave Desert.

**S**

2 Draw a sketch of the landscape in Figure **3** and add labels to identify the main landforms. Describe the landscape in a couple of sentences.

3 Locate the aerial photo (Figure **5**) on the map in Figure **4**. Now draw a simple sketch map to show the main characteristic features of the alluvial fan.

4 Assess the importance of time in the development of the landforms and landscape of the Mojave Desert.

5 Describe the processes (past and present) that have been responsible for the formation of Soda Lake.

6 Study Figure **8**. Describe the form and location of the Kelso Dunes. What is the evidence that the dunes in the photo are moving?

## STRETCH YOURSELF

Find out more about the landscapes and landforms of the Mojave Desert. The USGS website has extensive literature about the desert and the Mojave National Preserve. Focus on one landform that interests you and investigate its formation in detail.

# Field data: Dumont Dunes, Mojave Desert

## How does the prevailing wind affect the ridge top profile of sand dunes?

The Dumont Dunes are located in the north of the Mojave Desert near the southern tip of Death Valley (Figure **1**). Scientists from the California Institute of Technology chose to study a 50 m high sand dune to consider the impact of wind direction on the angles of the dune ridge. They were interested to see if the dune faces were symmetrical or whether windward and leeward sides were characterised by different angles. The sand dune was described as 'a barchanoid ridge with a distinct slip face' (Figure **9**). Its profile, measured by a laser rangefinder, is shown in Figure **10**.

North-facing Dumont dune

⬢ *Figure 9  The Dumont dune studied by the California Institute of Technology*

⬢ *Figure 10  Dumont dune profiles*

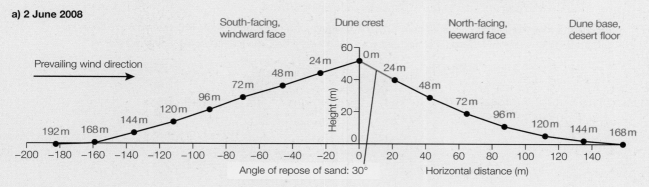

a) 2 June 2008

South-facing, windward face · Dune crest · North-facing, leeward face · Dune base, desert floor

Prevailing wind direction

Angle of repose of sand: 30°

Horizontal distance (m)

Height (m)

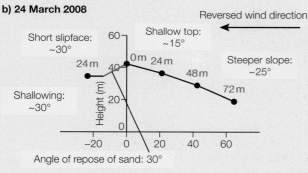

b) 24 March 2008

Reversed wind direction

Short slipface: ~30°
Shallow top: ~15°
Steeper slope: ~25°
Shallowing: ~30°

Angle of repose of sand: 30°

Height (m)

The prevailing winds are from the south. They carry sand grains up the windward side to the top (ridge) of the dune where they blow over and become deposited as *grainfall* on the sheltered leeward side. The angle on the leeward side builds up to reach the natural *angle of repose* (maximum angle before the slope starts to collapse, about 30° for sand), at which point local slope failure results in *grainflow*. The windward slope has firmer sand due to being *combed* by the prevailing wind, and the angle is consistently at about 20°.

Therefore, the study concluded that the sand dunes were asymmetrical and that strong prevailing winds were a controlling factor in preventing sand on the windward side achieving its natural angle of repose.

## ACTIVITIES

1 Summarise the purpose of the Dumont dune study.

2 Study Figure **9**. Suggest some of the possible reasons why scientists decided to choose this sand dune for study.

3 Study Figure **10a**.

   a Compare and contrast the windward and leeward faces of the sand dune.

   b What is meant by the *angle of repose*?

   c What is the difference between *grainfall* and *grainflow*?

   d The scientists suggested that 48 m down the leeward side marks the transition between grainfall and grainflow. What is the evidence for this judgement and why does it occur?

   e Why is the angle of the windward side consistent and lower than the angle of the leeward side?

4 Study Figure **10b**. Describe and suggest reasons for the changes to the ridge-top profile as a result of a change in wind direction.

Ⓢ 5 Critically evaluate the study, particularly the data collection and the conclusion. Could there be other controlling factors, such as moisture, that might explain the asymmetry?

The following are sample practice questions for the Hot desert systems and landscapes chapter. They have been written to reflect the assessment objectives of the Component 1: Physical geography Section B of your A Level.

These questions will test your ability to:

◆ demonstrate knowledge and understanding of places, environments, concepts, processes, interactions and change, at a variety of scales [AO1]

◆ apply knowledge and understanding in different contexts to interpret, analyse and evaluate geographical information and issues [AO2]

◆ use a variety of quantitative, qualitative and fieldwork skills [AO3].

---

**1** Outline the sources of energy in a desert which drive landscape change. (4)

**2** Study Figure **1** (below) and Figure **3** on page 67. Interpret the data presented, with reference to water balance in the region. (6)

🔽 **Figure 1** *Rainfall anomaly in the Middle East, 2016. Infographic developed by iMMAP as part of the Regional Food Security Analysis Network, a project between FAO and iMMAP, funded by USAID. All rights reserved*

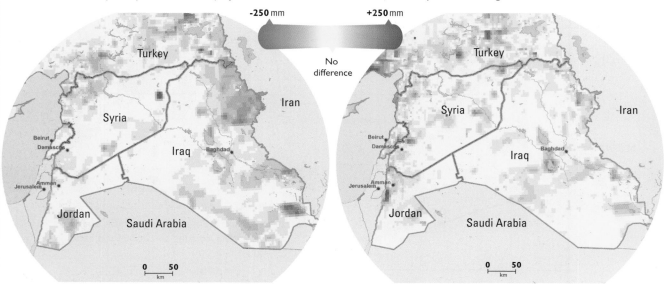

**1a** *Rainfall in December 2016 compared to December 2015*   **1b** *Rainfall in December 2016 compared to 15-year average*

**3** Study Figure **2** on page 60. Using Figure **2** and your own knowledge, explain the global distribution of mid- and low-latitude deserts with reference to the key natural causes of aridity. (6)

**4** 'Water from within and without creates desert landforms.' Assess the role of exogenous and endogenous sources of water in shaping characteristic desert features. (20)

> **Tip** 💡
>
> Geomorphology is the study of physical features of the surface of the Earth. Physical processes work in tandem to create a landscape or landform.

---

**5** Outline the role of wind in shaping desert landforms at different scales. (4)

**6** Using Figure **2**, analyse the challenges the desert environment of the Uluru-Kata Tjuta National Park poses for its sustainable development. (6)

🔽 **Figure 2** *Australia closes climb on sacred Uluru*

In October 2019 Parks Australia permanently closed the climbing route over Uluru in the Uluru-Kata Tjuta National Park. The rock Uluru is sacred to the indigenous population. Aboriginal people have lived in the area for at least 30,000 years, while Uluru's origins go back half a billion years.

Tourists began climbing the rock in the 1930s, retracing the steps of the ancestral people who first arrived at Uluru. A chain link handrail was installed in 1966 without consulting indigenous owners, who had implored people to respect their culture and not climb the rock. Climbers have worn away the outer oxidation surface of the rock on the route, discolouring it. Some scientists estimate it could take decades or even centuries for it to return to its natural colour, especially considering the dry climate of central Australia. Some climbers had used the top of the rock as a toilet, contaminating waterholes and other wildlife.

The safety of visitors had also proved a challenge for the authorities managing access to the site. At least 35 people are known to have died while climbing Uluru over the last 50 years, mostly from heart attacks and strokes brought on by exhaustion. For climbers' safety, park rangers conducted regular temperature, wind speed, and weather checks to assess whether it was safe for people to climb.

7   Using Figures **1** and **2** on page 74 and your own knowledge, assess the value of using a systems approach to understand the formation of deflation hollows in desert environments. (6)

8   To what extent are desert landscapes relics of the past? (20)

9   Outline the interactions between climate, soils and vegetation for a hot desert environment that you have studied. (4)

10  Figures **3** and **4** (right) provide information about a project designed to address desertification in the Tal Rimah Rangelands, eastern Jordan. Analyse the data presented about the success of water harvesting methods designed to retain water. (6)

| Shrub species | Number planted (2002) | Survival rate in June (5%) | | | |
|---|---|---|---|---|---|
| | | 2003 | 2004 | 2005 | 2006 |
| Atriplex nummularia (An) | 3,456 | 98 | 92 | 90 | 88 |
| Atriplex halimus (Ah) | 3,040 | 95 | 92 | 89 | 87 |
| Salsola vermuclatea (Sv) | 3,660 | 96 | 88 | 90 | 90 |

**Figure 3** *Survival of shrub seedlings at Tal Rimah (Tal Rimah Range Rehabilitation Report, 2008)*

11  Using Figure **1** on page 88, and your own knowledge, critically assess current causes of desertification. (6)

12  'Poverty is both a cause and a consequence of land degradation.' With reference to a specific community or communities affected by desertification, critically assess the validity of this statement. (20)

| Plant species | Harvesting method | 2003 | 2004 | 2005 | 2006 |
|---|---|---|---|---|---|
| Atriplex halimus | Contour furrows | 23 | 196 | 345 | 395 |
| | Crescent micro | 15 | 186 | 279 | 260 |
| | V-shaped micro | 15 | 133 | 258 | 332 |
| Atriplex nummularia | Contour furrows | 24 | 245 | 370 | 452 |
| | Crescent micro | 21 | 209 | 314 | 341 |
| | V-shaped micro | 24 | 151 | 306 | 319 |
| Salsola vermuclatea | Contour furrows | 6 | 22 | 45 | 77 |
| | Crescent micro | 5 | 20 | 61 | 60 |
| | V-shaped micro | 5 | 14 | 34 | 49 |

**Figure 4** *Estimate biomass of shrubs planted in different water harvesting structures from 2003–6. Values expressed in kilograms of dry matter per hectare (kg DM/ha) (Tal Rimah Range Rehabilitation Report, 2008)*

*Coasts are being increasingly hit by severe storms or long-term sea level rise associated with climate change. What short- and long-term impacts could these events have on a coastal landscape you know?*

## Specification key ideas

## Your exam

**AL** You must answer one question in Section B of Paper 1: Physical geography, from a choice of three: Hot desert systems and landscapes or Coastal systems and landscapes or Glacial systems and landscapes. Paper 1 makes up 40% of your A Level.

**AS** You must answer one question in Section A of Paper 1: Physical geography and people and the environment, from a choice of three: Water and carbon cycles or Coastal systems and landscapes or Glacial systems and landscapes. Paper 1 makes up 50% of your AS Level.

## Your key skills in this chapter

**S** In order to become a good geographer you need to develop key geographical skills. In this chapter you will practise a range of quantitative and relevant qualitative skills, within the theme of 'Coastal systems and landscapes'. Examples of these skills are:

- Using atlases and other map sources  3.7, 3.8, 3.10
- Use of overlays  3.10
- Drawing and annotating field sketches, annotating photographs  3.1, 3.6, 3.8, 3.9, 3.10, 3.11
- Measurement and geospatial mapping  3.10
- Presenting field data and interpreting graphs  3.3
- Analysing data, including applying statistical skills to field measurements  3.7
- Drawing and annotating diagrams of physical systems  3.2, 3.3, 3.6, 3.7

## Fieldwork opportunities

### 1. Investigating longshore drift

Whether or not the process of longshore drift is taking place along a beach, and in which direction, may be a simple focus for fieldwork and can also form the basis of a quick, effective field study. Readily available equipment (a metre rule, plus a camera, data recording sheet and a pen) can be used to collect data that can be presented and analysed, informed by accompanying photographs. Measurements taken from the top of a groyne to the bank of sand or sediment found on either side can be compared (for example, on the northward side v. the southward side) and the difference calculated. A bar graph of the differences in sediment height across a whole beach of groynes may be constructed. Data will appear, for example, above the x-axis where northward side is higher, and below the x-axis where southward side is higher to aid analysis of any pattern.

### 2. Investigating sand dunes and plant succession

Test theories of plant succession by mapping changing biodiversity along a designated transect, which crosses an area of coastal dunes. Your sampling technique will need some thought. A systematic approach, placing your quadrat every five or ten metres, may avoid the human bias of site selection but the differences between vegetation found on top of dune ridges and down below in dune slacks (the low-lying areas between dunes) may, alternatively, prompt you to use a stratified approach.

Coasts as natural systems

In this section you will learn about the concept of physical systems and their application to coastal landscapes

### The coast as an open system

The coast is an example of an open system. This means that it has inputs that originate from outside the system (such as sediment carried into the coastal zone by rivers) and outputs to other natural systems (such as eroded rock material transported offshore to the ocean). In Selwicks Bay (Figure **1**), a small stream discharges water and inputs sediment into the sea at Selwicks Bay. The waves that affect this stretch of coastline are often driven by North Atlantic storms that pass into the northern North Sea. Sediment eroded from cliffs is transported southwards along the coast and deposited in parts of the southern North Sea – an output from the system.

As an open system, the coast has important links with other natural systems such as the atmosphere (consider the importance of wind, for example, in generating waves), tectonics, ecosystems and oceanic systems. These natural systems are linked together by flows of energy and by the transfer of material. In Figure **1**, the chalk cliffs are vulnerable to the process of carbonation, which is an important aspect of the carbon cycle. Dissolved salts can be carried into the deep ocean well away from the coast.

🔻 **Figure 1**  *Selwicks Bay, Flamborough Head, UK*

# The coastal system

In common with other natural systems it is useful to apply systems terminology to help us to understand the connections between processes and landforms (Figure **3**). You will come across elements of the coastal system throughout this chapter, in particular the flows of energy and the transfers of sediment from place to place. A really clear example of the application of systems concepts to the coast is the sediment cell (see 3.3). There are eleven major sediment cells in England and Wales, which form the basis for coastal management (Figure **2**). Here, there are very clear inputs of sediment (e.g. from rivers and cliff erosion), transfers of sediment (e.g. longshore drift), stores (e.g. beaches and spits) and outputs (e.g. transfer to the deep ocean). This is probably the clearest and most straightforward way of understanding the coast using systems terminology.

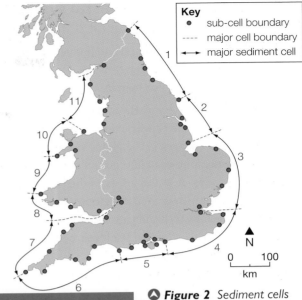

**Key**
● sub-cell boundary
---- major cell boundary
◄──► major sediment cell

▲ *Figure 2* *Sediment cells in England and Wales*

◄ *Figure 3* *The coastal system*

| Systems term | Definition | Coastal example |
|---|---|---|
| Input | Material or energy moving into the system from outside | Precipitation, wind |
| Output | Material or energy moving from the system to the outside | Ocean currents, rip tides, sediment transfer, evaporation |
| Energy | Power or driving force | Energy associated with flowing water, the effects of gravity on cliffs and moving air (wind energy transferred to wave energy) |
| Stores/ components | The individual elements or parts of a system | Beach, sand dunes, nearshore sediment |
| Flows/transfers | The links or relationships between the components | Wind-blown sand, mass movement processes, longshore drift |
| Positive feedback | Where a flow/ transfer leads to increase or growth | Coastal management can inadvertently lead to an increase in erosion elsewhere along the coast. Groynes trap sediment, depriving areas further down-drift of beach replenishment and this can exacerbate erosion. Seawalls can have the same effect by transferring high energy waves elsewhere along the coast. |
| Negative feedback | Where a flow/ transfer leads to decrease or decline | When the rate of weathering and mass movement exceeds the rate of cliff-foot erosion a scree slope is formed. Over time, this apron of material extends up the cliff face protecting the cliff face from subaerial processes. This leads to a reduction in the effectiveness of weathering and mass movement. |
| Dynamic equilibrium | This represents a state of balance within a constantly changing system | Constructive waves build up a beach, making it steeper. This encourages the formation of destructive waves that plunge rather than surge. Redistribution of sediment offshore by destructive waves reduces the beach gradient which, in turn, encourages the waves to become more constructive. This is a state of constant dynamic equilibrium between the type of wave and the angle of the beach (see 3.2). |

# Links between the coastal system and other natural systems

Coastal systems do not operate in isolation. They are interlinked with other physical and human systems, both affecting and being affected by change. Consider the impact of natural and also human-induced climate change.

During the Quaternary glacial and interglacial periods, sea levels rose and fell several times (as did the land) in response to changes in the global water cycle. The changing level of the sea affected the precise location of coastal processes at the edges of the land masses – several landforms owe their development to changes in the sea level. Recent changes in the global carbon cycle are indirectly affecting sea levels by causing global

warming and this, in turn, is affecting coastal (and terrestrial) ecosystems. Consider the impacts of these changes on human systems, with some coastal regions suffering more severe flooding and being at greater risk from storm surges. You can see, therefore, how the world's natural and human systems are inextricably linked together – if one changes, they are all subject to change.

Figure **4** shows the estuarine and coastal system in New Zealand, with particular reference to fisheries. It helps to show the linkages between various marine and terrestrial systems and between physical and human systems.

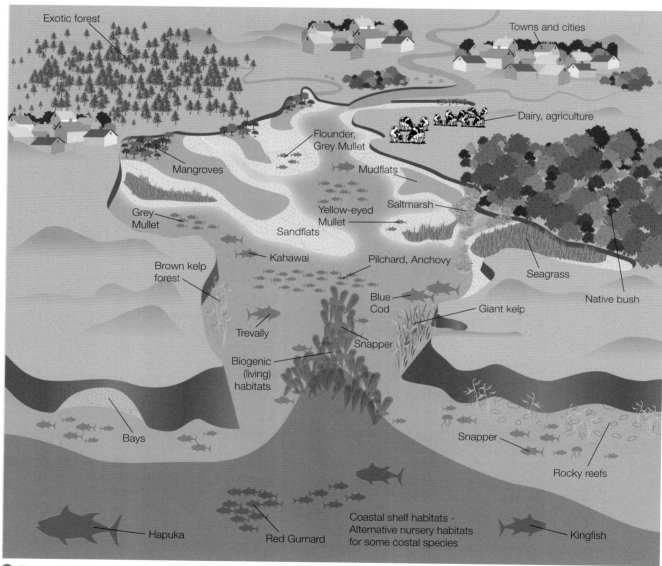

▲ **Figure 4** *The estuarine and coastal system in New Zealand*

## Dyrhólaey, Iceland

Dyrhólaey is a small peninsula on the south coast of Iceland famous for its dramatic coastal landscape and its puffins in the summer. Looking south, the next land mass is Antarctica! Waves driven over this vast area of ocean transfer a great deal of energy into the coastal system of southern Iceland and they are responsible for extremely active erosion and significant transfers of sediment. The landforms of erosion and deposition combine to create this very distinctive coastal landscape.

The black sand and pebbles in Figure **5** are volcanic basalt formed by tectonic activity over millions of years and transported to the coast by rivers, glaciers and the wind. Some has been eroded directly by the sea. The presence of this material clearly illustrates the linkages with other physical systems. Much of the landscape in Figure **5** is the result of sea level change triggered by long-term changes in the water cycle. In recent times, relative sea level has dropped, creating vast coastal plains and isolated remnants of coastal landforms (an isolated stack is shown in the photo). Vegetation has developed on these newly exposed surfaces and much of this area is renowned for its wildlife, in particular birds. This demonstrates the important connections between coastal systems and ecosystems, just one of many connections between natural systems illustrated at Dyrhólaey.

🔽 **Figure 5** *The coastal system at Dyrhólaey, southern Iceland*

volcanic basalt    coastal plain    'new' vegetation    isolated stack

## ACTIVITIES

1 Draw a sketch of Figure **1** and use annotated labels to identify components of the coastal system. Use Figure **3** to help you.

2 Study Figure **4**.
   **a** Describe and suggest reasons for the diverse range of fisheries at the coast.
   **b** Identify the physical and human systems (other than the coastal system) illustrated by Figure **4**.
   **c** Suggest how changes to one or more of these systems could impact on the coastal system.
   **d** Why is it important to understand the connections between natural and human systems in the management of coastal fisheries?
3 With reference to Figure **5** and your own knowledge, consider to what extent the coastal system is affected by the operation of other natural systems.

In this section you will learn about:
- sources of energy in coastal environments
- high-energy and low-energy coasts

St Nazaré in Portugal is renowned for its huge waves, which attracts surfers from all over the world (Figure **1**). In 2013, Garrett McNamara broke his own world record for the tallest wave ever surfed – the wave was 30 m high. So, how do waves such as this form and what are the factors that determine the energy of a wave?

## The Sun and wind – the energy behind the waves

The primary source of energy for all natural systems is the Sun. Heat and light from the Sun is converted by natural processes (such as photosynthesis in plants) to form energy. At the coast the main form of energy is derived from the sea in the form of waves. Although waves can be generated by tectonic activity or underwater landslides creating tsunami, they are mostly formed by the wind.

Wind is quite simply the movement of air from one place to another. In much the same way that air escapes from a punctured bicycle tyre, wind moves from high pressure to low pressure. Variations in atmospheric pressure primarily reflect differences in surface heating by the Sun. The greater the pressure difference between two places – called the pressure gradient – the faster (stronger) the wind.

In the UK, the prevailing (dominant) winds come from the south-west, the result from air moving from the subtropical high pressure belt at about 30 °N to the subpolar low pressure belt at about 60 °N. These winds blow over a broad expanse of the Atlantic Ocean and have the potential to transfer a great deal of energy to the waves that approach the UK.

A number of factors affect wave energy:

- The strength of the wind – determined by the pressure gradient.

- The duration of the wind – the longer the wind blows, the more powerful waves will become.

- The fetch – the distance of open water over which the wind blows. The longer the fetch, the more powerful the waves. As Figure **2** shows, the longest fetch in the UK extends for over 3000 miles across the Atlantic Ocean to Brazil. This coincides with the direction of the prevailing wind, accounting for the high energy waves that often affect south and west-facing coasts.

▲ **Figure 1** *Garrett McNamara surfing at St Nazaré in Portugal in 2015*

⊙ **Figure 2** *Fetch affecting wave energy in the UK*

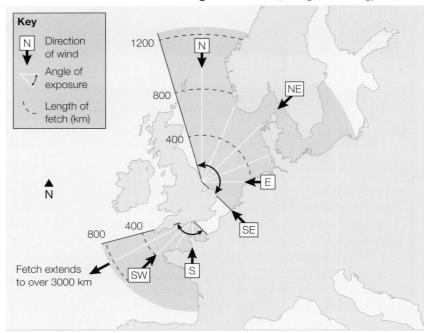

## How are waves formed?

As air moves across the water, frictional drag disturbs the surface and forms ripples or waves. In the open sea, there is little horizontal movement of water. Instead, there is an orbital motion of the water particles. Close to the coast, horizontal movement of water does occur as waves are driven onshore to break on the beach (Figure **3**).

⬇ *Figure 3* *Waves approaching and breaking onshore*

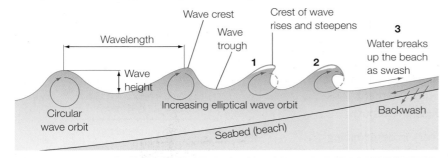

1   The water becomes shallower and the circular orbit of the water particles changes to an elliptical shape.
2   The *wavelength* (the distance between the *crests* of two waves) and the velocity both decrease, and the wave height increases – causing water to back up from behind and rise to a point where it starts to topple over (break).
3   The water rushes up the beach as *swash* and flows back as *backwash*.

## Different types of wave

Although waves vary, there are two main types: **constructive** and **destructive** (see Figure **4**). These two types of wave have significant impacts on the processes and landforms occurring at the coast.

### Beaches and waves: an example of a negative feedback

Constructive waves are usually associated with relatively gentle beach profiles, enabling waves to surge a long way up the beach. Over time, however, as more beach material is deposited, the profile steepens, working against the propagation of constructive waves. Instead, the waves become more destructive in their nature (plunging rather than surging), removing material from the beach and depositing it just offshore. This can result in the profile becoming less steep, encouraging constructive rather than destructive waves to form. This 'toing and froing' is a balancing act that will, all things being equal, result in a state of *dynamic equilibrium*. Of course, with wind strength and direction changing all the time and beach profiles responding accordingly, the state of balance may not exist in reality.

◀ *Figure 4* *Constructive and destructive waves*

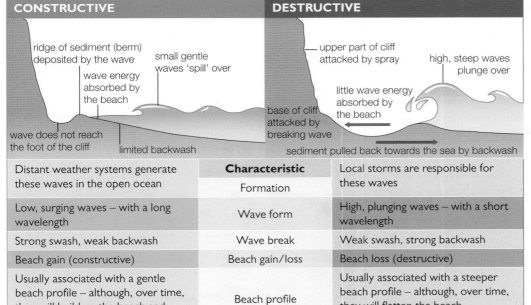

| CONSTRUCTIVE | Characteristic | DESTRUCTIVE |
| --- | --- | --- |
| Distant weather systems generate these waves in the open ocean | Formation | Local storms are responsible for these waves |
| Low, surging waves – with a long wavelength | Wave form | High, plunging waves – with a short wavelength |
| Strong swash, weak backwash | Wave break | Weak swash, strong backwash |
| Beach gain (constructive) | Beach gain/loss | Beach loss (destructive) |
| Usually associated with a gentle beach profile – although, over time, they will build up the beach and make it steeper | Beach profile | Usually associated with a steeper beach profile – although, over time, they will flatten the beach |

## Tides and currents

*Tides* are changes in the water level of seas and oceans caused by the gravitational pull (another source of energy) of the moon and, to a lesser extent, the Sun. The UK coastline experiences two high and two low tides each day. The relative difference in height between high and low tides is called the *tidal range*. The tidal range is also affected by the relative position of the sun and moon – highest during spring tides and lowest during neap tides (see Figure **5**). A high tidal range creates relatively powerful tidal currents (important sources of energy), as tides rise and fall, which can be particularly strong in estuaries and narrow channels. These currents are important transfer mechanisms in transporting sediment either within the coastal system or beyond (as an output).

The tides and tidal range are important factors in determining the precise height and duration of wave processes on a particular beach or a cliff. For example, at either side of high tide, often for a few hours, erosion of a cliff will be spatially concentrated, which results in the formation of a *wave-cut notch* (see 3.6). This will be exacerbated if there is a low tidal range as energy will be concentrated on a small section of cliff for a longer period of time. With a high tidal range, the waves will only break at a specific level on the cliff for a relatively short period of time.

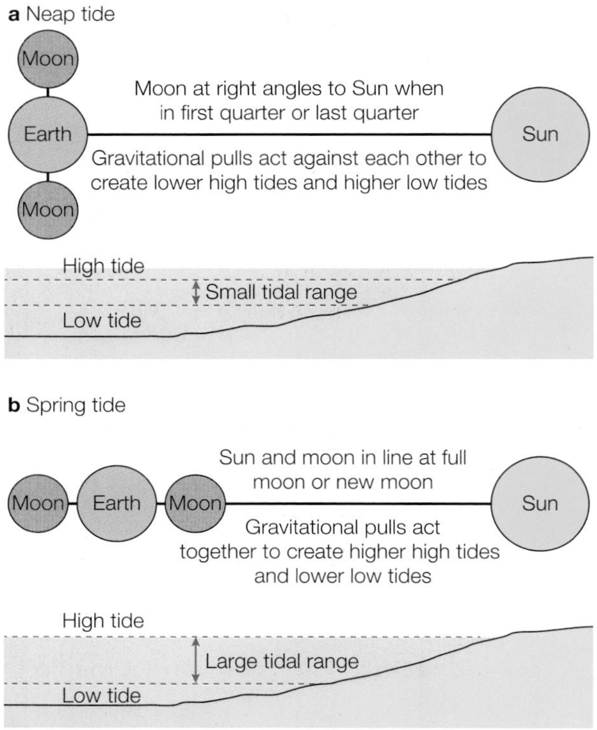

**a** Neap tide

Moon at right angles to Sun when in first quarter or last quarter

Gravitational pulls act against each other to create lower high tides and higher low tides

High tide

$\updownarrow$ Small tidal range

Low tide

**b** Spring tide

Sun and moon in line at full moon or new moon

Gravitational pulls act together to create higher high tides and lower low tides

High tide

$\updownarrow$ Large tidal range

Low tide

**Figure 5** *Spring and neap tides*

**Figure 6** *A rip current in Florida after Hurricane Jeanne*

### Rip currents

You may have heard of *rip currents* – strong localised underwater currents that occur on some beaches, posing a considerable danger to swimmers and surfers (Figure **6**). Rip currents are commonly formed when a series of plunging waves cause a temporary build-up of water at the top of the beach. Met with resistance from the breaking waves, water returning down the beach (the backwash) is forced just below the surface following troughs and small undulations in the beach profile. This fast-flowing offshore surge of water can drag people into deep water where they may drown. Figure **7** shows how to respond if caught in a rip current.

**RIP CURRENTS**

**Figure 7** *How to respond to a rip current*

## High-energy and low-energy coastlines

Rocky coasts are generally found in **high-energy environments**. In the UK, these tend to be:

◆ stretches of the Atlantic-facing coast, where the waves are powerful for much of the year (such as Cornwall or north-west Scotland)

◆ where the rate of erosion exceeds the rate of deposition.

Erosional landforms, such as *headlands*, *cliffs*, and *wave-cut platforms* (sometimes referred to as *abrasion* platforms, even though they are not!), tend to be found in these environments.

In contrast, sandy and estuarine coasts are generally indicative of **low-energy environments**. In the UK, these tend to be:

◆ stretches of the coast where the waves are less powerful, or where the coast is sheltered from large waves (such as the estuaries and bays of Lincolnshire)

◆ where the rate of deposition exceeds the rate of erosion.

Landforms such as *beaches*, *spits* and *coastal plains* tend to be found in these environments.

### Wave refraction

At a more localised scale, high and low-energy stretches of coast may result from *wave refraction* – the distortion of wave fronts as they approach an indented shoreline (Figure **8**). Wave refraction causes energy to be concentrated at headlands and dissipated in bays. This accounts for the presence of erosive features at headlands (cliffs, stacks) and deposition features in bays (beaches).

The concept of negative feedback can be seen to operate here. Variations in rock strength lead to the formation of headlands and bays. This causes wave refraction which, in turn, encourages erosion of the headlands and deposition in the bays – working against the erosion of the softer rock that formed the bay originally. If conditions remained stable for a long period of time (which they don't!), a state of equilibrium would be reached, where the shape of the coastline remains static due to a balance between the potential erodibility of the rocks and the effect of wave refraction.

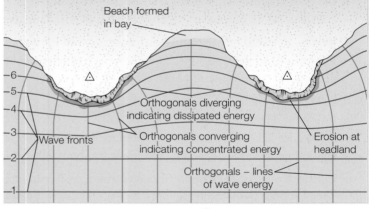

▲ **Figure 8** *Wave refraction*

---

### ACTIVITIES

1 Describe the formation of waves and identify the factors affecting the energy of waves.

2 What are the main differences between constructive and destructive waves?

**S**

3 With the aid of a diagram, explain what happens to waves when they approach the coast.

4 Use simple diagrams to explain how wave types and subsequent beach formation can form an example of a negative feedback.

5 Evaluate the role of tides and currents in the development of coastal landforms and landscapes.

**S**

6 With reference to Figure **2**, attempt to show potential high-energy and low-energy stretches of coastline on an outline map of the UK. Write a reasoned commentary to support your map – ensure you refer to both fetch and wind direction.

7 What is wave refraction and how does it demonstrate the concept of negative feedback?

In this section you will learn about sediment sources, cells and budgets in coastal systems

## Sources of sediment

Globally, most beach sediment comes from rivers, streams and coastal erosion. However, there are considerable local variations. The main sediment sources are as follows:

◆ Rivers – sediment that is transported in rivers often accounts for the vast majority of coastal sediment, especially in high-rainfall environments where active river erosion occurs. This sediment will be deposited in river mouths and estuaries where it will be reworked by waves, tides and currents.

◆ Cliff erosion – this can be extremely important locally in areas of relatively soft or unconsolidated rocks. The extensive till cliffs along the Holderness coast in Lincolnshire comprise sand and clay and rates of erosion can be as high as 10 m a year (Figure **1**). This contrasts with the tough, igneous granites in Cornwall that erode at very slow rates.

◆ Longshore drift – sediment is transported from one stretch of coastline (as an output) to another stretch of coastline (as an input).

◆ Wind – in glacial or hot arid environments, wind-blown sand can be deposited in coastal regions. Sand dunes are semi-dynamic features at the coast that represent both accumulations (sinks) of sand and potential sources.

◆ Glaciers – in some parts of the world, such as Alaska, Greenland and Antarctica, ice shelves *calve* (chunks of ice breaking off a glacier or ice sheet) into the sea, depositing sediment trapped within the ice (Figure **2**).

◆ Offshore – sediment from offshore can be transferred into the coastal (littoral) zone by waves, tides and currents. In the UK, sea levels rose at the end of the last glacial period, resulting in a considerable amount of coarse sediment being bulldozed onto the south coast of England to form landforms such as barrier beaches (e.g. Start Bay in Devon and Chesil Beach in Dorset). Storm surges associated with tropical cyclones and tsunami waves can also be responsible for inputs of sediment into the coastal system.

## Sediment cells

A *sediment cell* is a stretch of coastline, usually bordered by two prominent headlands, where the movement of sediment is more or less contained. Figure **3** shows the sediment cell in the form of a conceptual systems diagram, with inputs (sources), transfers (flows) and stores (sinks).

◆ Inputs (sources) – these are primarily derived from the river, coastal erosion and offshore sources, such as bars or banks.

◆ Transfers (flows) – these involve longshore (littoral) drift together with onshore and offshore processes such as rip currents.

◆ Stores (sinks) – these include the beach, sand dunes and offshore deposits (bands and bars).

⬆ **Figure 1** *Rapidly eroding Holderness coast*

⬇ **Figure 2** *Ice calving at Alaska Bay, Alaska, USA*

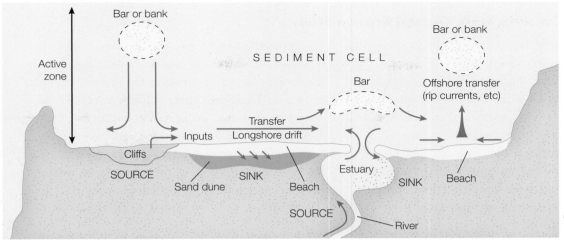

Figure 3 *The sediment cell system*

Some material within the cell may be swept out to sea to act as an output from the system. This may occur as a result of a severe storm event. The temporal nature of the sediment stores is interesting to consider; just how permanent are they? It is all a matter of time scale, as even long-term stores such as sand dunes, sandbanks and beaches may be destroyed by severe storms or long-term sea level rise associated with climate change. In terms of energy, the primary source of energy is the Sun, converted into wave energy by the wind. Tectonic energy is another source of energy in that it will generate tsunami waves.

## Sediment cells in England and Wales

Each of the eleven major cells in England and Wales (Figure **2**, 3.1) can be divided into several smaller subcells. One of these subcells lies in Christchurch Bay in Dorset (Figure **4**). Within the Christchurch Bay subcell, the interlinking marine processes of erosion, transportation and deposition can be mapped. It is possible to identify sediment sources (inputs) such as the cliffs to the west of Barton, areas of deposition (stores or sinks) such as The Shingles, and transfer mechanisms, such as longshore drift operating from west to east.

Figure 4 *Sediment movement in the Christchurch Bay subcell*

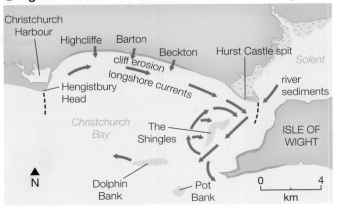

## Sediment budgets

Material in a sediment cell can be considered in the form of a *sediment budget*, with losses and gains. Losses from the system involve deposition in sediment sinks, whereas gains tend to involve coastal erosion or sediment brought into the system by rivers or from offshore sources. In principle, the sediment budget seeks to achieve a state of *dynamic equilibrium* where erosion and deposition are balanced. This balance can be upset by events, such as a surge in river discharge following floods introducing vast amounts of sediment into the system. This, in turn, leads to deposition in the river estuary. A severe storm might also upset the balance by eroding a beach and transferring sediment outside the system.

Figure **5** shows the main inputs from coastal erosion together with the transfers along the coast in East Anglia. By comparing the values of sediment movement, it is possible to identify losses and gains at points around the coast and to make assertions about the location of sediment sinks or alternative sources of sediment. This diagram illustrates the difficulty in obtaining data for all components of a sediment cell.

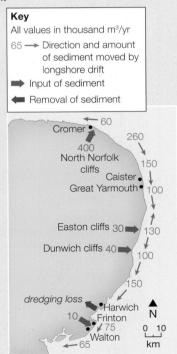

Figure **5** *Coastal sediment budget for East Anglia*

## The South Carolina, USA, Coastal Erosion Study

The US Geological Survey (USGS) conducts regional studies of coastal erosion to provide impartial scientific information necessary for the protection and management of valuable coastal resources. One such study involved the north-east coast of South Carolina. The main objective of the study was to determine the geologic and oceanographic processes that control sediment movement along the region's shoreline, thereby improving projections of coastal change, in particular increased coastal erosion resulting from climate change.

The research involved dredging and beach profiling together with the use of secondary data. Figure **6** shows the main components of the sediment budget and Figure **7** lists the results of the research. These results show that, in order to account for the losses as measured directly by dredging of sediment sinks, a huge amount of erosion is taking place, particularly in the inner shelf. Interestingly, relatively small amounts of sediment are gained from river deposition.

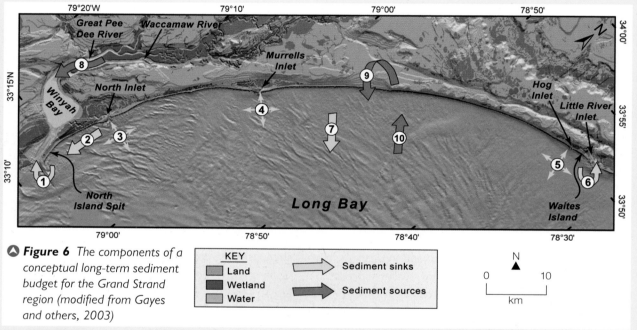

**Figure 6** *The components of a conceptual long-term sediment budget for the Grand Strand region (modified from Gayes and others, 2003)*

| KEY | |
|---|---|
| Land | |
| Wetland | Sediment sinks |
| Water | Sediment sources |

| Sediment sinks (losses) | Volume (m³/yr) | Basis of estimate |
|---|---|---|
| 1. Winyah Bay | 284 000+ | 1994–2002 dredging records |
| 2. North Island Spit | 79 000 | Historic spit growth |
| 3. North Inlet | unknown | |
| 4. Murrells Inlet | 75 000 | 1974–1978 dredging records |
| 5. Hog Inlet | unknown | |
| 6. Little River Inlet | 57 000 | 1982–1995 dredging records |
| 7. Loss offshore | unknown | |
| **Total Sinks** | **495 000** | |
| **Sediment sources (gains)** | **Volume (m³/yr)** | **Basis of estimate** |
| 8. Rivers | very small | Patchineelam and others (1999) |
| 9. Beach and shoreface erosion | 104 000 | Beach profile migration based on the long-term average erosion rate |
| 10. Inner shelf erosion | 391 000 | Difference between total sinks (1–7) and gains from other sources (8 and 9) |
| **Total Sources** | **495 000** | |

**Figure 7** *Estimated volumes of mobile sediment annually lost to sinks and gained from different sources (modified from Gayes and others, 2003)*

## The impact of coastal protection on sediment budgets and the sediment cell

Figure **8** shows a stretch of coastline before and after coastal protection. Notice that the coastal protection measures have the potential to significantly disrupt the operation of the sediment cell and affect the sediment budget (also see Figure **6**, 3.5).

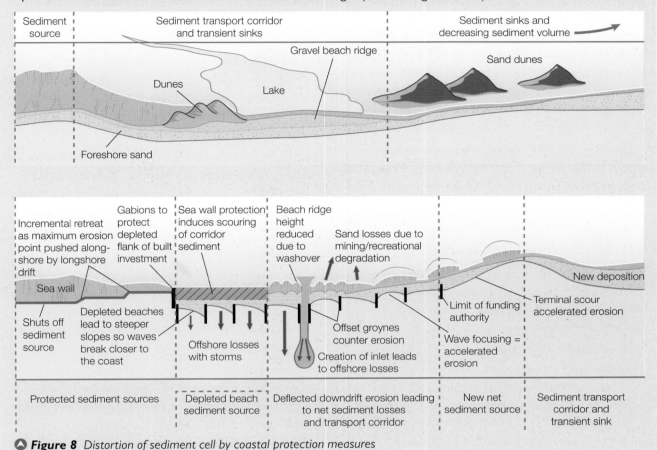

Sediment source

Sediment transport corridor and transient sinks

Sediment sinks and decreasing sediment volume

Gravel beach ridge

Sand dunes

Dunes

Lake

Foreshore sand

Incremental retreat as maximum erosion point pushed along-shore by longshore drift

Gabions to protect depleted flank of built investment

Sea wall protection induces scouring of corridor sediment

Beach ridge height reduced due to washover

Sand losses due to mining/recreational degradation

New deposition

Sea wall

Terminal scour accelerated erosion

Shuts off sediment source

Depleted beaches lead to steeper slopes so waves break closer to the coast

Offshore losses with storms

Offset groynes counter erosion

Creation of inlet leads to offshore losses

Limit of funding authority

Wave focusing = accelerated erosion

Protected sediment sources

Depleted beach sediment source

Deflected downdrift erosion leading to net sediment losses and transport corridor

New net sediment source

Sediment transport corridor and transient sink

▲ **Figure 8** *Distortion of sediment cell by coastal protection measures*

## ACTIVITIES

**(S)**

1  Draw a large conceptual diagram to show the various sources of sediment. Use the internet to expand your annotations and consider using a selection of photos to illustrate your diagram.

2  Make a careful copy of Figure **4** and use labels and/or symbols explained in a key to identify the inputs (sources), transfers (flows) and stores (sinks). Suggest energy flows and consider whether any feedback loops can be identified. Use Figure **2**, 3.1 to help you.

3  Study Figure **5**. Write a short discussion paper using the data to enable you to make assertions about the type and location of sediment sources and sediment sinks on this stretch of coastline. Include a simple sketch to support your discussion.

**(S)**

4  Write a brief report on the South Carolina coastal research project based on the stretch of coast at Grand Strand (Figures **6** and **7**). Your report should identify the purpose of the study, the sediment system in Long Bay, the results of the research and discussion of the implications for the future management of the coastline. Use an appropriate graph or diagram to present the data in Figure **7**. For additional information, access the website at http://pubs.usgs.gov/of/2008/1206/index.html

5  Study Figure **8**. To what extent can coastal protection measures disrupt the operation of a sediment cell and affect its budget?

In this section you will learn about geomorphological processes in coastal environments

## Weathering

*Weathering* is the breakdown or disintegration of rock in situ (its original place) at or close to the ground surface. Energy flows can be clearly demonstrated as most processes involve (directly or indirectly) energy transfer from the Sun, in the form of radiation, or rain. As a process, weathering leads to the transfer (flow) of material. There are also important links with other natural systems, such as the water cycle (e.g. freeze-thaw) and the carbon cycle (e.g. carbonation).

Weathering is active at the coast where rock faces are exposed to the elements and cliff faces kept fresh by the constant removal of debris by the sea.

◆ If the rate of debris removal exceeds the rate of weathering and *mass movement* then a positive feedback may operate, as the rate of weathering and mass movement could increase.

◆ If debris removal is slow and ineffective, this will lead to a build-up of an apron of debris (*scree*) that reduces the exposure of the cliff face as it extends up the cliff face. Weathering and mass movement rates will decrease – a negative feedback.

Weathering can be divided into three different types: mechanical, biological and chemical. By breaking rock down, weathering creates sediment that the sea can then use to help erode the coast.

### Mechanical (physical) weathering

Mechanical (or physical) weathering involves the break-up of rocks without any chemical changes taking place. There are several types of mechanical weathering processes that are active at the coast.

◆ *Frost shattering* (also known as *freeze-thaw*) occurs when water enters a crack or joint in the rock when it rains and then freezes in cold weather. When water freezes, it expands in volume by about 10 per cent. This expansion exerts pressure on the rock, which forces the crack to widen (see 2.6 and 4.6). With repeated freezing and thawing, fragments of rock break away and collect at the base of the cliff as scree. These angular rock fragments are then used by the sea as tools in marine erosion. Although the coast tends to be milder than inland, frost shattering is still important.

For example, in 2001 (following a very wet autumn and a cold February) frost shattering triggered several major rockfalls along the south coast of England (see Figure **1**). Chalk (a permeable and porous rock) was the main rock affected.

◆ *Salt crystallisation*. When salt water evaporates, it leaves salt crystals behind. These can grow over time and exert stresses in the rock, just as ice does, causing it to break up (see 2.6). Salt can also corrode rock, particularly if it contains traces of iron.

◆ *Wetting and drying*. Frequent cycles of wetting and drying are common on the coast. Rocks rich in clay (such as shale) expand when they get wet and contract as they dry. This can cause them to crack and break up.

▶ **Figure 1** *A major rockfall at the White Cliffs of Dover in 2012, caused by frost shattering*

## Biological weathering

The breakdown of rocks by organic activity is *biological weathering*. There are several ways in which this can operate at the coast.

◆ Thin plant roots grow into small cracks in a cliff face. These cracks widen as the roots grow, which breaks up the rock (Figure **2**).

◆ Water running through decaying vegetation becomes acidic, which leads to increased chemical weathering (see below).

◆ Birds (e.g. puffins and sand martins) and animals (e.g. rabbits) dig burrows into cliffs.

◆ Marine organisms are also capable of burrowing into rocks (e.g. piddocks, which are similar to clams), or of secreting acids (e.g. limpets).

▲ **Figure 2** *Biological weathering*

## Chemical weathering

Chemical weathering involves a chemical reaction where salts may be dissolved or a clay-like deposit may result which is then easily eroded.

◆ *Carbonation* – rainwater absorbs carbon dioxide from the air to form a weak carbonic acid. This reacts with calcium carbonate in rocks, such as limestone and chalk, to form calcium bicarbonate, which is easily dissolved. The cooler the temperature of the rainwater, the more carbon dioxide is absorbed (so carbonation is more effective in winter). Carbonation is an important part of the carbon cycle (see 1.13).

◆ *Oxidation* – the reaction of rock minerals with oxygen, for example iron, to form a rusty red powder leaving rocks more vulnerable to weathering.

◆ *Solution* – the dissolving of rock minerals, such as halite (rock salt).

## Mass movement

The downhill movement of material under the influence of gravity is known as **mass movement**. It can range from being extremely slow – less than 1 cm a year (e.g. soil creep) – to horrifyingly fast (e.g. rockfalls and landslides). Mass movement at the coast is common – the sheer weight of rainwater, combined with weak geology, is the major cause of cliff collapse.

In February 2014, following the wettest winter on record, the Jurassic Coast near Lyme Regis in Dorset was affected by a number of dramatic landslips, damaging holiday chalets (Figure **3**). This exposed stretch of coastline and is constantly being shaped and reshaped by processes of mass movement invigorated by undercutting by the sea.

Mass movement forms an important group of processes and flows within the coastal system, transferring both energy (in response to gravity) and sediment. The sediment forms an important input to shoreline processes, forming the 'tools' for erosion and providing material to be transported and deposited elsewhere along the coastline. Mass movement, along with cliff erosion, provides an important input to sediment cells (see 3.3).

▲ **Figure 3** *One of seven holiday chalets that had to be demolished and removed from this location at Lyme Regis in 2014 following a series of landslips*

## Types of mass movement

Mass movement can be classified into four main types – creep, flow, slide and fall (Figure **4**). Each process represents a flow or transfer of material and can be considered to be an output from one store (land) and an input to another store (beach/sea). The type of movement at any one place depends upon a range of factors – angle of the slope or cliff; rock type and structure; vegetation cover; how wet the ground is.

| Type of mass movement | Nature of movement | Rate of movement | Wet/dry |
|---|---|---|---|
| Soil creep Solifluction | Creep/flow | Imperceptible | Wet |
| Mudflow | Flow | Often quite rapid | Wet |
| Runoff | Flow | Rapid | Wet |
| Landslide/debris slide Slump/slip | Slide | Usually rapid | Dry Wet |
| Rockfall | Fall | Rapid | Dry |

⬆ **Figure 4** *Classifying different types of mass movement*

### Soil creep

As the name implies, *soil creep* is an extremely slow form of movement of individual soil particles downhill. The precise mechanism of movement often involves particles rising towards the ground surface due to wetting or freezing and then returning vertically to the surface in response to gravity as the soil dries out or thaws (see 4.7). This zigzag movement is similar to that of longshore drift. Soil creep cannot be seen operation but its action can be implied by the formation of shallow terracettes, the build-up of soil on the upslope side of walls and the bending of tree trunks (Figure **5**).

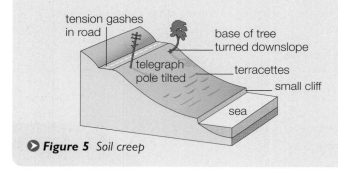

▶ **Figure 5** *Soil creep*

### Mudflows

A *mudflow* involves earth and mud flowing downhill, usually over unconsolidated or weak bedrock such as clay, often after heavy rainfall. Water gets trapped within the rock, increasing pore water pressure, which forces rock particles apart and leads to slope failure. Pore water pressure is a form of energy within the slope system and it is an extremely important factor in determining slope instability. Mudflows are often sudden and fast-flowing so can represent a significant natural hazard.

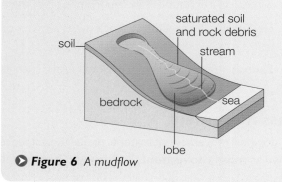

▶ **Figure 6** *A mudflow*

### Landslide

A *landslide* involves a block of rock moving very rapidly downhill along a planar surface (a slide plane), often a bedding plane that is roughly parallel to the ground surface (Figure **7**). Unlike a mudflow, where the moving material becomes mixed, the moving block of material in a landslide remains largely intact.

Landslides are frequently triggered by earthquakes or very heavy rainfall, when the slip surface becomes lubricated and friction is reduced. Landslides tend to be very rapid and pose a considerable threat to people and property. In 1993, 60 m of cliff slid onto the beach near Scarborough in North Yorkshire, taking with it part of the Holbeck Hall Hotel (Figure **8**).

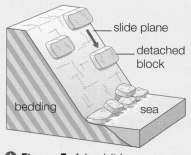

⬆ **Figure 7** *A landslide*

▼ **Figure 8** *Landslide in 1993 at Holbeck Hall, Scarborough*

## Rockfall

A *rockfall* involves the sudden collapse or breaking away of individual rock fragments (or a block of rock) at a cliff face. They are most commonly associated with steep or vertical cliffs in heavily jointed and often quite resistant rock. A rockfall is often triggered by mechanical weathering (particularly freeze-thaw) or an earthquake. Once broken away from the source, rocks fall or bounce down the slope to form scree (also known as talus) at the foot of the slope (Figure **9**). Scree often forms a temporary store within the coastal system, with material gradually being removed and transported elsewhere by the sea. When this occurs the scree forms an input into the sediment cell.

◀ *Figure 9  A rockfall*

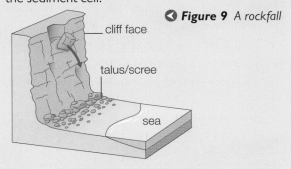

## Runoff

*Runoff* is a good illustration of the link between the water cycle and the coastal system. When overland flow occurs down a slope or cliff face, small particles are moved downslope to enter the littoral zone, potentially forming an input into the sediment cell. Runoff can be considered a type of flow that transfers both water and sediment from one store (the rock face) to another (a beach/the sea).

Toxic chemicals can contaminate stormwater and cause threats to coastal ecosystems, illustrating yet another link between natural systems (see 6.10).

## Solifluction

Essentially, *solifluction* is similar to soil creep but specific to cold periglacial environments (see 4.7 and 4.11). In the summer, the surface layer of soil thaws out and becomes extremely saturated because it lies on top of impermeable frozen ground (permafrost). Known as the active layer, this sodden soil with its blanket of vegetation slowly moves downhill by a combination of heave and flow. Solifluction characteristically forms features called *solifluction lobes* (see 4.11).

## Landslip or slump

A *landslip* or *slump* differs from a landslide in that its slide surface is curved rather than flat. Landslips commonly occur in weak and unconsolidated clays and sands, often when permeable rock overlies impermeable rock, which causes a build-up of pore water pressure. Landslips or slumps are characterised by a sharp break of slope and the formation of a scar (Figures **10** and **11**). Multiple landslips can result in a terraced appearance on the cliff face.

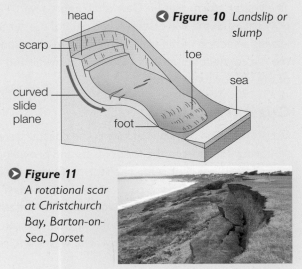

◀ *Figure 10  Landslip or slump*

▶ *Figure 11  A rotational scar at Christchurch Bay, Barton-on-Sea, Dorset*

### ACTIVITIES

1  In pairs, discuss and decide (**a**) the part played by weathering and its influence on the rate of coastal retreat, (**b**) which forms of weathering are likely to have the greatest impact in different parts of the UK.

2  Explain why evidence of past solifluction can be seen in some parts of the UK.

3  Explain the significance of weathering and mass movement in relation to the coastal system. Use the correct systems terminology in your answer and consider drawing a simple diagram to support your answer.

4  Explain how the methods by which sediment is supplied to the coastal system are likely to vary between areas of resistant and weak geology.

5  Distinguish between the following pairs of terms: (**a**) weathering and mass movement; (**b**) flows and slides; (**c**) landslides and landslips; (**d**) soil creep and solifluction; (**e**) rockfalls and slumping.

Ⓢ 6  In pairs, discuss and devise two flow diagrams to outline the sequence of processes that probably led to the rockfall at Dover (Figure **1**) and the landslip at Lyme Regis (Figure **3**).

**In this section you will learn about coastal processes**

## An unbelievable storm

The residents of Riviera Terrace in Dawlish are used to their homes shaking when a storm hits the South Devon coast. 'But this was different,' said one resident. 'It was like being in a car wash. The storm was unbelievable and waves were pounding against the terrace.'

The winter storm that hit Dawlish in February 2014 was so powerful that the waves destroyed part of the sea wall – leaving a section of rail track dangling in mid-air and cutting the rail connection between Devon, Cornwall and the rest of the UK for two months.

## Coastal erosion

Coastal erosion plays a vital role in the coastal system, removing debris from the foot of cliffs and providing an input into coastal sediment cells. The storm of February 2014 transformed the shape of parts of the Devon coast as huge quantities of sediment were swept away from beaches; a few miles from Dawlish, sand dunes at Dawlish Warren (spit) were severely eroded by the powerful waves and several groynes were damaged.

Coastal erosion is a manifestation of the energy of the Sun, converted by the power of the wind into waves capable of sculpting landforms and eroding sediment. While it is possible to identify several distinctive processes of marine erosion, in reality they will often work collaboratively to erode a stretch of coastline. Figure **2** shows undercutting of a cliff resulting from the processes of erosion.

### Coastal erosion processes
#### Hydraulic action

Look at Figure **3**. The sheer force of the water as it crashes against a coastline is called *hydraulic action*. When a wave advances, air can be trapped and compressed, either in joints in the rock or between the breaking wave and the cliff. When the wave retreats, the compressed air expands. This continuous process can weaken joints and cracks in the cliff, causing pieces of rock to break off. Simultaneously, bubbles formed in the water may implode under the high pressure. This generates tiny jets of water which will, over time, erode the rock. This process is specifically termed cavitation.

▲ **Figure 1** *The powerful storm of February 2014 – waves crash against the sea front and railway line in Dawlish*

▲ **Figure 2** *Cliff undercutting and weakening*

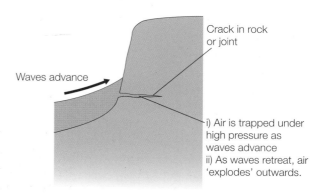

Crack in rock or joint

Waves advance

i) Air is trapped under high pressure as waves advance
ii) As waves retreat, air 'explodes' outwards.

▲ **Figure 3** *Cliff-foot erosion – hydraulic action*

## Wave quarrying

*Wave quarrying* is the action of waves breaking against unconsolidated material such as sands and gravels. Waves scoop out the loose material in a similar way to the action of a giant digger in a quarry on land.

## Corrasion

When waves advance, they pick up sand and pebbles from the seabed, a temporary store or sediment sink. When they break at the base of the cliff, the transported material is hurled at the cliff foot – chipping away at the rock. This is *corrasion* (Figure **4**). This is a good example of an energy flow in action within the coastal system. The size, shape and amount of sediment picked up by the waves, along with the type of wave, determines the relative importance of this erosive process.

## Abrasion

Corrasion hurls sediment at a cliff face. Abrasion involves more of a 'sandpapering effect' as sediment is dragged up and down or across the shoreline, eroding and smoothing rocky surfaces. Abrasion is particularly important in the formation of a *wave-cut platform* (see 3.6).

## Solution (corrosion)

Weak acids in seawater can dissolve alkaline rock (such as chalk or limestone), or the alkaline cement that bonds rock particles together. This is *solution*. The action of this type of erosion may be indistinguishable from the action of carbonation, a type of weathering (and an important link to the carbon cycle). This is a good example of processes (in this case erosion and weathering) working collaboratively and being hard to separate. Does it really matter?

In addition, the process of *attrition* also takes place at the coast but it is not directly responsible for the erosion of a coastline. Attrition refers to the gradual wearing down of rock particles by impact and abrasion, as the pieces of rock are moved by waves, tides and currents. This process gradually makes stones rounder and smoother.

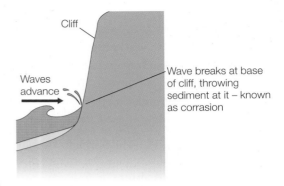

**Figure 4** *Cliff-foot erosion – corrasion*

**Figure 5** *Boulder clay cliffs on the south Cornwall coast*

## Factors affecting coastal erosion

There are several factors that affect the nature and rate of coastal erosion, including waves, rock type and geological structures, the presence or absence of a beach, subaerial processes and management.

◆ Waves – the rate and type of erosion experienced on a particular stretch of coast is primarily influenced by the size and type of waves that reach that coast. Most erosion happens during winter storms (such as the one that hit Dawlish in 2014), when destructive waves are at their largest and most powerful.

◆ Rock type (lithology) – rock lithology (its physical strength and chemistry) is important in determining the rate of erosion. Tough and resistant rocks such as granite erode at very slow rates compared to weaker clays and shales. In the UK, some of the fastest rates of erosion occur on the Holderness coast in Lincolnshire where unconsolidated glacial till deposits have been eroded by 120 m in the past century. During the same time period, the granite at Land's End has been eroded by just 10 cm!

- ◆ Geological structure – cracks, joints, bedding planes and faults create weaknesses in a cliff that can be exploited by erosive processes. On a large scale, variations in rock type and geological structure can lead to the formation of headlands and bays, as a result of the subsequent differential erosion.

- ◆ Presence or absence of a beach – beaches absorb wave energy and reduce the impact of waves on a cliff. If a beach is absent, following excessive erosion or as a result of management techniques elsewhere on the coast, a cliff may experience increased erosion as it is more vulnerable to wave attack.

- ◆ Subaerial processes – weathering and mass movement will weaken cliffs and create piles of debris that are easily eroded by the sea, potentially increasing the rate of erosion.

- ◆ Coastal management – the presence of structures such as groynes and sea walls will have an impact on sediment transfer (and the build-up of beaches) and patterns of wave energy along a coastline. In trapping sediment moved by longshore drift, groynes may deprive beaches further down-drift of sediment input and may decrease in extent. Sea walls may deflect wave energy elsewhere along the coast exacerbating erosion in those localities.

## Coastal transportation processes

Coastal transportation plays a major role in the coastal system, transferring sediment from one store to another, for example, from the foot of a cliff to a spit or an offshore bar. As with coastal erosion, it is a manifestation of an energy flow, governed, controlled and determined by the power of waves, tides and currents. It is possible to identify four methods of transportation:

- ◆ *Traction* – the rolling of coarse sediment along the sea bed that is too heavy to be picked up and carried by the sea

- ◆ *Saltation* – sediment 'bounced' along the seabed, light enough to be picked up or dislodged but too heavy to remain within the flow of the water

- ◆ *Suspension* – smaller (lighter) sediment picked up and carried within the flow of the water

- ◆ *Solution* (dissolved load) – chemicals dissolved in the water, transported and precipitated elsewhere. This form of transportation plays an important role in the carbon cycle, transferring and redepositing carbon in the oceans.

The key factors affecting the type of transportation are velocity (energy) and particle size (mass). In high-energy environments, larger particles will be able to be transported, whereas in low-energy environments only the finest particles (clays) will be transported.

## Longshore (littoral) drift

Most waves approach a beach at an angle – generally from the same direction as the prevailing wind. Along the coast of southern England, for example, winds blow onshore from the south-west throughout the year. As the waves advance, material is carried up the beach at an angle. The backwash then pulls material down the beach at right angles to the shore (due to the force of gravity, in effect a source of energy). The net effect of the zigzag movement of sediment up and down the beach is a process is known as *longshore (littoral) drift* (see Figure **6**).

Longshore drift is an important transfer (flow) mechanism as it is responsible for moving vast amounts of sediment along the coastline and eventually out to sea, for example, at the tip of a spit. It is a very important component in a sediment cell and, if interrupted by management strategies, can lead to distortions of natural patterns, depriving beaches of material and exacerbating erosion.

Stores, flows and sinks will all be affected if longshore drift is interrupted. For example, in West Africa, the prevailing south-west winds transport sediment from west to east, from Ghana eastwards to Togo and Benin. However, due to coastal management, not enough sand is reaching the eastern countries to replace that transported away by longshore drift. As a result of this, parts of Benin's coastline is being eroded at an astonishing 10 m per year due to the lack of a protective beach in front of the cliffs.

◆ **Figure 6** *The process of longshore drift*

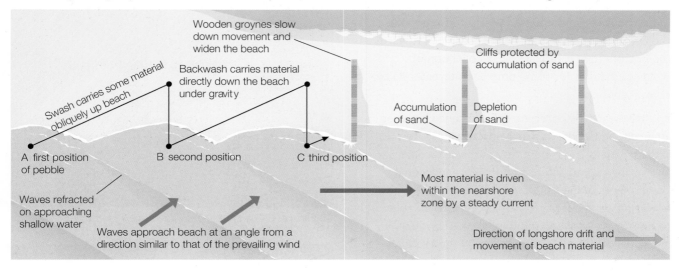

Wooden groynes slow down movement and widen the beach

Cliffs protected by accumulation of sand

Swash carries some material obliquely up beach

Backwash carries material directly down the beach under gravity

Accumulation of sand

Depletion of sand

A first position of pebble

B second position

C third position

Most material is driven within the nearshore zone by a steady current

Waves refracted on approaching shallow water

Waves approach beach at an angle from a direction similar to that of the prevailing wind

Direction of longshore drift and movement of beach material

## Coastal deposition

Deposition takes place when the velocity of the water (or wind) falls below a critical value for a particular size of particle and can no longer be transported. In high-energy environments, such as exposed parts of the south coast of England, clay and sand will be easily transported away leaving behind the larger, coarser pebbles to form the characteristic shingle beaches. In low-energy environments, such as river estuaries in the lee of spits, the very smallest clay particles will eventually drop to the seabed to form mudflats. Some sediment may be carried offshore to form underwater sandbanks.

Sediment deposition is an extremely important part of the overall coastal system and, more specifically, the sediment cell. Areas of deposition – beaches, spits, mudflats, sand dunes and offshore bar – are all sediment stores or 'sinks' (Figure **7**). While some sediment sinks may be considered to be outputs they can also act as important inputs to the coastal system. For example, offshore sediment deposits have been driven onto the south coast of England by rising sea levels following the end of the last glacial period.

▲ **Figure 7** *A sandbank at Ko Poda Island, Thailand – an example of a sediment sink*

## ACTIVITIES

**1 a** Study Figure **2**. What types of erosion are likely to have been dominant? Give reasons for your answer.

 **b** Suggest the factors responsible for the nature and extent of coastal erosion of this cliff.

**(S)**

 **2** With the aid of a simple diagram describe the process of longshore (littoral) drift.

**3** Under what circumstances and for what reasons does coastal deposition occur?

**4** Discuss the factors affecting the rate and nature of coastal transportation.

**5** How can coastal management schemes such as constructing sea walls and groynes result in a positive feedback loop by increasing the rate of erosion elsewhere at the coast?

## STRETCH YOURSELF

Find out more about the effect of the storm at Dawlish in 2014 and the subsequent rebuilding that was required. Consider the reasons for the severe erosion and summarise the impacts the storm had on this stretch of coastline. Explain how the magnitude of the different components of the coastal system changed during the storm.

In this section you will learn about landforms and landscapes of coastal erosion and factors and processes in their development

## Coastal landforms and landscapes

Figure **1** was taken from Beachy Head, to the west of Eastbourne, on the Sussex Heritage Coast, one of the UK's most iconic stretches of coastline. Towards the bottom right is Birling Gap, a collection of buildings that has for some time been threatened by coastal erosion.

It is important to appreciate the difference between *landscape* and *landform*. The landscape in Figure **1** is the big picture – the entirety of the sea, coast and rolling countryside. Landforms are individual components of the landscape – cliffs, beach and the emerging wave-cut platform. You need to be able to distinguish between the two; always consider individual landforms in the context of the broader landscape.

## Landforms of coastal erosion

### Cliffs and wave-cut platforms

When waves break against the foot of a *cliff*, erosion (hydraulic action and corrosion in particular) tends to be concentrated close to the high-tide line. This creates a *wave-cut notch* (Figure **2**). As the notch gets bigger, the cliff is undercut and the rock above it becomes unstable, eventually collapsing.

As these erosional processes are repeated, the notch migrates inland and the cliff retreats (Figure **3**), leaving behind a gently sloping *wave-cut platform* (Figure **4**), which is usually only completely exposed at low tide. Wave-cut platforms rarely extend for more than a few hundred metres, because a wave will break earlier and its energy will be dissipated before it reaches the cliff, thus reducing the rate of erosion, limiting the further growth of the platform. This is another excellent example of a negative feedback.

▲ **Figure 1**  The landscape of the Sussex Heritage Coast

▲ **Figure 2**  A wave-cut notch at Flamborough Head in Yorkshire

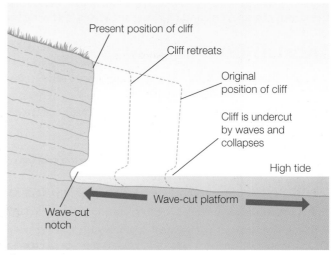

▲ **Figure 3**  Cliff retreat

## Cliff profile and rate of retreat

There are several factors that affect the cliff profile and its rate of retreat.

◆ Steep cliffs tend to occur where the rock is strong and resistant to erosion, such as most igneous and metamorphic rocks. Sedimentary rocks that are dipping steeply or even vertically tend to produce steep and dramatic cliffs (Figure **5**), as will the absence of a beach and an exposed orientation with a long fetch and high-energy waves that encourage erosion and undercutting by the sea.

◆ Gentle cliffs usually reflect weak or unconsolidated rocks that are prone to slumping. Rocks that are dipping towards the sea also tend to have low-angle cliffs. A sheltered location with low-energy waves and a short fetch will result in subaerial debris building up at the foot of the cliff, reducing its overall angle. A wide beach will absorb wave energy, preventing significant undercutting and steepening.

◆ The rate of retreat of a cliff very much depends on the balance between marine factors – such as wave energy, fetch, presence of a beach – and terrestrial factors – such as subaerial processes, rock geology and lithology (Figure **6**). The most rapidly retreating cliffs tend to be composed of very weak rock, such as the glacial till cliffs of the Holderness coast.

▲ *Figure 4* Cliffs and a wave-cut platform near Eastbourne, East Sussex

▲ *Figure 5* Vertical rock structure in cliffs near Tenby, South Wales

Other factors leading to rapid rates of cliff retreat include rising sea levels, and human activities such as coastal defences elsewhere leading to increased erosion.

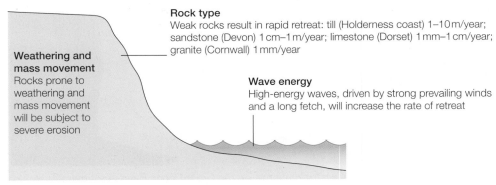

**Rock type**
Weak rocks result in rapid retreat: till (Holderness coast) 1–10m/year; sandstone (Devon) 1cm–1m/year; limestone (Dorset) 1mm–1cm/year; granite (Cornwall) 1mm/year

**Weathering and mass movement**
Rocks prone to weathering and mass movement will be subject to severe erosion

**Wave energy**
High-energy waves, driven by strong prevailing winds and a long fetch, will increase the rate of retreat

▲ *Figure 6* Factors affecting the rate of retreat

**Coastal morphology** is related not only to the underlying geology, or rock type, but also to its *lithology* – its physical composition. It includes any of the following characteristics:

◆ Strata – layers of rock

◆ Bedding planes – horizontal, natural breaks in the strata, caused by gaps in time during periods of rock formation

◆ Joints – vertical fractures caused either by contraction as sediments dry out, or by earth movements during uplift

◆ Folds – formed by pressure during tectonic activity, which makes rocks buckle and crumple (e.g. the Lulworth Crumple)

◆ Faults – formed when the stress or pressure to which a rock is subjected, exceeds its internal strength (causing it to fracture). The faults then slip or move along fault planes

◆ Dip – refers to the angle at which rock strata lie (horizontally, vertically, dipping towards the sea, or dipping inland).

The relief – or height and slope of land – is also affected by geology and geological structure. There is a direct relationship between rock type, lithology and cliff profiles. The five diagrams that make up Figure **7** help to illustrate this.

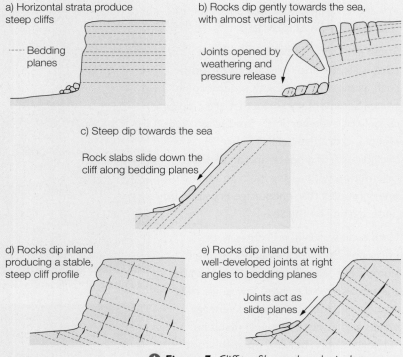

**Figure 7** *Cliff profiles and geological structure*

## Cliffs profile features – caves, arches and stacks

One of the world's most iconic road trips is the Great Ocean Road coastal route in Southern Australia between Melbourne and Adelaide. With its towering sandstone cliffs, isolated stacks and spectacular arches, this stretch of coastline is one of the most dramatic landscapes in Australia (Figure **8**). It demonstrates clearly how erosion on a high-energy coast will create several landforms as the cliffs – notice their steep profile – are steadily eroded.

**Figure 8** *Coastal landforms on the Great Ocean Road in southern Australia. The stacks in the foreground are called the Twelve Apostles and are a big tourist attraction.*

Caves, arches, stacks and stumps are all connected as part of a sequence of coastal landform development:

**Figure 9** *The formation of caves, arches, stacks and stumps at a headland*

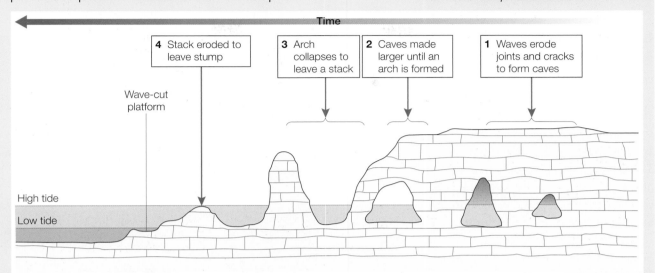

- The erosion of rocks like limestone and chalk tends to exploit any lines of weakness – joints, faults and cracks.

- When joints and faults are eroded by hydraulic action and abrasion, this can then create *caves*. If the overlying rock then collapses, a *blowhole* will develop. During storm high tides, seawater can be blown out of these blowholes with considerable and spectacular force.

- If two caves join up, or a single cave is eroded through a headland, an *arch* is formed. The base is widened and the gap is further enlarged by erosion and weathering.

- Eventually, the top of the arch collapses leaving an isolated pillar of rock called a *stack*. As it continues to be eroded by the sea, the stack collapses to leave a *stump*, which may only appear above the surface at low tide.

## ACTIVITIES

**S**

1  Draw a sketch of Figure **1** and add labels to identify aspects of the coastal system, such as inputs, outputs, stores and transfers. Can you suggest the operation of any feedback loops?

2  Study Figure **5**. To what extent do you think geological structure is responsible for the profile of these cliffs?

**S**

3  **a**  Draw an annotated field sketch of the coastal landforms shown in Figure **8**.

   **b**  What is the evidence that this stretch of coastline is being actively eroded?

   **c**  Suggest how this landscape might change over the next hundred years.

**S**

4  Produce a flow chart to show the development of wave-cut notches, wave-cut platforms and cliffs.

5  How might a negative feedback loop operate during the development of an extensive wave cut platform? Use simple diagrams to support your answer.

## STRETCH YOURSELF **S**

Using Google Maps/Earth imagery, identify two stretches of coast in the UK, and two stretches in the rest of the world, where features in this section can be seen. Add labels to identify the landforms and write a few sentences describing the overall landscape.

In this section you will learn about landforms and landscapes of coastal deposition and factors and processes in their development

## Landscapes of coastal deposition

Figures **1** and **2** show two contrasting landscapes of coastal deposition. Figure **1** is Chesil Beach in Dorset, a huge shingle ridge several metres in height and stretching for some 29 km. It is a high-energy, south-facing coastline with a long fetch across the Atlantic Ocean and is an important store of sediment. The relatively coarse sediment is a reflection of the high-energy waves (energy flow) that batter this coastline for much of the year. Although it is currently being reshaped by longshore drift (a transfer in the coastal system), most of the sediment was driven onshore by rising sea levels after the last glacial period. The direct source of this sediment was the English Channel, where it was originally dumped by meltwater some 8000 years ago, as the ice on the land melted. This illustrates the 'temporary' nature of some sediment stores as well as the importance of time; is 'temporary' measured in years or thousands of years?

In contrast, Figure **2** shows a sheltered bay on the south coast of Finland. Here, the low-energy waves (resulting from wave refraction) in the bay have created a large sandy beach, which is popular with visitors in the summer (what do you think the white structures are?). Some of the sand is blown onshore by winds (energy flow) to form sand dunes (a store), which are colonised by vegetation, tolerant of the salty and exposed conditions.

▲ *Figure 1* *High-energy, shingle coastline at Chesil Beach, Dorset, UK*

▲ *Figure 2* *Low-energy, sandy coastline at Hanko on the south coast of Finland*

## Landforms of coastal deposition

Deposition occurs along the coast when waves no longer have enough energy to transport sediment. Depending on how and where the sediment is deposited, a variety of landforms can be produced.

### Beaches

A beach can be described as a depositional landform extending from approximately the highest high tide to the lowest low tide. It is an extremely important temporary store in the coastal system. Beach **accretion** will take place during a prolonged period of constructive waves driven by storms many hundreds of miles away. Destructive waves, resulting from localised storms, may excavate the beach, removing vast quantities of sediment and even exposing previously covered wave-cut platforms.

## Swash-aligned and drift-aligned beaches

Beaches can be described as being *swash aligned* or *drift aligned* depending on their orientation relative to the prevailing wind (and wave) direction (see Figure **3**).

◆ **Swash-aligned beaches** tend to form in low-energy environments such as bays that are affected by waves arriving roughly parallel to the shore. You can see the effect of wave refraction in Figure **3**. The bayhead beach may consist of either sand or shingle, depending on factors like the nature of the sediment and the power of the waves. High-energy waves will transport sand leaving behind coarser shingle (Figure **1**) whereas low-energy waves will deposit sand (Figure **2**) or mud.

◆ **Drift-aligned beaches** form where the waves approach the coast at an angle. Longshore drift (a transfer process) moves sediment along the beach, often culminating in the formation of a spit, essentially a sediment sink or store. Sediment may be graded along a drift-aligned beach. Finer shingle particles are likely to be carried further by longshore drift and also to become increasingly rounded as they move.

## Beach forms

Beaches often form part of a much broader area of deposition extending into the offshore zone (Figure **5**). Several features characterise beaches, including **berms** (ridges), **cusps** (Figure **4**) and runnels. Notice in Figure **5** that there are often several berms on a beach representing different tidal levels – there may also be a storm berm at the highest point on a beach. Berms can be made of sand or pebbles.

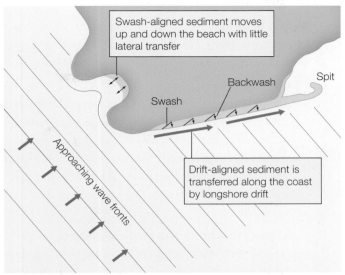

▲ *Figure 3* The formation of swash-aligned and drift-aligned beaches

▲ *Figure 4* Berms and cusps, Gulf of Salerno, Italy

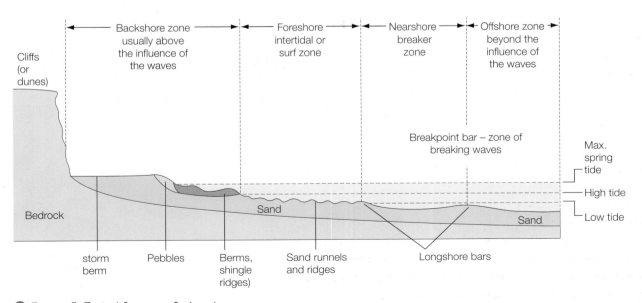

▲ *Figure 5* Typical features of a beach

**Beach profiles**

The material along a beach profile also varies in size and angularity, depending on distance from the shoreline (as Figure **6** shows).

♦ Larger pebbles tend to be near the top of the beach. Constructive waves will carry a range of sediment sizes up a beach due to the strong swash but, due to water percolating into the beach, the weaker backwash will only be able to drag back the smaller pebbles. Over time, this leads to the pebbles being sorted with large at the top through to smaller at the bottom.

♦ Pebbles at the bottom of the beach tend to be more rounded due to the constant action of the waves causing abrasion and attrition. Scree falling off a cliff face can also explain the presence of mostly angular pebbles near the top of the beach.

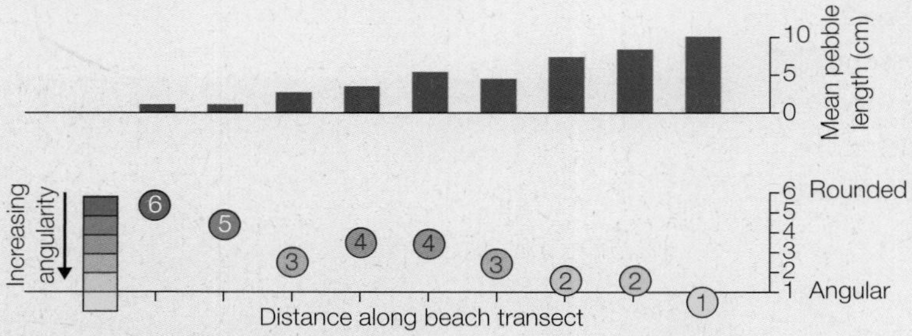

▲ **Figure 6** *Pebble size and shape along a beach transect*

Seasonal changes in wave type create summer and winter profiles – sediment is dragged offshore by destructive waves in winter and returned by constructive waves in summer (Figure **7**).

♦ Beach profiles are steeper in summer, when waves are more constructive than destructive. Constructive waves are less frequent and have a longer wavelength (6–9 per minute), so wave energy dissipates and deposits over a wide area (weakening the backwash).

♦ In winter, destructive waves occur at a higher frequency (11–16 per minute). Berms may be eroded by plunging waves and high-energy swash crashing down onto the beach. Strong backwash transports sediment offshore (depositing it as *offshore bars*). Sometimes, the backwash exerts a *rip current*, or undertow – dragging sediment back as the next wave arrives over the top (see 3.2).

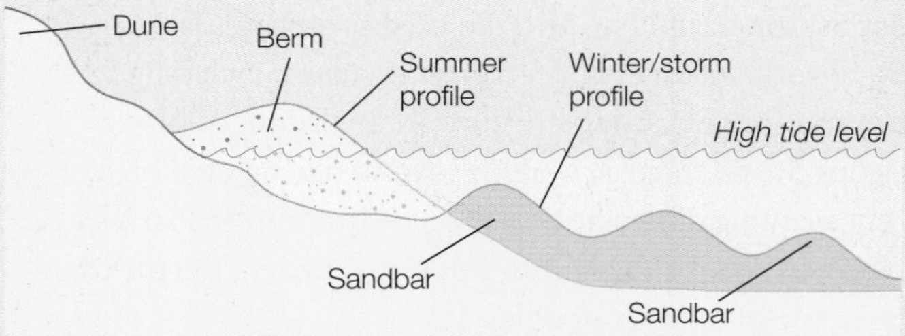

▶ **Figure 7** *Typical summer and winter beach and dune profiles*

## Spits

A *spit* is a long, narrow feature, made of sand or shingle, that extends from the land into the sea (or part of the way across an estuary). Spits form on drift-aligned beaches (see Figure **3**). Sand or shingle is moved along the coast by longshore drift, but if the coastline suddenly changes direction (e.g. because of a river estuary), sediment begins to build up across the estuary mouth and a spit will form (see Figure **8**). The outward flow of the river will prevent the spit from extending right across the estuary mouth. The end of the spit will also begin to curve round, as wave refraction carries material round into the more sheltered water behind the spit. This is known as a *recurved tip*.

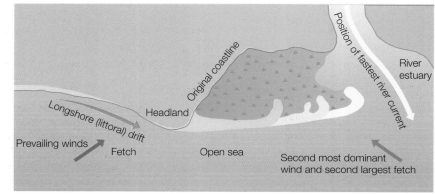

▲ **Figure 8** *The formation of a spit*

The entrance to Poole Harbour (Figure **9**) is unusual in that it has two spits extending from both the northern and southern sides of the bay – forming a *double spit*. A saltmarsh may develop behind a spit, where finer sediment settles and begins to be colonised by salt-tolerant plants (such as the one behind the left-hand spit in Figure **9**). A *compound spit* occurs where the transport processes are variable over time, which produces a series of 'barbs' along the spit (Figure **8**).

## Tombolo

A **tombolo** is a beach (or ridge of sand and shingle) that has formed between a small island and the mainland. Deposition occurs where waves lose their energy and the tombolo begins to build up. Tombolos may be covered at high tide, for example, at St Ninian's in the Shetland Islands (Figure **10**) and at Lindisfarne in Northumberland. Chesil Beach (Figure **1**) is also an example of a tombolo, linking the Isle of Portland with Weymouth on the mainland.

## Offshore bars

Also known as sandbars, **offshore bars** are submerged (or partly exposed) ridges of sand or coarse sediment created by waves offshore from the coast. Destructive waves erode sand from the beach with their strong backwash and deposit it offshore (Figure **11**). Offshore bars act as both sediment sinks and, potentially, sediment input stores. They can absorb wave energy thereby reducing the impacts of waves on the coastline.

**Figure 9** *The double spit at the entrance to Poole Harbour, Dorset*

**Figure 10** *The tombolo linking St Ninian's Isle to Mainland, Shetland is the largest active sand tombolo in the UK*

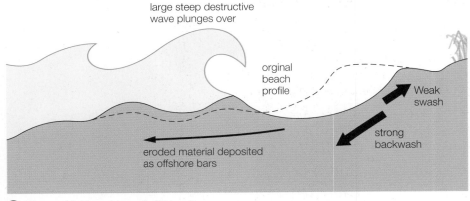

**Figure 11** *Formation of offshore bars*

large steep destructive wave plunges over

orginal beach profile

Weak swash

strong backwash

eroded material deposited as offshore bars

## Barrier beaches (bars)

Where a beach or spit extends across a bay to join two headlands, it forms a *barrier beach* or *bar*. The barrier beach at Start Bay in Devon (Figure **12**) is 9 km long and is formed from rounded shingle deposits (consisting mostly of flint and quartz gravel). Barrier beaches and bars can also trap water behind them to form lagoons, such as Slapton Ley in Figure **12**.

Barrier beaches and bars on the south coast of England are believed to have been deposited following rising sea levels after the last glacial period. Sediment deposited by meltwater in what is now the English Channel was bulldozed onshore to form the present-day barrier beach or bar. Subsequently, longshore drift has added more material and has reworked the sediment. This is a good illustration of energy flows and it also demonstrates the importance of time in the formation of present-day landforms. We cannot always assume that landforms are the result solely of processes operating at the present time.

Where a beach becomes separated from the mainland, it is referred to as a *barrier island*. Barrier islands vary in scale and form – are usually sand or shingle features – and are common in areas with low tidal ranges, where the offshore coastline is gently sloping. Large-scale barrier islands can be found along the coast of the Netherlands, and in North America along the South Texas coast.

▲ **Figure 12** *Start Bay barrier beach and Slapton Ley lagoon*

## Sand dunes

Many depositional landforms consist of sand and shingle – loose, unstable sediment that can be easily eroded and transported. Sandy beaches may be backed by *sand dunes*, like those at Studland in Dorset (see Figure 13), which consist of sand that has been blown off the beach by the onshore winds. In order for sand dunes to form a number of prerequisites are required:

◆ large quantities of available sand, washed onshore by constructive waves (an offshore sand bar is an ideal source of sand)

◆ large tidal range, creating a large exposure of sand that can dry out at low tide

◆ dominant onshore winds, that will blow dried sand to the back of the beach.

▲ **Figure 13** *Sand dunes at Studland in Dorset*

Dunes develop where sand is initially trapped by debris towards the back of the beach. Vegetation helps to stabilise the sand and gradually dunes develop. Over a period of several hundred years a transformation takes place known as a *vegetation succession* (see 1.9 and 6.9).

◆ The first colonising plants are called **pioneer species**, which have special adaptations to help them survive in hostile conditions. Plants such as sea rocket and couch grass are able to cope with the very dry, salty and exposed conditions. When they die, the plants add important organic matter to the developing soil.

◆ As the pioneer plants take hold, they help to bind the sand and form low sand dunes called *fore dunes*. *Marram grass* is a typical species found in this zone. It is extremely well adapted with long tap roots to seek water. The growth of marram grass is stimulated by burial and its tangle of lateral roots is perfect for binding the sand.

◆ As the environment changes over time, different species colonise the sand dunes until they become stable. The final community will be adjusted to the climatic conditions of the area, and is known as the *climatic climax community*. In Figure **15**, this is represented by trees such as oaks and pines.

**Figure 14** *European sea rocket has adapted to hostile conditions*

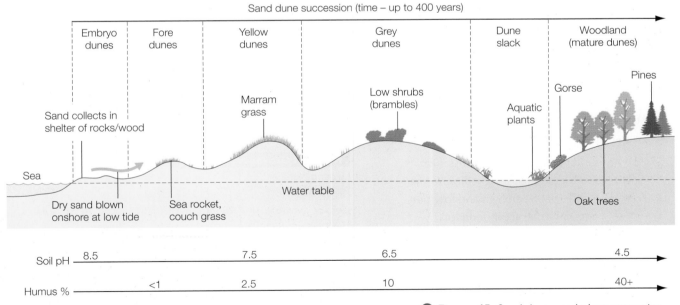

**Figure 15** *Sand dunes and plant succession*

- Embryo dunes are the first dunes to develop.
- As embryo dunes develop, they grow into bigger fore dunes – which are initially yellow in colour, but darken to grey as decaying plants add humus.
- Depressions between dunes can develop into dune slacks – damper areas where the water table is closer to, or at, the surface.

## Estuarine mudflats and saltmarshes

River estuaries are important sediment stores (sinks) where huge quantities of river sediment is deposited in water close to the edges of the river, away from the faster tidal currents that scour the channels. Rising tides create a buffer to the river flow, slowing velocity and leading to considerable deposition. Most of the sediment that accumulates here is mud, due to the low velocities and, over time, expansive *mudflats* can form that then develop into **saltmarshes**.

Saltmarshes are areas of flat, silty sediments that accumulate around estuaries or lagoons. They develop in three types of environment:

◆ in sheltered areas where deposition occurs (e.g. in the lee of a spit)

◆ where salt and freshwater meet (e.g. estuaries)

◆ where there are no strong tides or currents to prevent sediment deposition and accumulation.

Saltmarshes are covered at high tide and exposed at low tide. They are common around the coast of Britain and, as with sand dunes, develop over time, exhibiting a clear vegetation succession (Figure **16**).

◆ To begin with, mud is deposited close to the high-tide line, dropping out of the water by a process known as *flocculation*. This involves the tiny individual particles of clay (mud) sticking together such that their combined mass enables them to sink to the seabed.

◆ As with sand dunes, pioneer plants such as *eelgrass* and *cordgrass* start to colonise the transition zone between high and low tide. These plants can tolerate inundation by salty water and they also help to trap further deposits of mud.

◆ Gradually, the mud level rises above high tide and a lower saltmarsh develops with a wider range of plants that no longer need to be so well adapted to salty conditions.

◆ Soil conditions improve and the vegetation succession continues to form a meadow.

◆ Eventually, shrubs and trees will colonise the area as the succession reaches its climatic climax.

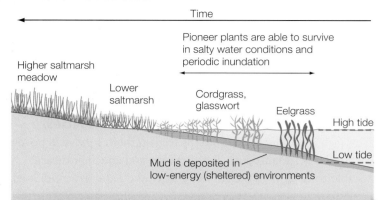

- As mud flats develop, salt-tolerant plants (such as *eelgrass*) begin to colonise and stabilise them.
- *Halophytes* (salt-tolerant species), such as *glasswort* and *cordgrass*, help to slow tidal flow and trap more mud and silt.
- As sediment accumulates, the surface becomes drier. Different plants begin to colonise (such as *sea asters* and *meadow grass*).
- Creeks (created by water flowing across the estuary at low tide) divide up the saltmarsh.

▲ **Figure 16** *Saltmarsh formation and plant succession*

## ACTIVITIES

**S**

1 The data in Figure **17** was obtained at Hornsea on the Holderness coast. It shows beach profile data on either side of a groyne. If longshore drift was operating along the coastline they would expect to find sediment piling up on one side of the groyne.

a Use the angles and section lengths to construct two profiles to show the beach profiles to the north and south of the groyne. Ensure that you use the same scale for each profile. Section 1 is at the bottom of the beach closest to the sea.

b Use your completed profiles, together with other information in Figure **17**, to suggest and justify the likely direction of longshore drift along this stretch of coastline. Use a simple sketch map to support your answer.

c What other information could be used to support your conclusion?

| North of Groyne | | | | | | | |
| --- | --- | --- | --- | --- | --- | --- | --- |
| Section Number | 1 | 2 | 3 | 4 | 5 | 6 | 7 |
| Beach Angle (°) | 9 | 9 | 8 | 7 | 8 | 9 | −1 |
| Section Length (m) | 1.76 | 8.2 | 1.5 | 3.14 | 6.27 | 4.2 | 6.74 |

Drop in height from top of groyne to beach (at bottom end of beach): 0.7 cm
Total beach length: 34.2 m

| South of Groyne | | | | | | |
| --- | --- | --- | --- | --- | --- | --- |
| Section Number | 1 | 2 | 3 | 4 | 5 | 6 |
| Beach Angle (°) | 5 | 6 | 5 | 8 | 12 | 10 |
| Section Length (m) | 0.76 | 3.09 | 3.54 | 4.32 | 6.29 | 8.9 |

Drop in height from top of groyne to beach (at bottom end of beach): 75 cm
Total beach length: 27.5 m

Data courtesy of Cranedale Centre www.cranedale.com

▲ **Figure 17** *Beach profile data for groynes at Hornsea, East Riding of Yorkshire*

2 Study Figures **1** and **2**. Use the internet to find your own photos to illustrate deposition on a high-energy and low-energy stretch of coast. Add detailed labels describing the main landscape features and try to use systems terminology as appropriate.

3 Devise a labelled diagram to explain the difference between swash-aligned and drift-aligned beaches.

4 How would you expect beach characteristics (pebble size, angularity, beach forms) to differ between swash-aligned and drift-aligned beaches?

5 Do you agree that beaches and spits are 'temporary stores' of sediment within the coastal system?

6 **a** Draw an annotated diagram of (i) a summer beach, and (ii) a winter beach.

**b** Identify on each diagram where the largest sediment will be found, and why the sediment size will vary from the top of the beach to the shoreline.

7 Study Figure **12**. To what extent is this landscape the result of past and present coastal processes?

8 For either sand dunes or saltmarshes, design an information poster describing the characteristics, formation and vegetation succession. Consider how your chosen depositional landform fits into the coastal system concept and ensure that you refer to systems terminology in your work. Make use of annotated photos to illustrate your poster. Remember that this is A Level so your poster should reflect this academic level.

9 **a** Using Figure **18**, create a dispersion diagram for wave frequency for each of columns **A** and **B**, and then calculate the range for each set of data.

**b** Calculate the mean, median and mode for wave frequency in each of columns **A** and **B**.

**c** Write a brief comparison of the wave data for each period.

**d** Explain the likely impact of these waves on Hornsea's beaches during 17–18 January 2016.

| A 17 Jan midnight – noon | Waves per minute | B 17 Jan 13:30 – 18 Jan 01:30 | Waves per minute |
|---|---|---|---|
| 00:00 | 9.1 | 13:30 | 14.0 |
| 00:30 | 9.0 | 14:00 | 14.3 |
| 01:00 | 8.7 | 14:30 | 14.3 |
| 01:30 | 8.6 | 15:00 | 15.4 |
| 02:00 | 8.8 | 15:30 | 16.2 |
| 02:30 | 8.2 | 16:00 | 16.2 |
| 03:00 | 8.2 | 16:30 | 15.8 |
| 03:30 | 8.6 | 17:00 | 15.8 |
| 04:00 | 8.5 | 17:30 | 16.2 |
| 04:30 | 8.8 | 18:00 | 15.8 |
| 05:00 | 8.8 | 18:30 | 16.2 |
| 05:30 | 9.2 | 19:00 | 15.8 |
| 06:00 | 9.7 | 19:30 | 15.0 |
| 06:30 | 9.5 | 20:00 | 15.0 |
| 07:00 | 9.5 | 20:30 | 15.0 |
| 07:30 | 9.8 | 21:00 | 14.6 |
| 08:00 | 10.0 | 21:30 | 14.0 |
| 08:30 | 10.2 | 22:00 | 14.6 |
| 09:00 | 10.2 | 22:30 | 14.0 |
| 09:30 | 10.5 | 23:00 | 15.0 |
| 10:00 | 10.9 | 23:30 | 14.6 |
| 10:30 | 10.5 | 00:00 | 15.0 |
| 11:00 | 10.3 | 00:30 | 15.8 |
| 11:30 | 10.0 | 01:00 | 15.0 |
| 12:00 | 10.0 | 01:30 | 15.8 |

**Figure 18** *Wave frequency data at Hornsea, 17–18 January 2016*

## STRETCH YOURSELF

1 Use Google Maps satellite imagery together with internet research to study the depositional landforms in Christchurch Bay (Figure **4**, 3.3). Consider the main characteristics of these landforms and the processes responsible for their formation. Referring to the sediment cell, describe how these landforms are interlinked by transfer processes and assess the extent to which they are dependent on each other.

2 Carry out further research on the origin and evolution of Chesil Beach and report on your findings. Include annotated diagrams and maps in your report. (You may find www.chesilbeach.org/fsg/mbfrmtn.pdf useful)

In this section you will learn about eustatic, isostatic and tectonic sea level change

## A stack out of water?

Look at Figure **1**. At first sight the isolated outcrop of rock looks like a typical stack. But look again – its position in terms of the current coastline does not make sense. It should be permanently surrounded by water and some distance offshore, as it represents a remnant of a retreating section of cliff. The stack in the photo is no longer being actively eroded by the sea – it is crumbling and being broken down by subaerial processes only. Here, the sea level has fallen by a few metres leaving the stack, as well as the beach and the cliffs beyond, high and dry above the current sea level.

▲ **Figure 1** *Features associated with sea level fall, Antrim, Northern Ireland*

## Evidence for long-term sea level change

The term 'sea level' is essentially used to identify the boundary between land and sea. It is the level to which the sea reaches and defines the coastline. The daily tides establish high and low-tide levels separated, in most cases, by a tidal range of a few metres, which can become more extreme at spring tides and during storms (see 3.2). However, the action of the sea is essentially restricted to a zone of just a few vertical metres from day to day.

In the past, over much longer time scales, sea levels have varied enormously compared to their current position. During the Quaternary period, there were several alternating cold (glacial) periods, during which sea levels fell, and several warm (interglacial) periods, during which sea levels rose, in response to the nature of precipitation.

## Eustatic and isostatic change

There are two main causes of sea level change:

◆ **eustatic change** – when the sea level itself rises or falls

◆ **isostatic change** – when the land rises or falls, relative to the sea.

Eustatic change is global. In cold glacial periods, precipitation falls as snow (rather than rain) and forms huge ice sheets that store water that is usually held in the oceans. As a result, sea levels fall. As temperatures rise at the end of glacial periods, the ice sheets begin to melt and retreat. Their stored water then flows into the rivers and the sea, and sea levels rise. This is shown clearly in Figure **3**.

Isostatic change occurs locally. During glacial periods, the enormous weight of the ice sheets (which can be several kilometres thick) makes the land sink – *isostatic subsidence*. As the ice begins to melt at the end of a glacial period, the reduced weight of the ice causes the land to readjust and rise – *isostatic recovery*. This is the explanation for the landscape in Figure **1**, with the various coastal landforms stranded above the current high-tide line.

### Think about

Note the link between sea level change in the context of the coast and both the water and carbon cycles (Chapter **1**), and also glaciation (Chapter **4**). Consider the synoptic links and be prepared to make these connections in an answer to a question on this topic. Ensure you use systems terminology.

## Isostatic change in the UK

Eustatic change occurs relatively quickly, but isostatic change takes place over a much longer period of time. By the end of the last glacial period in Europe (about 8000 years ago), glacial meltwater had caused a relatively rapid rise in sea levels (Figure **3**), leading to the formation of the English Channel and the North Sea (turning Britain into an island). Despite the melting of a huge amount of ice, the land only started to rise very slowly.

In the UK there appears to be a pivotal motion due to isostatic change (Figure **2**).

◆ Land in the north and west (which was covered by ice sheets during the last Ice Age) is still rising as a result of isostatic recovery. It is along these coastlines that features of falling sea level are most evident (Figure **1**).

◆ Land in the south and east (which the ice sheets never covered) is sinking. Rivers pour water and sediment into the Thames Estuary and the English Channel. The weight of this sediment causes the crust to sink and relative sea levels to rise. Therefore, south-east England faces increased flood risks not only as a result of isostatic change, but also because of a rising sea level caused by global warming. It is along parts of the south of England where there are features associated with rising sea levels.

It is interesting to ponder whether the same kind of pivotal motion occurred during a glacial period, with the north-west sinking under the weight of ice and the south-east rising, rather like a seesaw!

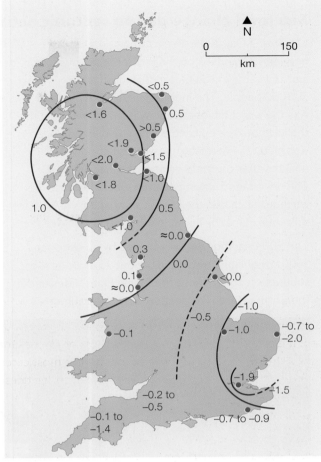

⊘ **Figure 2** *Isostatic change. The lines show how much parts of Great Britain are either rising (positive numbers) or falling (the negative numbers) in metres. Scotland is rising while most of south-east England is sinking.*

Figure **3** shows sea level rise since the last glacial period. Notice the steady but rapid rise in sea levels until about 8000 years ago – when the UK and much of the rest of Europe and North America became ice free (Figure **4**, 4.4). Amazingly, at the height of the last glacial period, sea levels were some 120 m below their current levels! Global changes in the last 8000 years are less easy to detect and the margin of error in observations often exceeds the changes themselves. Scientists now believe that sea levels have risen globally as a result of climate change and global warming.

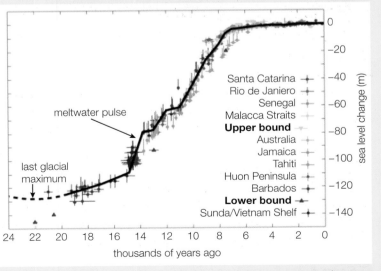

⊘ **Figure 3** *Sea level change from 24 000 years ago to present day*

## Sea level change due to tectonic activity

On Boxing Day 2004, an earthquake measuring between 9.0 and 9.3 on the Richter scale caused a tsunami in the Indian Ocean that killed approximately 300000 people. The Indonesian island of Sumatra was the worst hit, because it was the closest land to the earthquake's epicentre. The city of Banda Aceh was hit by 15-metre-high waves and flooded just 15 minutes after the initial earthquake (see Figure **4**). The devastation was made even worse because the earthquake caused the Earth's crust at Banda Aceh to sink – permanently flooding some parts of the city.

The 2004 earthquake was caused by an estimated 1600 km of fault line slipping about 15 m along the subduction zone where the Indian Plate slides under the Burma Plate. The seabed rose several metres – displacing an estimated 30 km³ of water and triggering the tsunami. Not only that, but the raising of the seabed reduced the capacity of the entire Indian Ocean – producing a permanent rise in sea level of an estimated 0.1 mm (see 5.11).

Tectonic activity generates a huge amount of energy that is transferred by the waves to the coast. The devastation and erosion of the coastline caused by the 2004 tsunami pays testament to the immense amount of energy harnessed by the waves. These one-off energy flows are very significant in causing coastal change and they exemplify the principle that most change in natural systems involves high-magnitude, low-frequency events.

⬆ **Figure 4** *Before and after the tsunami – the impact on Banda Aceh (images courtesy of DigitalGlobe)*

## Past tectonic activity

Past tectonic activity has had a direct impact on some coasts across the world, as well as on sea levels, due to:

- the uplift of mountain ranges and coastal land at destructive and collision plate margins that resulted in a relative fall in sea level in some parts of the world

- local tilting of land at destructive margins, for example, some ancient Mediterranean ports have been submerged and others have been stranded above the current sea level.

▶ **Figure 5** *The Roman Macellum of Pozzuoli near Napoli has remnants of marine molluscs on its three columns. This shows that the land has been submerged by the sea and then subsequently re-emerged.*

## Landforms caused by changing sea level

Changes in sea level affect the shape of the coastline and the formation of new landforms. A fall in sea level exposes land previously covered by the sea, creating an *emergent coastline*. A rise in sea level floods the coast and creates a *submergent coastline*.

### Emergent coastal landforms

As the land rose as a result of isostatic recovery, former wave-cut platforms and their beaches were raised above the present sea level. **Raised beaches** are common on the west coast of Scotland, where the remains of eroded cliff lines (called *relic cliffs*) can often be found behind the raised beach, with wave-cut notches and caves as evidence of past marine erosion. You have already seen a relic stack and a raised beach in Figure **1**. Wave-cut (marine) platforms can also be exposed if sea levels fall sufficiently, although these are often hidden beneath fresh beach deposits.

On the Isle of Arran (see Figure **6**), raised beaches represent separate changes in sea level. Notice the fresh beach deposits at a lower level than the raised beaches. These are probably blanketing ancient wave cut platforms below.

### Submergent coastal landforms

Essentially, submergence results in the flooding of coastlines. **Rias** (sheltered winding inlets with irregular shorelines) are one of the most distinctive features associated with a rise in sea level. They form when valleys in a dissected upland area are flooded. Rias are common in south-west England, where sea levels rose after the last Ice Age – the lower parts of many rivers and their tributaries were drowned to form rias. The Kingsbridge Estuary in Devon is one of these (Figure **7**). It provides a natural harbour with the deepest water at its mouth.

**Fjords** are formed when deep glacial troughs (see 4.8) are flooded by a rise in sea level. They are long and steep-sided, with a U-shaped cross-section and hanging valleys. Unlike rias, fjords are much deeper inland than they are at the coast. The shallower entrance marks where the glacier left the valley. Fjords can be found in Norway, Chile and New Zealand (Figure **8**).

⬆ *Figure 6* A raised beach on the Isle of Arran

⬆ *Figure 7* The Kingsbridge Estuary, Devon

⬇ *Figure 8* Milford Sound (a fjord) on New Zealand's South Island

**Dalmatian coasts** are distinctive submergent coastlines that form in a landscape of ridges and valleys running parallel to the coast. When the sea level rises, the valleys flood, although the tops of the ridges remain exposed, forming a series of offshore islands running parallel to the coast. The best example of a Dalmatian coastline is the one that gives this feature its name – the Dalmatian coast in Croatia (Figure **9**). Dalmatian coasts are also known as Pacific coasts, for example, in southern Chile.

## Contemporary sea level change

According to the Intergovernmental Panel for Climate Change (IPCC), sea levels stabilised about 3000 years ago and they have changed little since that time until very recently. From the late nineteenth century to the late twentieth century, sea levels rose globally by about 1.7 mm per year, although this has increased to about 3.2 mm per year in the period 1993–2010. The IPCC estimates that by 2100 sea levels could rise by between 30 cm and 1 m from current levels, although there is likely to be considerable variation from place to place.

Sea level rise is the result primarily of thermal expansion of water, due to heating, and the melting of *freshwater* ice, such as the Greenland and Antarctic ice sheets and mountain glaciers. Between 1880 and 2010, global temperatures rose by an average of 0.85 °C.

Notice the link here between coastal systems and the water and carbon cycles (Chapter **1**), and glacial systems (Chapter **4**).

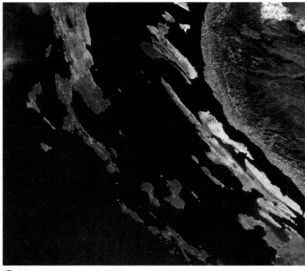

▲ **Figure 9** *A satellite image of the Dalmatian coast in Cro*

### The impact of climate change in Kiribati

The nation of Kiribati consists of 33 islands in the Pacific Ocean, that stretch across an area almost as wide as the USA (Figure **10**). Kiribati's islands are very low-lying sand and mangrove atolls – in most places only a metre or less above sea level. To visitors, Kiribati can seem like paradise, but it has been predicted that many of its islands could disappear under the sea in the next 50 years. In places, the sea level is rising by 1.2 cm a year (four times faster than the global average).

#### What next for Kiribati?

In 2014, President Anote Tong of Kiribati finalised the purchase of 20 km² of land on one of the islands of Fiji – 2000 km from Kiribati. Rising sea levels in Kiribati are contaminating its groundwater sources and affecting its ability to grow crops. So the land in Fiji will be used in the immediate future for agriculture and fish-farming projects, to guarantee the nation's food security. In the future, people could move there from Kiribati. The government has launched a 'migration with dignity' policy to allow people to apply for jobs in neighbouring countries such as New Zealand. If the islands are submerged, Kiribati's population will become *environmental refugees* – people forced to migrate as a result of changes to the environment.

▲ **Figure 10** *Tarawa Atoll, Kiribati – a vulnerable speck in the ocean*

**Figure 11** 1:50 000 Ordnance Survey map of Kingsbridge Estuary, Devon

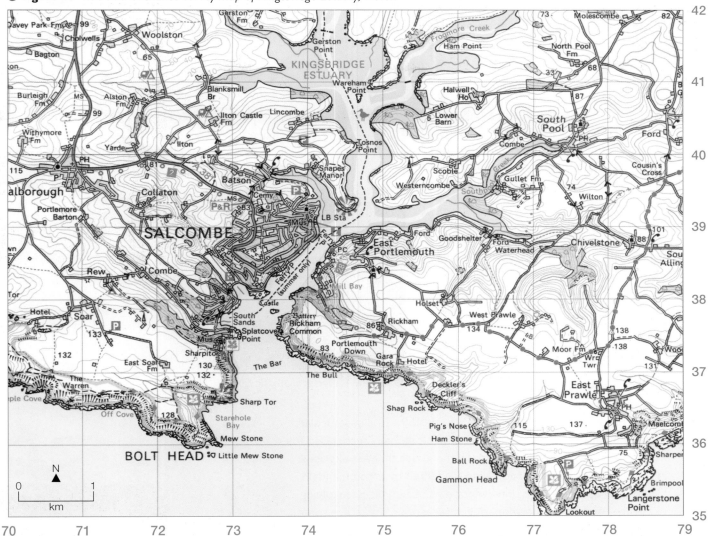

## ACTIVITIES

**(S)**

**1** Using an atlas, in pairs discuss and identify which European coastlines are at risk from contemporary sea level rises. Give examples.

**2 a** Look at the Ordnance Survey map in Figure **11**. Identify the features of a ria coastline on the map.

   **b** Research one other ria and its features in south-west England. You may find the following web sites useful: http://data.ordnancesurvey.co.uk and www.ordnancesurvey.co.uk/opendata/viewer

**3** In pairs, devise a virtual online field trip that focuses on 'Isostatic change in the UK landscape'. Use Google Maps and an internet search to identify the best UK locations to see and photograph **(a)** raised beaches, **(b)** relic cliffs, **(c)** any other evidence of eustatic change.

**4** To what extent are the risks from sea level rise in south-east England **(a)** similar to, **(b)** different from the risks to Kiribati?

**5** Distinguish between the following pairs of terms: **(a)** eustatic and isostatic change, **(b)** emergent and submergent coastlines, **(c)** rias and fjords, **(d)** relic cliffs and raised beaches.

**6 a** Summarise the problems facing Kiribati due to changes in sea level.

   **b** Use internet research to update how well Kiribati is coping with climate change.

## STRETCH YOURSELF (S)

Search Google Maps for satellite images of Croatia's Dalmatian coast, and then annotate them to show evidence of past changes in sea level.

In this section you will learn about coastal management, including sustainable approaches to flood risk and erosion management

## Human intervention at the coast

According to the United Nations Environment Programme (UNEP), about half the world's population live within 60 km of a coast and three-quarters of all large cities are at the coast. Coastal areas are under immense pressure for development resulting in environmental damage and habitat destruction (e.g. mangrove forests, sand dune systems, coral reefs) and pollution of coastal waters. Large numbers of people living at the coast are at risk of flooding and coastal erosion and this risk is likely to increase as global temperatures increase and sea levels rise.

▲ **Figure 1** *Dorset's Jurassic Coast*

Figure **1** shows part of Dorset's Jurassic Coast, England's first natural UNESCO World Heritage Site. The site covers some 150 km of spectacular scenery from East Devon to Dorset and it is globally recognised for its geological significance, particularly the fossils that are exposed in the cliffs. This stretch of coastline is affected by erosion and, in places such as Lyme Regis, intervention has taken place, involving extensive hard and soft engineering. Intervention may solve some problems but scientists are concerned that the lack of coastal erosion will result in fossils no longer being exposed and uncovered in the cliffs. Key questions are raised: should we intervene or let nature take its course? Should we control natural processes of learn to adapt to them?

Human intervention to address these coastal issues can take many forms. Traditional approaches have involved direct action – hard and soft engineering that focuses on relatively short stretches of coastline intended to stop or slow down erosion or reduce the risk of flooding (Figure **2**) Increasingly, a more holistic approach to coastal management is taking shape, involving a more sustainable and long-term approach that considers the entire coastal zone and seeks to achieve a balance between natural processes and human needs and concerns. Adaptation to coastal change is now seen as a viable alternative to the previously held position of trying to prevent and control natural processes.

▼ **Figure 2** *Sea wall and beach nourishment at West Bay, Dorset*

## Traditional approaches

Coastal erosion can be prevented – up to a point – but it is an expensive business. The cost of protecting the coast is often controversial. Until the 1990s, it was usual for local councils to tackle coastal erosion by designing *hard-engineering* structures. However, most of those structures are very expensive to build (see Figure **3**), so now the use of *soft-engineering* techniques is more popular. Each method has advantages and disadvantages.

### Hard engineering

Many stretches of coast in the UK exhibit a mixture of nineteenth-century hard-engineering structures, which have been since upgraded, extended and altered. Some of the most common types of hard-engineering structures are described in Figure **3**.

### Cost-benefit analysis (CBA)

A cost-benefit analysis is carried out before a coastal-management project is given the go-ahead. Costs are forecast (e.g. a sea wall – its design, building costs, maintenance, etc.) and then compared with the expected benefits (e.g. value of land saved, housing protected, savings in relocating people, etc.). Costs and benefits are of two types:

◆ tangible – where costs and benefits are known and can be given a monetary value (e.g. building costs)

◆ intangible – where costs may be difficult to assess but are important (e.g. the visual impact of a revetment).

A project where costs exceed benefits is unlikely to be given permission to go ahead.

**Figure 3** *Hard engineering – advantages, disadvantages and costs*

| Type of structure | Description | Advantages | Disadvantages | Cost |
|---|---|---|---|---|
| Groynes | Timber or rock structures built at right angles to the coast. They trap sediment being moved along the coast by longshore drift – building up the beach. | Work with natural processes to build up the beach, which increases tourist potential and protects the land behind it. Not too expensive. | Starve beaches further along the coast of fresh sediment (because they interrupt longshore drift), often leading to increased erosion elsewhere. Unnatural and can be unattractive. | £5000 to £10 000 each (at 200-metre intervals). |
| Sea walls | Stone or concrete walls at the foot of a cliff, or at the top of a beach. They usually have a curved face to reflect waves back into the sea. | Effective prevention of erosion. They often have a promenade for people to walk along. | They reflect wave energy, rather than absorbing it. They can be intrusive and unnatural looking. They are very expensive to build and maintain. | £6000/m |
| Rip rap (rock armour) | Large rocks placed at the foot of a cliff, or at the top of a beach. It forms a permeable barrier to the sea – breaking up the waves, but allowing some water to pass through. | Relatively cheap and easy to construct and maintain. Often used for recreation – fishing, sunbathing. | Can be very intrusive. The rocks used are usually not local and can look out of place with local geology. Can be dangerous for people clambering over them. | £1000 to £3000/m |
| Revetments | Sloping wooden, concrete or rock structures placed at the foot of a cliff or the top of a beach. They break up the waves' energy. | They are relatively inexpensive to build. | Intrusive and very unnatural looking. They can need high levels of maintenance. | Up to £4500/m |
| Offshore breakwater | A partly submerged rock barrier, designed to break up the waves before they reach the coast. | An effective permeable barrier. | Visually unappealing and a potential navigation hazard. | Similar to rock armour – depending on the materials used. |

## Soft engineering

A range of soft-engineering techniques attempt to work with natural processes to protect coasts (see Figure **4**). They can also be used to manage changes in sea level, for example, by allowing low-lying coastal areas to flood (creating marshes).

⯆ **Figure 4** *Soft engineering – advantages, disadvantages and costs*

| Method | Description | Advantages | Disadvantages | Cost |
|---|---|---|---|---|
| Beach nourishment | The addition of sand or pebbles to an existing beach to make it higher or wider. The sediment is usually dredged from the nearby seabed. | Relatively cheap and easy to maintain. It looks natural and blends in with the existing beach. It increases tourist potential by creating a bigger beach. | Needs constant maintenance because of the natural processes of erosion and longshore drift. | £3000/m |
| Cliff regrading and drainage | Cliff regrading reduces the angle of the cliff to help stabilise it. Drainage removes water to prevent landslides and slumping. | Can be effective on clay or loose rock where other methods will not work. Drainage is cost-effective. | Regrading effectively causes the cliff to retreat. Drained cliffs can dry out and lead to collapse (rock falls). | Cost is variable |
| Dune stabilisation | Marram grass can be planted to stabilise dunes. Areas can be fenced in to keep people off newly planted dunes. | Maintains a natural coastal environment. Provides important wildlife habitats. Relatively cheap and sustainable. | Time consuming to plant marram grass. People may respond negatively to being kept off certain areas. | £2 to £20/m |
| Marsh creation | A form of managed retreat, by allowing low-lying coastal areas to be flooded by the sea. The land then becomes a salt marsh. | Relatively cheap (land reverts to its original state before management). Creates a natural buffer to powerful waves. Creates an important wildlife habitat. | Agricultural land is lost. Farmers or landowners need to be compensated. | The cost is variable – depending on the size of the area left to the sea. |

## Sustainable integrated approaches

Soft-engineering schemes adopt sustainable principles (they are intended to work with nature over a long period of time) but they tend to be very focused on a particular stretch of coastline to solve an issue at a specific locality. Increasingly, management authorities – both political and non-political – are looking at holistic management plans for significant stretches of coastline.

⯅ **Figure 5** *Managed retreat at Alkborough Flats, Lincolnshire*

## ACTIVITIES

**S**

1   Using diagrams, explain how hard-engineering methods alter physical systems and processes.

2   To what extent do soft-engineering techniques represent sustainable approaches to coastal management?

3   Describe the coastal management measures shown in Figure **2** and suggest the advantages of combining hard and soft engineering.

4   Assess the advantages of integrated management plans such as Shoreline Management Plans (SMP) and Integrated Coastal Zone Management (ICZM).

5   Is there ever a case for deliberately *not* protecting a stretch of coastline from erosion?

3 Coastal systems and landscapes

## Shoreline Management Plans (SMP)

You may remember that there are 11 sediment cells in England and Wales (see 3.1 and 3.3). Each sediment cell defines a distinct management zone for which a *Shoreline Management Plan (SMP)* has been written, which identifies the natural processes, human activities and management decisions. SMPs are extremely detailed, comprehensive documents and are based on the sediment cell principle that intervention will be largely self-contained within each cell, having little or no knock-on effects elsewhere. Each sediment cell is treated (from a management point of view) as a 'closed cell', even though in practice we know that not to be absolutely the case.

SMPs are recommended for all sections of the coastline in England and Wales by Defra (the Department for Environment, Food and Rural Affairs). Four options are considered for any stretch of coastline:

1 *Hold the line* – maintaining the current position of the coastline (often using hard-engineering methods).

2 *Advance the line* – extending the coastline out to sea (by encouraging the build-up of a wider beach, using beach-nourishment methods and groyne construction).

3 *Managed retreat/strategic realignment* – allowing the coastline to retreat in a managed way (e.g. creating salt-marsh environments by deliberately breaching flood banks that protect low-quality farmland. (Figure **5**)).

4 *Do nothing/no active intervention* – letting nature take its course and allowing the sea to erode cliffs and flood low-lying land and allowing existing defences to collapse.

## STRETCH YOURSELF Ⓢ

Use the satellite imagery on Google Maps and the internet to investigate a coastal defence scheme, if possible, close to your home or school. Annotate the imagery to describe and explain the defence measures that have been adopted and to suggest how these measures affect the natural coastal system (e.g. sediment movement).

## Integrated Coastal Zone Management (ICZM)

The process that brings together all of those involved in the development, management and use of the coast. The aim is to establish sustainable levels of economic and social activity; resolve environmental, social and economic challenges and conflicts; and protect the coastal environment.

The move to adopt an ICZM strategy means that complete sections of coast are now being managed as a whole, rather than by individual towns or villages. This is because human actions in one place affect other places further along the coast. Essentially, this is because of the transfers (flows) within the sediment cell, moving sediment from one place to another. What is eroded in one location eventually becomes a protective beach somewhere else. Stop the erosion in one place and sediment is starved somewhere else; the problem is not so much solved as shifted.

ICZM and SMPs provide a more holistic overview of the coast, enabling sustainable measures to be implemented based on a fundamental understanding of natural process and the coastal system.

In 2013 the EU adopted a new initiative on Maritime Spatial Planning and Integrated Coastal Management. It stated that: 'Integrated coastal management aims for the coordinated application of the different policies affecting the coastal zone and related to activities such as nature protection, aquaculture, fisheries, agriculture, industry, off shore wind energy, shipping, tourism, development of infrastructure and mitigation and adaptation to climate change. It will contribute to sustainable development of coastal zones by the application of an approach that respects the limits of natural resources and ecosystems, the so-called 'ecosystem-based approach'.

Integrated coastal management covers the full cycle of information collection, planning, decision-making, management and monitoring of implementation. It is important to involve all stakeholders across the different sectors to ensure broad support for the implementation of management strategies.'

http://ec.europa.eu/environment/iczm/home.htm

In this case study you will apply coastal processes and landscape outcomes to a coastal environment at a local scale and also engage with field data

Case study

## The Holderness coastal system

The Holderness coast is a well-known stretch of coastline in eastern England. It forms a subcell in Sediment Cell 2 (Figure **2**, 3.1) and essentially comprises three distinct coastal units (Figure **1**):

◆ Flamborough Head in the north, a chalk promontory that exhibits many typical landforms associated with coastal erosion

◆ Bridlington Bay to Spurn Head, an extensive zone of erosion and sediment transfer characterised by a very rapid rate of cliff retreat

◆ Spurn Head, a classic spit formed at the estuary of the River Humber.

Within this subcell, the main input is erosion of the weak and unconsolidated till cliffs. Some of the finer sediment is washed offshore to form an output from the system while the slightly coarser material is moved southwards as a transfer involving longshore drift. Some sediment is deposited to form Spurn Head, while a significant amount continues south towards the Wash and East Anglia.

## Factors affecting the coastal system

Geology is an important factor affecting the processes and landforms of the Holderness coast. Chalk, a relatively resistant rock, forms a broad arc in the region, stretching from the Lincolnshire Wolds in the south to the coast at Flamborough Head. Notice in Figure **2** that the eastern edge of the chalk outcrop formed the preglacial coastline and that the great sweep of the present day coastal zone is the result of sediment carried and dumped by ice sheets originating from Scandinavia. As sea levels rose at the end of the last glacial period, the North Sea took shape and started to erode the thick till deposits to help form the present-day cliffs.

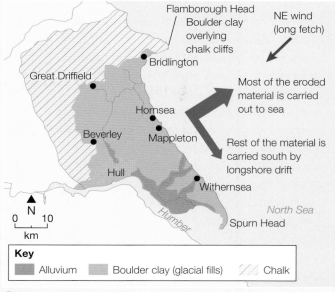

**Figure 1** *The Holderness coastal system*

**Figure 2** *The Holderness coast before and after glaciations*

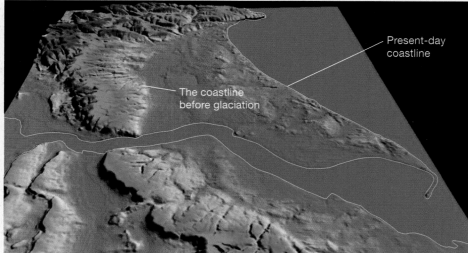

When the wind is blowing from the north-east (direction of greatest fetch), it can drive powerful waves towards the Holderness coast. Occasionally, areas of extremely low pressure move down the North Sea, funnelling water and creating storm surges several metres high. These low-frequency, high-magnitude events can lead to significant erosion and flooding – in 1953 more than 300 people lost their lives along the east coast of England during such an event. As a result of these powerful north-east waves, longshore drift operates from north to south along the Holderness coast.

In response to the rapid rate of erosion and the threat to settlement and infrastructure, parts of the coastline have been protected with hard-engineering structures such as sea walls, rock armour and groynes (see 3.9). Although these interventions have helped to protect specific localities, such as Hornsea and Mappleton, they have deprived areas further south of sediment, thereby exacerbating coastal erosion. The lack of a beach renders cliffs much more vulnerable to undercutting and collapse (Figure **3**).

## Flamborough Head

Jutting into the North Sea from the east coast of England, Flamborough Head is one of the most recognisable features on a map of the UK (Figures **1**, **4** and **5**). The main reason for the formation of the headland is because it is made of chalk – a resistant, sedimentary rock.

Chalk has a very distinctive white colour, as can be seen in Figure **4**. The layers or beds of the chalk are clearly visible and are roughly horizontal (Figure **5**). Vertical cracks run through the chalk (*joints*). In some places whole sections of chalk have been displaced along lines called *faults*. These joints and faults are weaknesses in the chalk, which are readily exploited by the processes of weathering and erosion to form narrow clefts in the coastline. One major faultline has been exploited to form Selwick's Bay (Figure **4**).

The sea is actively eroding and undercutting the base of the cliffs leading to frequent rockfalls. The high tide line is clearly shown by the dark staining at the foot of the cliffs in Figure **5**. Over time the cliff retreats, forming wave-cut platforms and stacks.

When waves approach the coastline they are bent or refracted by the shape of the coast. The waves are therefore curved and have low energy in the bay, resulting in a deposited beach. The more exposed headlands bear the full force of the incoming waves – this is why they are often characterised by steep cliffs and other features of coastal erosion.

⬆ **Figure 3** *Recent erosion threatening the settlement of Skipsea, where rates of erosion are 2 metres a year*

⬆ **Figure 4** *Distinctive chalk cliffs at Selwicks Bay, Flamborough Head*

⬇ **Figure 5** *Landforms of coastal erosion at Flamborough Head*

## Bridlington Bay to Spurn Head

This stretch of coast has retreated by up to 5 km since Roman times, which accounts for the loss of several settlements and ports by erosion. With rates of erosion in excess of 1 m per year (and up to 10 m per year in places), the Holderness coast has one of the most rapid rates of erosion in Europe.

Although erosion creates threats, it does generate a vast amount of sediment that feeds the sediment cell. Much of the finer sediment is carried offshore but a great deal of coarse sediment is transferred by longshore drift to the south, building up beaches and reducing erosion (a negative feedback for the system).

Spurn Head itself is nourished by this sediment transfer and it has an important role in protecting the towns and land bordering the River Humber from the effects of storm waves and flooding. Further south, the Wash is an important sediment sink protecting towns such as King's Lynn. There is even evidence that some sediment from Holderness ends up on the coast of the Netherlands, helping to protect it from flooding. This is a massive planning conundrum – should erosion be halted to protect a few houses and agricultural land or should it be allowed to continue because it is a vital source of sediment?

## Spurn Head

Spurn Head spit represents a temporary sediment store or sink (Figure **7**). Much of the material that forms the spit is derived from the Holderness coast and is transferred south by longshore drift. On reaching the River Humber estuary, the deposited sediment grows out to form a narrow finger of new land. Notice its curved tip resulting primarily from direct wave action.

Spurn Head is extremely narrow for much of its length and has frequently been breached and destroyed by major storms (hence its classification as a temporary store). It first formed some 8000 years ago at the end of the last glacial period and evidence suggests that the feature has gone through a number of cycles of growth and decline, lasting on average about 250 years.

Following a massive breach in 1849, groynes and revetments (wooden barriers) were erected to stabilise the spit. In subsequent years, when military forts were established at Spurn Point, the Royal Engineers took over the task of maintaining coastal defences. In the 1950s the military left and, in 1960, the spit was bought by the Yorkshire Naturalists' Trust. Unable to afford the maintenance costs of the spit, the Trust had to allow some of the sea defences to fall into disrepair. When the largest tidal surge in 60 years hit Spurn Point in 2013, the defences could not cope – buildings were destroyed and the access road swept away.

△ **Figure 6**  *Rapidly eroding cliffs on the Holderness coast*

Several factors account for the rapid rate of erosion of the Holderness coast:

- long fetch and powerful waves from the north-east
- weak and unconsolidated till cliffs
- extensive mass movement, especially slumping, caused by undercutting and saturation of clay within the cliffs
- narrow beach making the cliffs vulnerable to wave attack and undercutting
- lack of coastal defences.

△ **Figure 7**  *Spurn Head from the air*

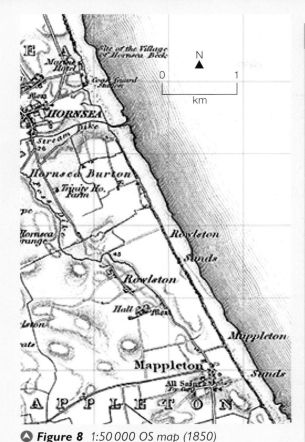

▲ **Figure 8** 1:50 000 OS map (1850)

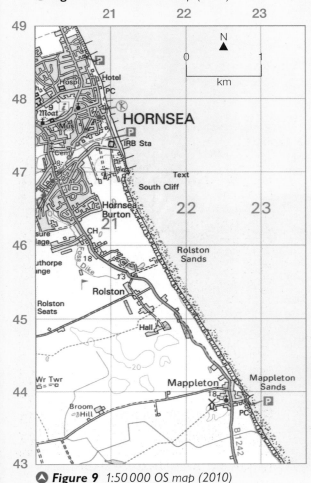

▲ **Figure 9** 1:50 000 OS map (2010)

## ACTIVITIES

**(S)**

**1** Design a large (A3) information poster to summarise the process and landforms of the Holderness coast. Use Figure **1** as a base map and begin by showing the operation of the sediment cell, inputs, outputs, transfers and stores. Are there any feedbacks in operation? Add text boxes to describe processes and landforms and try to incorporate some labelled photos. This should provide you with a comprehensive overview of this case study.

**2** What are the factors affecting the cliff profiles and rates of cliff retreat along the Holderness coast?

**(S)**

**3** Study Figure **5**.
  **a** Was the photo taken at high tide or low tide? Justify your answer.
  **b** Draw a sketch of the photo and label the main landforms and evidence of coastal processes
  **c** How might this landscape change in the next century?

**4** Consider the possible impacts of sea level rise on coastal process and landforms along the Holderness coast. How might this impact on the sediment cell?

**5** With specific reference to the coastal system, what are the arguments for and against protecting the Spurn Head coastline?

**(S)**

**6** Figure **8** is a 1:50 000 OS map (1850) of part of the Holderness coast. Figure **9** is a 2010 map of the same stretch of coastline. The 1850 map has been scaled so that it is the same scale as the modern map and can therefore be readily compared. Investigate the changes to the coastline between 1850 and 2010.

  • Create your own grid or use a sheet of tracing paper and draw the extent of both the past and present coastlines. Follow the high tide line on both maps. Add information onto your map, such as settlements, roads and coastal defences. Indicate the direction of maximum fetch and the direction of longshore drift.
  • Use two colours to show accretion and erosion.
  • Use the scale to work out the rates (metres per year) of accretion and erosion for selected locations.
  • Write a brief commentary describing the variations in accretion and erosion. What is the evidence to show that the coastal defences may have been responsible for increased erosion to the south of Hornsea?

## STRETCH YOURSELF

Use Google Earth to investigate the downdrift impacts of coastal defences on beaches and cliff erosion. You could focus on Mappleton and see if there is any evidence that the coastal defences have interfered with longshore drift causing beach depletion and excessive cliff erosion further south. Use annotations to document your evidence.

In this case study you will:
- learn how a contrasting coastal landscape can present risks and opportunities for human occupation
- evaluate human responses of resilience, mitigation and adaptation

*Case study*

## Odisha: a distinctive and contrasting coastal landscape

Odisha is a state on the eastern coast of India bordering the Bay of Bengal (Figure **1**). It is India's 9th largest state by area and 11th by population. Odisha has a relatively straight coastline (about 480 km long) with few natural inlets or harbours. The narrow, level coastal strip known as the Odisha Coastal Plains supports the bulk of the state's population.

The coastal plain mostly comprises depositional landforms of recent origin, geologically belonging to the post-Tertiary Period. There are six major deltas on the Odisha coast, which explains why the coastal plain is known locally as the 'Hexadeltaic region' or the 'Gift of Six Rivers' (the Subarnarekha, Budhabalanga, Baitarani, Brahmani, Mahanadi, and the Rushikulya).

The Odisha coast has a wide range of coastal and marine flora and fauna (including 1435 km² of mangrove forest). Figure **2** identifies the three major coastal ecological environments – it also gives a good impression of its relief. One of these environments is Chilika Lake, a brackish salty lagoon, well-renowned for its birdlife. During the monsoon season the lake becomes less saline, being diluted by the freshwater rainfall, and occupies a larger area than during the rest of the year. Chilika Lake is a good example of a temporary store in the water cycle; the beach that has created the lake is an important store within the coastal system.

In summary, the Odisha coast is essentially one of deposition, comprising several major deltas. It therefore represents a significant sediment store, providing a source (and sink) of sediment for this part of the Bay of Bengal. Rivers provide important transfers of sediment into the region in forming deltaic deposits.

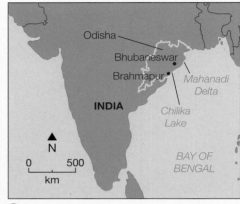

**Figure 1** *Location of Odisha, India*

**Figure 2** *The Odisha coast*

## Opportunities for human occupation and development

In addition to providing relatively flat land for settlement, the coastal plain provides several economic and environmental opportunities.

◆ Odisha's coastal zone has a wide variety of marine and coastal flora and fauna including mangroves, sea grasses, salt marshes, sand dunes, estuaries and lagoons.

◆ There are large stocks of fish, marine mammals, reptiles and Olive Ridley turtles, seagrass meadows, and abundant seaweeds.

◆ There area has huge potential for offshore wind, tidal and wave power.

◆ Thirty-five per cent of the coastal stretch is laden with substantial placer (sediment) minerals and heavy metal deposits. There are important clay and limestone resources in the north of the state.

◆ There are opportunities for offshore oil and natural gas, as well as seabed mining.

◆ Many local people are employed in coastal fishing and increasingly in aquaculture such as shrimps.

◆ Cultural and archaeological sites also dot the coast, drawing visitors from around the world.

◆ Tourism is important, with the coastal beaches and wildlife sanctuaries being major attractions.

◆ The Chilika Lake Bird Sanctuary boasts over 150 migratory and resident species of birds (Figure **3**).

⊙ **Figure 3** *Chilika Lake Bird Sanctuary*

## Risks for human occupation and development
### Coastal erosion

In 2011 the Ministry of Environment and Forests released its latest Assessment of Shoreline Change for the state of Odisha, focusing on mapping the areas of greatest erosion along the coast. Rates of erosion have increased in recent decades (Figure **4**), partly through natural processes but also as a consequence of human intervention methods, which have been used to protect infrastructures. With the majority of the state's population living on the coastal plain, the Indian government has become concerned about the increased vulnerability of coastal communities to storm surges and tsunami as well as the longer-term threats posed by climate change and rising sea levels.

The Odisha coastline is a naturally changing environment. Erosion provides important inputs of sediment that, once transferred along the coast by waves, tides and currents, is deposited to form beaches, dunes and barrier beaches, which characterise this stretch of India's coast. Natural seasonal variations occur along the coast, with accretion occurring in the summer during relatively low-energy wave conditions and erosion in the winter when high-energy destructive waves remove and deposit sediment offshore.

'Attempting to halt natural coastal process with seawalls and other hard structures, only shifts the problem, subjecting downdrift coastal areas to similar losses. Also, without the sediment transport, some of the beaches, dunes, barrier beaches, salt marshes, and estuaries are threatened and would disappear as the sand sources that feed and sustain them are eliminated.'

(Shoreline Change Assessment for Odisha Coast, 2011)

⊙ **Figure 4** *Coastal erosion, Odisha*

**Case study**

The key findings of the Assessment of Shoreline Change included the following:

◆ The coast of Odisha is largely accreting (46.8 per cent), with 36.8 per cent eroding and 14.4 per cent stable.

◆ Most accretion is in the north, focused on the major deltas.

◆ Most of the erosion is in the south. Here, there are major structures (sea walls, breakwaters and rock armour) protecting infrastructures (mainly ports), which have interfered with natural processes exacerbating rates of erosion.

◆ Due to the presence of dense mangrove vegetation, the coastal districts of Kendrapara, Bhadrak and Baleshwar show levels of accretion that are more than 50 per cent higher than anywhere else on the coast (Figure **5**).

◆ Shoreline change is extremely dynamic along the mouths of rivers, suggesting that the inflow pattern determines the nature of the shoreline – rates of accretion and erosion were found to vary considerable either side of the major river mouths.

In summary, the research has found the coastline to be a very dynamic coastal system (only 14.4 per cent of the coast is 'stable') and subject to considerable change. It is affected by significant seasonal variation in wave energy and sediment input via the region's major rivers. Human intervention has had a major impact on the system, interfering with sediment transfer and destabilising patterns of wave energy, resulting in severe erosion in certain localities. These changes are of concern to the state authorities given the enormous economic value of the coastal strip and possible increased threats from storm surges, tsunami and sea level rise.

▲ **Figure 5** *Mangrove in Odisha*

▲ **Figure 6** *The impact of Cyclone Phailin on the Odisha coast*

### Tropical cyclones – resilience, mitigation and adaptation

The Odisha coast is at risk from tropical cyclones and the associated storm surges. There is some evidence to suggest that the frequency and intensity of these storms may increase in the future as a result of climate change. With sea levels rising, this represents a significant threat to the coast.

In October 2013 Cyclone Phailin struck the Odisha coast near Gopalpur (Figure **6**). Windspeeds touched 200 km/h, tearing down power lines and uprooting trees. Over one million people were evacuated from those areas deemed to be at greatest risk. Forty-four people died in Odisha and many thousands were affected, with buildings damaged and economic activity disrupted. The coastal district of Ganjam was most severely affected by the storm. In total, some 500 000 ha of agricultural crops were destroyed and economic losses were close to US$700 000. Chilika Lake suffered from a storm surge

that may take the ecosystem years to recover and, along the coast, thousands of mangrove trees were destroyed, temporarily rendering this stretch of coastline more vulnerable to storm surges.

Odisha is very vulnerable to tropical cyclones and its people have developed considerable resilience in adapting to the threat. In 1999, the infamous Odisha Cyclone, the strongest cyclone ever recorded in the northern Indian Ocean, brought massive destruction to the region, killing over 10 000 people. The authorities now employ a number of mitigation strategies – providing relief supplies ahead of an approaching storm, broadcasting warnings and conducting staged evacuations away from the most vulnerable areas. The relatively small death toll in 2013 pays testament to the mitigation strategies as well as the resilience of the people and their ability to adapt to changing circumstances.

# Managing the Odisha coast

In the light of the 2011 Assessment of Shoreline Change report, it is clear that piecemeal and highly localised management strategies are likely to cause more long-term harm than good – they are certainly likely to upset any existing coastal system balance (less than 15 per cent of the coastline was found to be stable).

A recent ICZM project has coordinated the activities of the various stakeholders and promoted the sustainable use of the natural resources of the Odisha coast while maintaining the natural environment (Figure **7**). The ICZM aims to:

| Central (federal) government | State and local government | Stakeholders in the local economy |
| --- | --- | --- |
| Archaeology Department of Culture | Odisha State Disaster Management Authority | Odisha Tourism Development Corporation |
| Water Resource Department | Odisha State Pollution Control Board | Handicraft and cottage Industries |
| Fisheries Department | Wildlife Wing of Forest and Environment Department (State) | |

▲ **Figure 7** *ICZM Project Odisha, players and stakeholders*

◆ establish sustainable levels of economic and social activity

◆ resolve environmental, social and economic challenges and conflicts

◆ protect the coastal environment.

The project is a joint venture involving the Ministry of Forest and Environment, the Indian government, the World Bank and the government of Odisha. The major issues that have been identified include:

◆ coastal erosion and associated oceanographic processes

◆ assessing vulnerability to disaster (particularly tropical cyclones)

◆ biodiversity conservation

◆ livelihood security (e.g. fishing)

◆ pollution and environmental quality management

◆ conservation of cultural/archaeological assets.

Many different organisations have an interest in managing the coast and these have been consulted, along with others who have a stake in its future. In addition to the inter-organisational consultations, a wide range of public consultations have also been held, including with individual villages about issues such as:

◆ the assessment and control of coastal erosion

◆ the development of ecotourism (Figure **8**)

◆ planting or replanting mangroves

◆ building cyclone shelters.

Greenpeace India (an environmental pressure group) has also been involved in meetings about income generation and the management of marine resources, acting with some of the villages included in the ICZM project.

◆ **Figure 8** *Godwit Eco Cottage is a collection of holiday cottages close to Chilika Lake. They are made of bamboo and mud, with thatched roofs.*

**Case study**

## Management of the Mahanadi Delta

The Mahanadi Delta (Figures **2** and **9**) is an important ecological zone, providing important natural habitats for a wide variety of wildlife. However, in recent decades there has been a considerable loss of mangroves, largely due to the development of fisheries and other economic demands. Fifty years ago, coastal villages in Odisha had an average width of 5.1 km of mangroves protecting them. Today, that figure is an average of 1.2 km. In 1999, during 'super-cyclone' Kalina, villages that still had four or more kilometres of mangroves, recorded no deaths. However, in areas where the protective belt was less than 3 km wide, death rates rose sharply. With sea levels rising and tropical cyclones expected to become more frequent and intense, integrated management of this stretch of the Odisha coast is essential.

The NGO, Wetlands International, along with the Indian government and Odisha's ICZM project, is now trying to reverse decades of mangrove destruction. They are helping villagers to cultivate and plant mangroves along the coastline, and also on the banks of all tidal rivers along Odisha's coast. This is a good illustration of a sustainable and highly appropriate coastal management (mitigation) scheme that is followed, not just in India, but in many parts of the world that have similar flooding issues (Figure **10**).

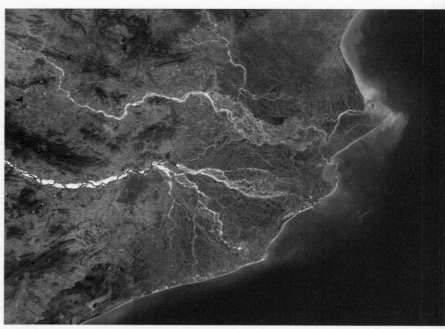

**Figure 9** *Satellite image of the Mahanadi Delta, India*

**Figure 10** *Planting mangrove seedlings, Andaman Coast, Thailand. According to Daniel Alongi of the Australian Institute of Marine Science, the faster sea levels rise, the faster mangroves accumulate sediment (sediment store) in their roots. They can keep up with a rise of 25 mm a year – eight times the current global rate. No sea wall can do that.*

### STRETCH YOURSELF

Access the Odisha ICZM website at www.iczmpodisha.org/aim_and_objective.htm to find out more about the project. Evaluate the work done so far in terms of its sustainability. What are the challenges for the future?

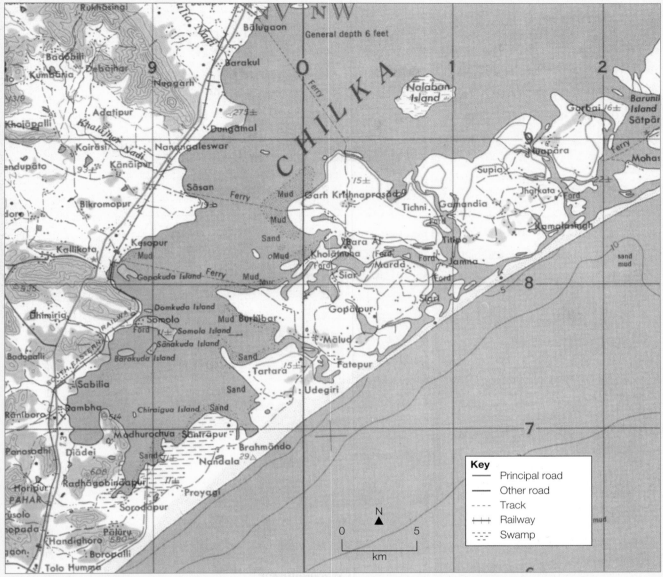

⬥ **Figure 11** *A 1:250 000 topographic map around Chilika Lake, Odisha*

## ACTIVITIES

**1** Study Figure **2**. What are the distinctive aspects of Odisha's coast. How does it compare with named stretches of coastline in the UK?

**S**

**2** Use Figure **2** to draw a large sketch of the Odisha coast. Use text boxes to identify some of the issues associated with this stretch of coastline. Consider both opportunities and risks. Use the internet to provide additional information and illustrate your work with photos.

**3 a** What are the aims of the ICZM project?

　 **b** Suggest the criteria that will be used to judge whether it has been successful in achieving its aims.

**4 a** Examine the role of mangroves in reducing the threat from flooding.

　 **b** To what extent is the replanting of mangroves a sustainable solution to the problem of flooding in India and Bangladesh?

**5** Consider the ways in which resilience, mitigation and adaptation have been evident in the approaches to coastal management adopted by organisations and local communities.

**6** 'An understanding of coastal systems is essential if coastal management is to be successful.' Do you agree? Explain your answer.

**7** Look at Figure **11**. Describe in detail the coastal landscape and the human uses of this area of coastline.

The following are sample practice questions for the Coastal systems and landscapes chapter. They have been written to reflect the assessment objectives of the Component 1: Physical geography Section B of your A Level.

These questions will test your ability to:

◆ demonstrate knowledge and understanding of places, environments, concepts, processes, interactions and change, at a variety of scales [AO1]

◆ apply knowledge and understanding in different contexts to interpret, analyse and evaluate geographical information and issues [AO2]

◆ use a variety of quantitative, qualitative and fieldwork skills [AO3].

**1** Outline how the coast functions as an open system. (4)

**2** Study Figure **1** below and Figure **9** on page 147. Analyse the data shown. (6)

| Erosion profile | Location | Cliff erosion data | | | | Maximum cliff loss between profiles | | |
| | | Cliff lost at profile (m) | | Erosion rate (m/yr) | | Height of cliff | Maximum recorded individual loss (m) | Date of maximum cliff loss |
| | | Nov 14–Apr 15 | Apr 15–Sept 15 | 1852–1989 | 1989–Sept 2015 | | | |
| 34 | At north end of Long Lane, Atwick | 0.00 | 0.00 | 1.11 | 0.88 | 14.9 | 8.91 | Sept 2004 |
| 35 | Opposite Long Lane, Atwick | 0.00 | 0.00 | 1.06 | 1.38 | 17.4 | 8.13 | Mar 2005 |
| 36 | Opposite Cliff Road, Atwick | 0.00 | 0.00 | 1.01 | 1.09 | 14.8 | 7.79 | Sept 2005 |
| 37 | South of Atwick | 0.00 | 0.00 | 1.06 | 1.02 | 20.2 | 10.97 | April 2013 |
| 38 | Just north of Atwick Gap boat club, Hornsea | 1.20 | 0.00 | 0.95 | 0.70 | 17.2 | 13.99 | Mar 2008 |
| 39 | Within campsite north end of Cliff Road, Hornsea | 0.00 | 0.00 | 0.82 | 0.48 | 19.2 | 10.03 | Apr 2011 |
| 40 | Just south of Nutana Avenue, north Hornsea | 0.00 | 0.00 | 0.65 | 0.17 | 16.6 | 5.82 | Sept 2005 |
| 41 | North end of Hornsea frontage | 0.00 | 0.00 | 0.56 | 0.11 | 15.2 | 1.60 | 1853 to 1890 |
| 42 to 44 | Sea defences: concrete seawalls, groynes and rock armouring Hornsea frontage | | | | | | | |
| 45 | Within caravan park to south of defences | 0.00 | 0.00 | 1.62 | 2.31 | 17.6 | 8.76 | Mar 2006 |
| 46 | South of Hornsea | 0.58 | 0.00 | 1.86 | 2.98 | 17.5 | 10.83 | Sept 2004 |
| 47 | Within Rolston firing range | 0.00 | 0.00 | 1.77 | 2.89 | 17.1 | 9.79 | Sept 2012 |
| 48 | Opposite Rolston | 1.16 | 0.00 | 1.77 | 2.71 | 16.8 | 9.88 | Mar 2004 |
| 49 | South end of old children's camp site, Rolston | 3.04 | 0.00 | 1.67 | 2.34 | 17.1 | 7.47 | Sept 2009 |
| 50 | North of Mappleton | 0.00 | 0.00 | 1.58 | 1.20 | 17.3 | 7.89 | Sept 2012 |
| 51 | North of Mappleton defences | 0.00 | 0.00 | 1.56 | 0.26 | 17.5 | 4.10 | Spring 2001 |

▲ **Figure 1** *Cliff erosion data for erosion profile points at 500-metre intervals, located north and south of Hornsea, Sept 2015 update*

www.coastalexplorer.eastriding.gov.uk

**3** With reference to Figure **2**, page 106, and your own knowledge of both high- and low-energy stretches of coastline, explain the factors affecting wave energy. (6)

**4** 'The coastal system is as much shaped by its links with other physical systems as it is by flows and transfers within itself.'
To what extent do you agree with this statement? (20)

**Tip**

**To what extent** as a command requires you to decide how much you agree with the statement, having thought about the evidence both supporting and contradicting this point of view. In your explanation you must give the reasons (and evidence for) why you have come to your conclusion.

5  Explain the distinction between swash-aligned and drift-aligned beaches. (4)

6  Using Figure **1**, page 102, analyse the natural processes changing the coast in this chalk location. (6)

7  Using Figure **6**, page 137, evaluate the evidence for, and processes of, sea level change. (6)

8  Assess the view that 'the coastal environment is a classic illustration of the complicated interaction of natural processes involving different degrees of energy, acting on different timescales'. (20)

9  Explain the distinction between hard engineering and soft engineering approaches to protecting the coast. (4)

10 Using Figures **8** and **9**, page 147, analyse the map evidence that this section of the Holderness coast is both vulnerable to erosion and managed. (6)

11 Study Figure **2**. Using evidence relating to the protection of the Holderness coast and other named places you have studied, assess the sustainability of different shoreline management plans. (6)

### Repair and upgrade history for main defence structures

Following their construction, coastal structures soon require maintenance and upgrading. Inappropriate initial construction can increase this workload and leave many coastal authorities with an extensive programme of costly works. For the East Riding, this maintenance commitment is complicated further by the continued erosion of the clay foreshore and adjacent coastlines.

| Date | Description of works | Cost (£) |
|---|---|---|
| Prior to 1970s | No maintenance records exist for this time, however, works are likely to minimal groyne repairs as the original sea walls remained intact and the newly constructed North and South revetments would still be relatively maintenance free. | |
| 1970s | Underpinning to North, Central and North Gate Promenade seawalls<br>General groyne repairs | 1 800 000<br>352 000 |
| 1980s | General sea wall repairs<br>Major upgrading and repair of groyne field<br>Outflanking rock protection to South Revetment | 200 000<br>1 413 500<br>93 620 |
| 1990s | General sea wall repairs – note mainly South and Seathorne revetments which began to fail in the mid 90s<br>General groyne repairs<br>Upgrading original sea walls:<br>1. Pier Towers to North Gate Promenade: Replace upper seawall with RC recurved seawall, plus rock armour protection to toe<br>2. Pier Towers to South Cliff Rd: Replace upper seawall with RC recurved seawall, plus rock armour protection to its toe to breakwater and outflanking protection to south of South Revetment<br>3. North Gate Promenade: Replace upper seawall with a new RC seawall and encase upper seawalls with RC<br>Extend and reprofile existing rock armour outflanking to south of South Revetment | 265 000<br>307 000<br><br><br>2 890 000<br><br><br>3 101 500<br><br><br>325 000<br>271 700 |
| 2000–2003 | General seawall repairs; note mainly South and Seathorne revetments<br>General groyne repairs | 130 000<br>58 800 |

▲ **Figure 2**  *Repair and upgrade history of main defence structures in Withernsea on the Holderness coast*

*Crown copyright. All rights reserved. East Riding of Yorkshire Council www.coastalexplorer.eastriding.gov.uk*

12 'An understanding of coastal systems is essential if coastal management is to be successful.' To what extent do you agree with this statement? (20)

# 4 Glacial systems and landscapes

The Perito Moreno Glacier in Argentina is one of the most accessible glaciers in the world and is a UNESCO World Heritage site – look at how it towers above the tourists on the viewing platform

## Your exam

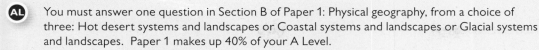

You must answer one question in Section B of Paper 1: Physical geography, from a choice of three: Hot desert systems and landscapes or Coastal systems and landscapes or Glacial systems and landscapes. Paper 1 makes up 40% of your A Level.

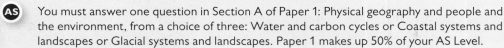

You must answer one question in Section A of Paper 1: Physical geography and people and the environment, from a choice of three: Water and carbon cycles or Coastal systems and landscapes or Glacial systems and landscapes. Paper 1 makes up 50% of your AS Level.

## Your key skills in this chapter

In order to become a good geographer you need to develop key geographical skills. In this chapter, whenever you see the skills icon you will practise a range of quantitative and relevant qualitative skills, within the theme of 'Glacial systems and landscapes'. Examples of these skills are:

- Interpreting photographs and remotely-sensed images  4.1, 4.6, 4.8
- Using atlases and other map sources  4.2, 4.8, 4.10, 4.15
- Using electronic databases  4.4
- Drawing and annotating sketch maps  4.8
- Measurement and geospatial mapping  4.3, 4.4, 4.9
- Presenting data and interpreting graphs  4.3, 4.4, 4.10, 4.14
- Data manipulation, including applying statistical skills to data  4.13
- Understanding and calculating mass balance or glacial budget  4.13
- Drawing and annotating diagrams  4.2, 4.4, 4.5, 4.7, 4.11

## Fieldwork opportunities

In the UK, fieldwork investigating glaciated landscapes necessarily involves fossil landforms – there aren't any glaciers in the UK today. Fossil landforms are likely to have been shaped by ice during the last glacial period, referred to in Britain as the Devensian glaciation. This glacial period was at its maximum extent about 20 000 years ago (see 4.2).

*1. Observation in the field: evidence of past climate*

Identifying erosional or depositional features of a glaciated landscape, for example, in the Lake District, creates a good opportunity to develop your field-sketching skills (take care with scale). Use photography to compliment your drawings; you can use photo-editing software to crop, enhance and label your photographs later. Keep a notebook to hand to make a detailed record of where the features are located in the landscape and map them on your return to the classroom, or use GIS software on a hand-held device in the field. From your evidence can you estimate the direction of flow or identify different periods of advance and retreat?

*2. The impact of subsequent physical processes on glaciated landscapes and beyond*

Geomorphological processes don't take place in isolation, and over several thousand years fluvial processes have shaped and reshaped glacial deposits. Before that, fluvioglacial processes associated with the action of meltwater, eroded, transported and deposited material beyond those areas of Britain that were ice covered. Observations about the way more recent processes may have altered or removed clues about glaciation is as relevant a focus for investigation. Sediment analysis, for example, comparing sediment in glacial till against fluvial or fluvioglacial deposits, may take the form of the classification of sediment samples by texture, roundness, size and stone orientation as well as the respective level of sorting of material.

In this section you will learn about glaciated landscapes and the glacial system

## What is a glaciated landscape?

Look at Figure **1**. It shows a glaciated landscape in the French Alps close to Mont Blanc, Europe's highest mountain. In common with all landscapes, it comprises many colours, textures and features. Yet, it is a landscape that is very distinctive in that its characteristics are linked to glaciation.

In the foreground is an active glacier, the Mer de Glace, set amid magnificent peaks, known locally as *aiguilles* (meaning 'needles' in French), that have been carved and shattered by ice processes. In places the mountainsides are strewn with rocks, dumped by the melting ice or piled up beneath frost-shattered cliffs. Elsewhere, exposed rock surfaces bear evidence of glacial erosion, with deep scratches or polished surfaces.

▲ **Figure 1** *A glaciated landscape – the Mer de Glace, France*

There are trees in the foreground and also some traces of vegetation beginning to take hold on the valley sides. There is no evidence of human activity, yet it is possible to infer that such a landscape would be popular with climbers and walkers. With its glaciers, snowfields, steep mountainsides and deep valleys, this is a distinctive glaciated landscape.

## What is the glacial system?

In common with many aspects of geography, systems concepts can be applied to glaciers to assist in understanding how they operate. A glacier can be viewed as an open system – with inputs from and outputs to external systems, such as atmospheric and fluvial systems. In common with all systems in geography, it is possible to identify particular characteristics and components such as inputs, outputs, flows and stores (Figure **3**).

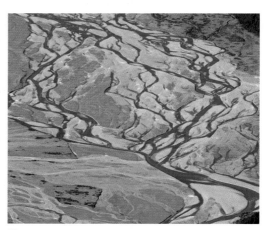

▲ **Figure 2** *Braided river, Hopkins River, South Island, New Zealand*

▼ **Figure 3** *The glacial system*

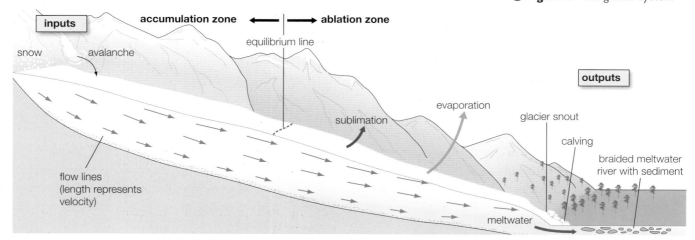

inputs

accumulation zone ← | → ablation zone

equilibrium line

snow     avalanche

outputs

evaporation

sublimation

glacier snout

calving

braided meltwater river with sediment

flow lines (length represents velocity)

meltwater

## Inputs

The main *input* is of course snow. As it becomes increasingly compacted over many years, it gradually turns from low-density 'fluffy' white ice crystals (snowflakes) to high-density clear glacial ice. Avalanches from mountainsides can also be an input to the glacial system.

## Outputs

The main *output* from a glacial system is liquid water resulting from the melting of ice close to the snout, where temperatures are higher. Where the ice front extends over water, such as the ice shelves in Antarctica, huge chunks of ice may break off to form icebergs. This process is called *calving* (Figures **3** and **4**). The processes of evaporation and *sublimation* also act as outputs from the glacial system.

## Energy

A glacier's mass combines with the force of gravity to generate potential energy. As the glacier moves, this potential energy is converted into kinetic energy, enabling the glacier to carry out the processes of *erosion*, *transportation* and ultimately *deposition*. The presence of meltwater facilitates this conversion of potential energy to kinetic energy (work).

## Stores/components

The main *stores* in the glacial system are snow and ice. There may be seasonal variations in the magnitude of these stores particularly in more temperate regions where there can be significant winter snowfall and summer melting. Over the last 30 years or so the stores of many of the world's glaciers have shown a decline in mass, attributed to recent trends in climate change and global warming.

## Flows/transfers

There are many flows and transfers of energy and material in a glacial system. These include processes such as evaporation, sublimation, meltwater flow and the processes of glacial movement (**internal deformation** and **basal sliding**). Flows and transfers are more pronounced and active in warmer environments where there are significant seasonal variations in temperature. In the world's coldest environments, such as Greenland and Antarctica, glacial systems are less active.

▲ *Figure 4* *Calving, Hubbard Glacier, Alaska USA*

## Feedback loops

Positive and negative *feedback loops* are highly significant aspects of all geomorphic systems. Negative feedbacks perform regulatory functions, working to establish balance and equilibrium. Positive feedbacks enhance and speed up processes promoting rapid change. There are examples of both positive and negative feedbacks in glacial systems, as you will see during this chapter (see 4.13).

## Dynamic equilibrium

Many physical systems move towards a state of **dynamic equilibrium**, where landforms and processes are in a state of balance. In a glacial system, an equilibrium line can be drawn to mark the boundary between the accumulation zone (glacial inputs) and the ablation zone (glacial outputs). If the glacier is in a state of balance where inputs equal outputs, the equilibrium line will remain in the same place. As this balance shifts, the equilibrium line will move up or down the glacier, hence the term 'dynamic' equilibrium.

### ACTIVITIES

1. Obtain a photo of a glaciated landscape, such as in South America, New Zealand or the Himalayas. Write a description of your chosen landscape in the form of detailed annotations. Consider physical landforms and processes, vegetation and human activity.

2. Using a copy of the same photo or selecting a different one, use Figure **3** to help you identify and label systems concepts such as inputs, outputs and flows. Try to identify or infer as many concepts as possible in your photo.

3. Explain why a glacier can be considered to be an *open* system.

In this section you will learn about the distribution and physical characteristics of cold environments

## What was the past distribution of cold environments?

The distribution of cold environments has changed significantly over time. During the last two million years – a blink of the eye in geological terms – temperatures have fluctuated considerably. This period of time is known as the Pleistocene, a geological period that lasted from about 1.8 million years ago until about 11 700 years ago (Figure **1**). During this time there was a pattern of alternating cold periods (glacials) and warm periods (interglacials); in the last one million years there may have been as many as ten glacial periods.

During the cold glacial periods, the climate cooled sufficiently for precipitation to fall as snow rather than rain. This resulted in the formation and growth of huge ice masses, which in the northern hemisphere spread south over large parts of Europe, Asia and North America. During the warmer interglacial periods – some being considerably warmer than the conditions we experience today – much of the ice melted and the ice sheets and glaciers retreated.

Look at Figure **2**. It shows the approximate distribution of ice at its maximum extent about 20 000 years ago. Notice that vast ice sheets covered much of North America and Europe. There were also extensive glaciers and ice caps in South America and in mountainous regions. With so much water locked up as ice on the land, the water cycle was significantly distorted and sea levels fell by 120 m.

These temperature graphs were derived from the measurement of *deuterium* isotopes in ice cores extracted from two sites in Antarctica. The third trace is derived from microscopic plants buried in deep ocean sediments. Remember that at these times, there were no thermometers to measure temperatures!

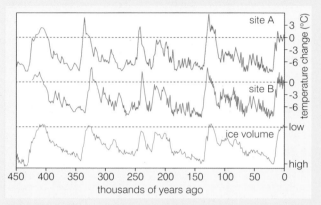

**Figure 1** *Antarctic temperature change and reconstruction of global ice volume during the last 450 000 years.*

**Figure 2** *The extent of ice during the last glacial maximum about 20 000 years ago*

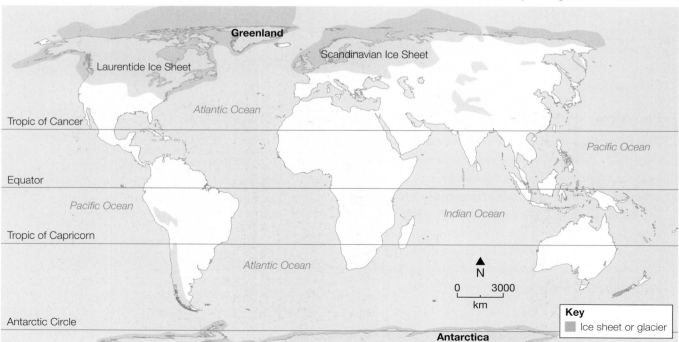

At this small scale, the finer detail of the actual ice extent needs to be treated with caution as evidence may be absent or unclear. For example, Figure **2** shows the entire land mass of the UK to be covered by ice. Yet Figure **3** shows this was not the case. Ice extended over much of the UK but the southern-most regions stayed ice-free despite being frozen and experiencing periglacial conditions.

## What is the present-day distribution of cold environments?

Look at Figure **4**. It shows the present-day distribution of cold environments. Notice that, with the exception of Antarctica and the far south of South America and the Andes, the world's cold environments are located in the far north. This is a reflection of the latitudinal position of the land masses – in the southern hemisphere, the equivalent latitudes coincide with ocean rather than land.

**Figure 3** *The extent of ice cover over the British Isles about 18 000 years ago*

**Key**
← ice movement
▢ tundra

It is possible to identify four types of cold environment:

◆ Polar – areas of permanent ice. Essentially the vast ice sheets of Antarctica and Greenland.

◆ Periglacial (tundra) – literally-speaking, at the 'edge' of permanent ice. Periglacial environments are characterised by permanently frozen ground (permafrost) and include large tracts of northern Canada, Alaska, Scandinavia and Russia.

◆ Alpine – mountain areas. For example the European Alps and the Southern Alps in New Zealand, where the high altitudes result in cold conditions particularly in winter.

◆ Glacial – glaciers are found at the edges of the ice sheets and, in particular, in mountainous regions such as the Himalayas and Andes.

Notice in Figure **4** that there is a clear latitudinal sequence of cold environments – polar environments in the far north are followed by continuous and then discontinuous permafrost as conditions become warmer further south. Alpine cold environments and most glacial environments exist in the mountainous areas.

**Figure 4** *Present-day global distribution of cold environments*

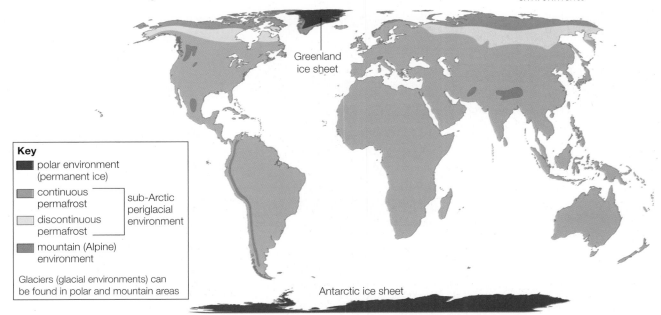

**Key**
■ polar environment (permanent ice)
▨ continuous permafrost
▢ discontinuous permafrost
■ mountain (Alpine) environment

sub-Arctic periglacial environment

Glaciers (glacial environments) can be found in polar and mountain areas

# Physical characteristics of cold environments

Cold environments include some of the world's last great wilderness areas. The vast expanses of ice and snow, combined with freezing temperatures and strong penetrating winds, make these some of the most inhospitable places on Earth.

## Soil development

Soil is a mixture of weathered rock, rotted organic matter, living organisms (biota), gases (particularly oxygen) and water. In cold environments, weathering is limited due to the lack of liquid water, the lack of vegetation means there is little organic matter and there are few decomposers – fungi and bacteria thrive in warm and humid conditions. Soil formation is, therefore, extremely slow, and any soils that do develop will be thin, acidic, sometimes waterlogged and mostly frozen.

Look at Figure **5**. It shows deep permafrost in northern Alaska. Notice that there is a thin layer of light grey unfrozen soil above the permafrost. This is called the **active layer**. With downward drainage prevented by the permafrost, the active layer will often become saturated and boggy.

## Climate

All cold environments experience significant periods of time when temperatures are close to or significantly below freezing (Figure **6**). Liquid water is limited to certain times of year and, in the most extreme environments, totally absent. Snowfall amounts vary enormously, from very little in polar environments to potentially huge amounts in coastal and mountain areas. Frequent strong winds add to the wind chill and also absorb precious moisture from plants. These extreme climatic conditions have a severely limiting effect on the development of soils and vegetation.

Svalbard is an archipelago (group of islands) in the northern Arctic. Most of Svalbard experiences a polar climate, although some stretches of coastline experience slightly less severe periglacial conditions. Its capital, Longyearbyen, is home to about 2000 people and is one of the most northerly permanent settlements in the world.

For much of the year it is dark both day and night. Temperatures can fall to below −20 °C in January and strong winds and heavy snowfall can make it extremely hazardous just to step outside. Despite the challenges, there is tremendous beauty and great potential for tourism, with opportunities to explore the natural environment (see 4.15).

▼ **Figure 5** *Permafrost in periglacial soils in northern Alaska, USA*

Polar: Arctic Bay, NW Territories, Canada

|  | J | F | M | A | M | J | J | A | S | O | N | D |
|---|---|---|---|---|---|---|---|---|---|---|---|---|
| Temp. (min) °C | −33 | −36 | −32 | −26 | −11 | −1 | 2 | 2 | −3 | −12 | −23 | −30 |
| Temp. (max) °C | −26 | −28 | −22 | −14 | −3 | 6 | 11 | 8 | 1 | −7 | −17 | −23 |
| Precipitation | 8 | 5 | 8 | 5 | 8 | 13 | 18 | 33 | 23 | 18 | 8 | 5 |

Periglacial: Verkhoyansk, Siberia, Russia

|  | J | F | M | A | M | J | J | A | S | O | N | D |
|---|---|---|---|---|---|---|---|---|---|---|---|---|
| Temp. (min) °C | −53 | −49 | −39 | −23 | −5 | 9 | 8 | 4 | −3 | −19 | −40 | −49 |
| Temp. (max) °C | −48 | −41 | −25 | −7 | 6 | 16 | 19 | 14 | 6 | −11 | −35 | −47 |
| Precipitation | 5 | 5 | 3 | 5 | 8 | 23 | 28 | 25 | 13 | 8 | 8 | 5 |

Alpine/Glacial: Santis, European Alps, Switzerland

|  | J | F | M | A | M | J | J | A | S | O | N | D |
|---|---|---|---|---|---|---|---|---|---|---|---|---|
| Temp. (min) °C | −11 | −11 | −9 | −6 | −2 | 1 | 3 | 3 | 1 | −3 | −7 | −10 |
| Temp. (max) °C | −7 | −7 | −4 | −2 | 3 | 6 | 8 | 8 | 6 | 2 | −3 | −6 |
| Precipitation | 202 | 180 | 164 | 166 | 197 | 249 | 302 | 278 | 209 | 183 | 190 | 169 |

▶ **Figure 6** *Climatic data for selected weather stations*

## Climate, soils and vegetation interactions – nutrient cycling

Figure **7** shows a typical nutrient cycle for a cold environment. Nutrients are plant foods that are constantly being recycled. The small nutrient stores indicate that nutrient availability is very limited in a tundra environment. Notice also that the transfer arrows are generally thin, indicating a very limited transfer of nutrients between the components. The only sizeable transfer is the *fallout pathway* – fallen leaves and other organic matter such as dead animals, which contribute nutrients to the litter store. Figure **7** shows the interaction between soils and vegetation. Climate is a major limiting factor in their development.

In contrast, Alpine environments enjoy warm, wet summers providing good conditions for both soil formation and plant growth. High Alpine meadows can exhibit tremendous biodiversity particularly in the summer.

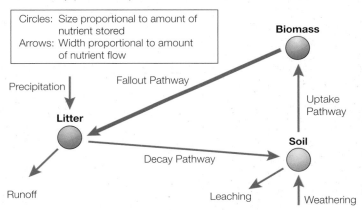

Circles: Size proportional to amount of nutrient stored
Arrows: Width proportional to amount of nutrient flow

Biomass

Precipitation        Fallout Pathway

Litter        Uptake Pathway

Decay Pathway        Soil

Runoff        Leaching        Weathering

## Vegetation development

In order to grow, plants require nutrients, liquid water and temperatures above 6 °C. Most plants require a soil to anchor them and from which they can obtain nutrients and water. The interaction between soils, vegetation and, inevitably, climate is very important. Soils need organic matter from rotting vegetation and plants need soils – many cold environments have neither.

Vegetation is largely absent from polar and glacial environments, apart from at the very periphery where bare rock surfaces can be colonised by lichens. If a rudimentary soil has formed then a few species of extremely tough plants may develop.

Periglacial environments are characterised by tundra vegetation consisting of low-growing plants, including mosses, lichens, grasses, sedges and dwarf shrubs. Their small, waxy leaves are well adapted to retain warmth and reduce the water loss caused by exposure to strong winds. They maximise the short, warmer summers by flowering and setting seed in just a few weeks. Tundra plants also have to cope with thin soils and potentially severe waterlogging on flat, poorly drained land.

◀ **Figure 7** *Nutrient cycle in a periglacial (tundra) environment*

## ACTIVITIES

1   Quoting evidence from Figure **1**, write a brief summary comparing glacial and interglacial periods during the last 450 000 years.

**(S)** 2   Study Figure **3**.
  a   Locate your home town or city. Describe the conditions that you would have experienced 18 000 years ago.
  b   Which parts of the British Isles never had a covering of ice?
  c   Use an atlas to identify the sources of ice flow in the British Isles.
  d   Why is it important to understand past climates when trying to explain the formation of present-day landscapes?
3   a   On a blank world outline map, copy Figure **4**.
  b   Use the internet to find a photograph to illustrate each of the environments.
  c   Using an atlas map, label the mountain ranges identified in the key as 'mountain (Alpine) environment'.
  d   Using your atlas, identify and label some regions/countries experiencing polar and periglacial conditions.

**(S)** 4   Construct a large summary table describing the distribution, climate, soil, vegetation and general characteristics of each of the four cold environments in Figure **4**. Make use of diagrams and photos to support your work.

5   Suggest some of the challenges people face when living and working in a cold environment, especially during the winter. Research and include additional information about Svalbard.

**(S)** 6   Copy Figure **7** and add detailed annotations to describe and explain its main characteristics. Focus on why the stores hold few nutrients and why there is very limited transfer of nutrients between the stores.

In this section you will learn about the glacial budget

## What is the glacial budget?

Just as a financial budget involves credits (payments in) and debits (withdrawals), a glacier also has a budget – the **glacial budget**. Essentially, the glacial budget considers the balance between the inputs and outputs – the term *mass balance* may also be used in glacial studies to effectively mean the same thing, although technically it is a more generic term that could be applied to other systems too. Look back at Figure **3** in 4.1. Notice that the glacier has been divided into two zones:

◆ The accumulation zone, where there is a net gain of ice over the course of a year. Here the inputs (snow and avalanches) exceed the outputs.

◆ The ablation zone, where there is a net loss of ice during a year. The losses (meltwater, calving, etc.) exceed the gains.

The boundary where gains and losses are balanced is called the *equilibrium line*. Over a period of several years, variations in mass balance (glacial budget) may result in the equilibrium line moving either up or down the glacier. This can often be linked to the advance or retreat of the glacier snout.

Mass balance varies during the course of a year (Figure **1**). In the summer, ablation will be at its highest due to rapid melting of the ice. During the winter, higher amounts of snowfall and limited melting will result in accumulation being greater than ablation.

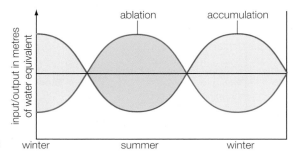

> **Figure 1** *Seasonal variations in mass balance*

## Gulkana Glacier, Alaska

The Gulkana Glacier is one of two 'benchmark' glaciers in Alaska that have been studied by the United States Geological Survey (USGS) since the 1960s. The purpose of this study has been to understand glacier dynamics and hydrology, and assess the glaciers' response to climate change.

The Gulkana Glacier is located on the southern flank of the eastern Alaska Range (Figure **2**) and has an elevation range from 1160 m to 2470 m. In 2016 the glacier occupied a total area of 16.0 km². The climate is continental, characterised by a large range in temperature and irregular, lighter precipitation as compared to more coastal areas.

Fixed points were identified on the glacier and, together with meteorological and runoff data, scientists have been able to plot trends and calculate the glacier's mass balance (Figure **2**). The data collected forms the longest continuous set of mass balance studies in North America.

Field visits to measure and maintain stakes at three index sites are made each spring, at the onset of the melt season, and again in early autumn, near its completion. The density of the material gained or lost is

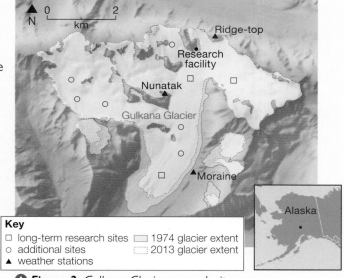

**Key**
□ long-term research sites ▢ 1974 glacier extent
○ additional sites ⋯ 2013 glacier extent
▲ weather stations

> **Figure 2** *Gulkana Glacier research sites and changes in extent (1974–2013)*

measured by digging a snow-pit or extracting an ice core. By collecting data near the balance maxima (spring) and minima (autumn), direct measurements closely reflect maximum winter accumulation and the annual balances at each location (Figures **3** and **4**). Since 1975 both the stakes and the glacier surface elevations at the actual index sites have been surveyed to allow calculations of velocity and surface elevation change.

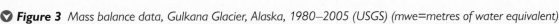

**Figure 3** *Mass balance data, Gulkana Glacier, Alaska, 1980–2005 (USGS) (mwe=metres of water equivalent)*

| Year | Winter balance (mwe) | Summer balance (mwe) | Net balance (mwe) | Cumulative balance (mwe) | Equilibrium line altitude (m) | Year | Winter balance (mwe) | Summer balance (mwe) | Net balance (mwe) | Cumulative balance (mwe) | Equilibrium line altitude (m) |
|---|---|---|---|---|---|---|---|---|---|---|---|
| 1980 | 1.13 | −1.20 | −0.07 | −0.07 | 1738 | 1993 | 0.82 | −2.49 | | | 1880 |
| 1981 | 0.96 | −0.93 | +0.03 | −0.04 | 1687 | 1994 | 1.37 | −1.96 | | | 1777 |
| 1982 | 1.55 | −1.67 | −0.12 | −0.16 | 1746 | 1995 | 0.94 | −1.65 | | | 1806 |
| 1983 | 1.14 | −1.11 | +0.03 | −0.13 | 1751 | 1996 | 0.87 | −1.39 | | | 1768 |
| 1984 | 1.30 | −1.61 | −0.31 | −0.44 | 1768 | 1997 | 0.99 | −2.68 | | | 1865 |
| 1985 | 1.41 | −0.73 | +0.68 | +0.24 | 1650 | 1998 | 0.79 | −1.43 | | | 1793 |
| 1986 | 1.09 | −1.03 | | | 1682 | 1999 | 1.04 | −2.15 | | | 1842 |
| 1987 | 1.24 | −1.37 | | | 1737 | 2000 | 1.44 | −1.49 | | | 1704 |
| 1988 | 1.26 | −1.48 | | | 1759 | 2001 | 1.40 | −2.08 | | | 1790 |
| 1989 | – | – | −0.70 | | 1791 | 2002 | 0.76 | −1.83 | | | 1833 |
| 1990 | 1.36 | −2.04 | | | 1794 | 2003 | 1.79 | −1.80 | | | 1718 |
| 1991 | 1.31 | −1.37 | | | 1704 | 2004 | 0.93 | −3.22 | | | 1851 |
| 1992 | 0.98 | −1.22 | | | 1758 | 2005 | 1.73 | −1.99 | | | 1758 |

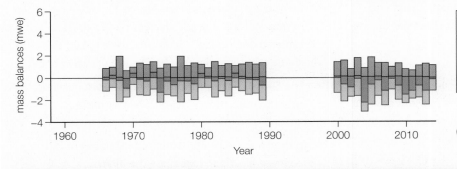

**Figure 4** *Gulkana Glacier mass balance 1966–2014 (USGS)*

## ACTIVITIES

**1** Why do you think the Gulkana Glacier was chosen as a 'benchmark' glacier?

**2** How was data collected for the study of the glacier's mass balance?

**3 a** Study Figure **2**. Describe how the extent of the glacier changed between 1974 and 2013.

  **b** Why might inferences based on evidence from Figure **2** alone possibly lead to inaccurate conclusions about the glacier's changing mass balance?

**(S)**

**4 a** Study Figure **3**. How can you explain the positive mass balance values in the winter and negative mass balance values in the summer?

  **b** Complete the two blank columns. The first few calculations have been done for you.

  **c** Represent the cumulative balance data in the form of a line graph. Be careful to choose an appropriate vertical scale. Describe the trends.

**(S)**

  **d** Now draw a similar graph to show the changes in the altitude of the equilibrium line. Are there any discernible trends? How do changes in the equilibrium line link to changes in mass balance?

**5** Figure **4** shows the trends in mass balance for the Gulkana Glacier between 1966 and 2014. Notice that the green shading indicates the annual (net) mass balance. The data for the 1990s are missing.

  **a** Use the data for 1990–1999 in Figure **3** to plot your own graph in the same style as that in Figure **4**.

  **b** Use your graph together with Figure **4** to describe the changes in mass balance between 1966 and 2014.

**6** With reference to all the evidence gathered and using your answers to the earlier activities, consider to what extent the Gulkana Glacier is showing signs of decline.

In this section you will learn about historic patterns of ice advance and retreat

## Why do glaciers advance and retreat?

Glaciers respond to long-term trends in mass balance, the balance between accumulation and ablation.

◆ If *accumulation* exceeds *ablation* (positive mass balance), the glacier's mass increases and it will probably advance.

◆ If *ablation* exceeds *accumulation* (negative mass balance), the glacier's mass decreases and it will probably retreat.

A glacier's mass can increase/decrease without any advance/retreat taking place. One is not always the consequence of the other.

Glacier advance or retreat is usually extremely slow by everyday standards. It is usually recorded annually or even over decades by the position of the glacier snout. The snout is often marked by a terminal moraine – a high ridge of sediment – that is pushed ahead of the glacier in much the same way as earth pushed by a bulldozer. If the glacier is retreating, the terminal moraine is left abandoned.

## Historic patterns of ice advance and retreat

You have already learnt how alternating cold and warm periods led to ice advance and retreat during the last 800 000 years (see 1.10).

A more recent cold period lasted from around 1500 to 1850 – the 'Little Ice Age' – when global temperatures were cooler than the present day. Ice advanced, particularly in mountain areas, even the UK was affected – frost fairs on the River Thames, winter food shortages and the formation of sea ice around the coastline.

Over the next hundred years the climate warmed and many of the world's glaciers retreated. A slight global cooling between 1950 and 1980 halted this trend and some glaciers advanced (see Figure **1**). However, since 1980 many of the world's glaciers have experienced considerable shrinkage. Scientists have linked this dramatic trend to climate change and global warming.

◆ The Swiss Glacier Monitoring Network currently notes that 86 of 95 glaciers observed are retreating, 6 are stationary and 3 advancing.

◆ In the Himalayas, an estimated 10 per cent of the ice mass has been lost since the 1970s.

◆ In the Cascades in North America, all 47 glaciers monitored by scientists are retreating and some have disappeared completely.

🔽 **Figure 1** *Worldwide changes in glacier length*

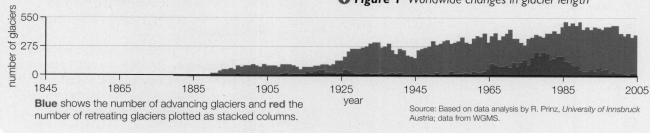

**Blue** shows the number of advancing glaciers and **red** the number of retreating glaciers plotted as stacked columns.

Source: Based on data analysis by R. Prinz, *University of Innsbruck Austria*; data from WGMS.

## ACTIVITIES

1 Study Figure **1**. Summarise the trends illustrated by the graph. Do you think there is strong evidence for climate change and global warming?

2 Design a summary diagram to describe the historic advances and retreats of the Mer de Glace. Base your diagram on a timeline. Identify the periods of advance and retreat and support your descriptions by using statistics and illustrations.

3 Study Figure **4**.
   a What is the evidence of the Little Ice Age?
   b Suggest a value on the *y*-axis that approximates with the pre-Little Ice Age situation. Explain your chosen value.
   c Is there evidence of current global warming (since about 1980)?

4 'Visit the Mer de Glace before it disappears!' Is this good advice?

## Ice advance and retreat: the Mer de Glace, France

The Mer de Glace is located on the north side of Mont Blanc, close to the resort of Chamonix. The 12 km-long glacier is one of the largest in this part of the Alps. Its elevation ranges from 1500 m to 4200 m. The glacier has been stagnant and thinning for some years, and a recent report in 2014 suggested that it may retreat by 1400 m by 2040.

During the Little Ice Age, the Mer de Glace extended to the floor of the Valle de Chamonix at an altitude of 1000 m and threatened to engulf outer parts of the town. Since then the glacier has retreated by 2300 m and thinned markedly (Figure 2). At the famous Montenvers Station, where tourists disembark to walk down to the glacier (Figure 3), it has thinned by 150 m since 1820.

Retreat has not been continuous during this period. During the 1970s and 1980s, a slight cooling led to the glacier advancing by 150 m. Figure 1 shows how this cooling period impacted on glaciers worldwide. However, since the 1980s the glacier has retreated by 500 m and thinned (at Montenvers Station) by 70 m. Figure 4 shows the advance and retreat of the Mer de Glace since 1570.

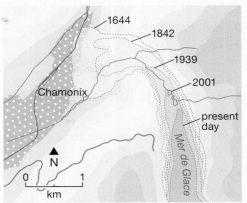
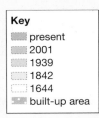
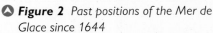

▲ **Figure 2** *Past positions of the Mer de Glace since 1644*

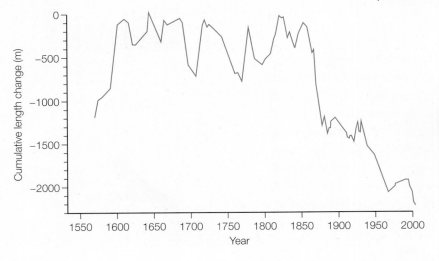

▲ **Figure 3** *Walkway from Montveners Station down to the human-made grotto that is built into the Mer de Glace. The grotto is rebuilt twice a year due to the movement of the ice (about 1 cm per hour).*

⊘ **Figure 4** *Fluctuations in the Mer de Glace (1570–2003)*

### STRETCH YOURSELF

Conduct a more detailed study of one or more glaciers in a region that interests you. Look at the website www.grid. unep.ch/glaciers which may be of use.

167

In this section you will learn about the characteristics and development of warm- and cold-based glaciers

## What are warm- and cold-based glaciers?

You might imagine that a glacier has the same temperature throughout its mass – extremely cold! However, a glacier's temperature profile – its *thermal regime* – often shows considerable variations with depth. The increased pressure of overlying ice combines with geothermal heating from below to cause temperatures to increase with depth.

At the ground surface, under normal atmospheric pressure, the melting point of ice is about 0°C. Under pressure, the melting point is lower. This is called the *pressure melting point* (pmp).

Look at Figure **1**. Notice that the pressure melting point, indicated by the broken red line, decreases slightly with depth. If, at any point, the actual temperature of the ice is the same as the pmp, then melting will take place. It is possible to identify two types of glacier according to their temperature profile, **warm-based glaciers** and **cold-based glaciers**.

### Warm-based glaciers

A warm-based glacier (Figure **1a**) is typically associated with a temperate environment, for example, the European Alps. For much of its profile, the temperature is barely below freezing. However, notice that there are significant variations in surface temperature between summer and winter. In summer, as temperatures exceed 0°C, melting will occur and the surface of the glacier will be awash with water. Notice also that at the glacier's base, the temperature of the ice reaches pmp during the year. This means that some meltwater will be produced beneath the glacier, acting as a lubricant to promote glacier movement by basal sliding (see 4.1). This results in erosion and scouring of the bedrock. Debris will be entrained and later deposited. These temperate, warm-based glaciers can erode and transport large volumes of sediment.

### Cold-based glaciers

Cold-based glaciers (Figure **1b**) are associated with polar environments, such as Antarctica or central Greenland. Here the ice is very much colder, with both winter and summer temperatures being well below 0°C, with little or no surface melting. Despite a marked increase in temperature with depth, it never reaches the pressure melting point and therefore the ice remains frozen throughout its profile. In the absence of meltwater at its base, the glacier is effectively frozen to the bedrock and basal sliding will not occur. Any movement will be solely by internal deformation (see 4.1). As a result, there will be limited entrainment of debris and little erosion or deposition.

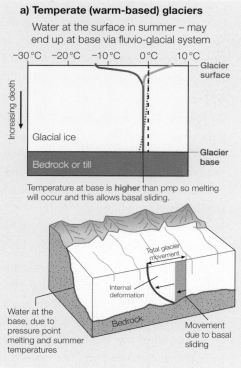

**a) Temperate (warm-based) glaciers**

Water at the surface in summer – may end up at base via fluvio-glacial system

Temperature at base is **higher** than pmp so melting will occur and this allows basal sliding.

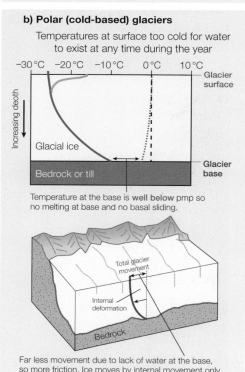

**b) Polar (cold-based) glaciers**

Temperatures at surface too cold for water to exist at any time during the year

Temperature at the base is **well below** pmp so no melting at base and no basal sliding.

Far less movement due to lack of water at the base, so more friction. Ice moves by internal movement only.

▶ **Figure 1** *Temperature profiles for warm- and cold-based glaciers*

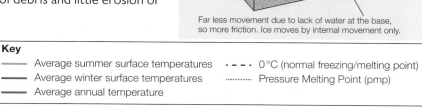

**Key**

| | |
|---|---|
| —— Average summer surface temperatures | - - - 0°C (normal freezing/melting point) |
| —— Average winter surface temperatures | ·········· Pressure Melting Point (pmp) |
| —— Average annual temperature | |

## EXTENSION

### Polythermal glaciers in Svalbard

Warm-based and cold-based glaciers represent either end of a spectrum. Between these extremes are individual glaciers that exhibit variability in their base temperature. Such glaciers are termed *polythermal*.

In Svalbard, a Norwegian archipelago in the Arctic, many of the smaller outlet glaciers are polythermal (Figure **2**). The very cold temperatures mean that high basal pressures must be reached if pressure melting point is to be attained. The outer edges of glaciers, where the ice is thinner, tend to have cold bases and therefore experience limited movement. At higher altitudes, the thicker ice may have a warm base due to the increased pressure. The pressure melting point may be reached and faster glacial movement can occur. A large number of glaciers across the world exhibit polythermal characteristics both in cross and long profile

▲ *Figure 2* *Midnight sun reflecting off the glaciers at Longyearbyen, Svalbard, Norway*

### The Dry Valleys, Antarctica

The Dry Valleys in Southern Victoria Land is one of the most extreme polar deserts in the world and has been likened to conditions on Mars (Figure **3**). Much of the area is rocky and ice-free, apart from a number of glaciers that flow into the area from external sources. There is only 10 mm of snow fall (water equivalent) per year and the mean annual air temperature is around −19.8 °C. The majority of the glaciers that flow into the area are cold throughout. They have basal temperatures of around −17 °C and no free-running water.

Yet, despite these glaciers being the coldest glaciers on Earth, research by Hambrey and Fitzsimons (2010) suggests that they are capable of erosion and deposition. Debris entrainment encompasses the detachment of frozen blocks of sediment that are then folded and thrusted by the movement of the ice. The geomorphological features that are created include sandy ridges and aprons draped with a veneer of windblown sand.

▲ *Figure 3* *Taylor Valley Glacier, Dry Valleys, Antarctica*

## ACTIVITIES

Ⓢ 1 Study Figure **1**. Draw your own annotated diagrams describing the characteristics of warm- and cold-based glaciers. Make sure you define the key terms such as pressure melting point.

2 What are the factors that determine the development of a glacier's thermal regime?

3 How does base temperature determine glacier movement and the effectiveness of glacial processes?

## STRETCH YOURSELF

Locate the Dry Valleys and describe the climate and landscape features of the area. Find out about the characteristics of the glaciers and see what you can discover about their movement and processes.

In this section you will learn about the wide variety of geomorphological processes that operate in cold environments

## Geomorphological processes in cold environments

*Geomorphological processes* are natural processes that result in the modification of landforms on the Earth's surface – they 'shape the Earth' ('geo' means 'earth' and 'morph' means 'shape'). While cold environments may appear somewhat inactive and sterile, the geomorphological processes that act upon them can be surprisingly vigorous and effective.

Geomorphological processes are most active at the margins of cold environments where precipitation amounts are high, liquid water is readily available and temperatures hover above and below freezing. In the most intensely cold environments where precipitation is low and temperatures remain well below freezing throughout the year, geomorphological processes are limited and extremely slow.

▼ **Figure 1** *Active geomorphological processes at the snout of Gigjökull Glacier, Iceland*

Look at Figure **1**. This photo shows proglacial lakes at the snout of Gigjökull Glacier in Iceland where, incidentally, the floodwater emerged from the Eyjafjallajökull ice cap following the eruption of 2010. Notice that the vast deposits of glacial debris in the foreground have formed hills and ridges. Rivers flowing from the lakes actively erode the landscape in the summer. The exposed mountains in the distance, with their covering of snow, will be gradually broken down and shaped by the processes of weathering and mass movement. This peripherally cold landscape is clearly being actively shaped by geomorphological processes.

## Weathering in a cold environment

*Weathering* is defined as the breakdown or disintegration of rock in its original position (in situ) at or just below the ground surface. There are two major weathering processes that operate in cold environments – frost shattering and carbonation.

### Frost shattering

*Frost shattering* (also known as freeze-thaw) commonly affects bare rocky outcrops high up on a mountainside (Figure **2**). The process begins when water (rainwater or meltwater) seeps into cracks and holes (pores) within a rock. When the temperature falls to 0 °C or below, the water turns to ice, expanding in volume by about 9 per cent. This exerts stresses within the rock, enlarging cracks and pores. As the process of freezing and thawing is repeated many times, so the cracks become enlarged until chunks of rock break away and pile up as scree at the foot of the slope (Figure **2**).

Frost-shattered rocks are very sharp and angular. As they become trapped under the ice they form extremely effective abrasive tools, rather like the sand on a sheet of sandpaper. Imagine how ineffective sandpaper would be if it had no sand! Frost shattering often prepares a landscape for glacial erosion by breaking up a rocky surface, enabling erosion to take place.

▼ **Figure 2** *The effects of frost shattering in the Alps*

## Carbonation

Carbon dioxide dissolved in water forms a weak carbonic acid. This reacts with and dissolves calcium carbonate in some rocks, particularly limestone, to form calcium bicarbonate. This is **carbonation**, a process of chemical weathering. Carbon dioxide is more soluble at low temperatures. This means that water can dissolve more carbon dioxide when it is cold, making carbonation an important process in cold environments.

### The effects of frost action on the landscape

When water turns to ice it expands, breaking apart rocks and sediments and forming a rock-strewn landscape called a **blockfield** or felsenmeer (Figure **3**). It will also result in the accumulation of scree at the base of cliffs. *Frost action* is most effective when there are diurnal (daily) fluctuations in temperature; however it can also be effective over a much longer time period when freezing and thawing only occurs seasonally (see 4.7).

During a period of hard frosts in winter you might have noticed that soils and lawns become very bumpy and irregular. This is the result of a process called **frost heave**, where freezing water just below the surface expands and pushes up the ground above. The growth of individual ice crystals can raise individual soil particles to create a spiky surface.

⬥ **Figure 3** *Blockfield in High Tatras mountains, Slovakia*

### Nivation

**Nivation** is an umbrella term used to cover a range of processes associated with patches of snow (Figure **5**). These processes are most active around the edges of snow patches (Figure **4**).

- Fluctuating temperatures and the presence of meltwater promote frost shattering.
- Meltwater in the summer will carry away any weathered rock debris to reveal an ever-enlarging *nivation hollow*.

- Slumping may also take place during the summer, as saturated debris collapses due to the force of gravity.

As long as the freshly weathered material is removed by meltwater or mass movement processes, the nivation hollow will continue to be enlarged. If climatic cooling takes place, the hollow will eventually be occupied by glacial ice and the nivation hollow might become enlarged to form a corrie (see 4.8).

⬇ **Figure 4** *Nivation processes resulting in the formation of a nivation hollow*

⬇ **Figure 5** *Nivation snow patches, Parângul Mare, Romania*

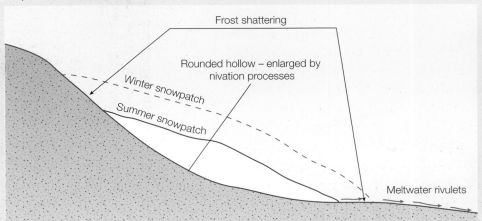

# Ice movement

There are two broad types of ice flow – internal deformation and basal sliding.

## Internal deformation

Internal deformation takes place in both cold- and warm-based glaciers. The rate of movement is estimated to be up to 1–2 cm a day.

There are two mechanisms of **internal deformation**.

- *Intergranular movement.* Individual ice crystals slip and slide over each other in much the same way as individual grains of sugar move when heaped in a pile. The ice crystals within the glacier tend to orientate themselves in the direction of ice movement and this allows them to slide past one another.

- *Intragranular movement.* Individual ice crystals become deformed or fractured due to the intense stresses within the ice that are exerted by the glacier's mass under the influence of gravity. Gradually, the mass of ice deforms and moves downhill in response to gravity.

## Basal sliding

Basal sliding involves the movement of a large body of ice, usually in a series of short jerks. It occurs in warm-based glaciers where meltwater is present to help lubricate the base of the ice. As it moves (up to 2–3 m a day) the ice will pick up and transport debris, using it to erode the underlying bedrock (Figure **7**).

When a glacier encounters an obstacle such as an outcrop of tough rock, the additional resistance to movement on the upslope side causes an increase in stress, and therefore pressure, which may result in melting of the base – *pressure melting*. This facilitates the movement of the glacier over the obstacle, although the meltwater often refreezes on the downslope side where pressure is reduced. This phenomenon of melting and freezing depending on pressure is called *regelation*.

The movement of the ice downhill can raise the temperature of the base ice through increased pressure and friction. This positive feedback may lead to further melting of the basal ice allowing the glacier to slip more easily over its bed. Figure **6** shows basal ice beneath a glacier in Antarctica. The dark-striped layer beneath the white glacier ice is the basal ice, the product of regelation. Notice that it is strongly layered, sheared and incorporates debris. Beneath the basal ice is lighter-coloured glacial debris called *till*.

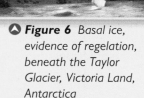

▲ *Figure 6 Basal ice, evidence of regelation, beneath the Taylor Glacier, Victoria Land, Antarctica*

▼ *Figure 7 Internal deformation and basal sliding*

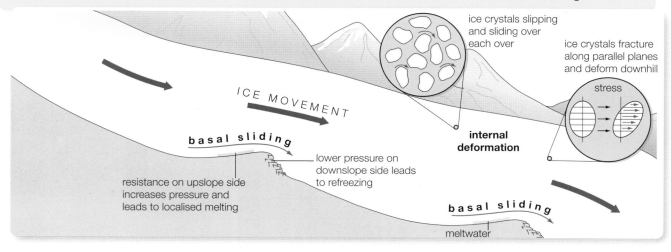

ice crystals slipping and sliding over each over

ice crystals fracture along parallel planes and deform downhill

ICE MOVEMENT

stress

**basal sliding**

**internal deformation**

lower pressure on downslope side leads to refreezing

resistance on upslope side increases pressure and leads to localised melting

**basal sliding**

meltwater

## Variations in the rate of ice movement

The movement of ice is affected by many controlling factors, variations in which can significantly affect the rate of flow from place to place. For example, when there is a sudden increase in the gradient, the ice will flow faster and, through internal deformation, the ice will become 'stretched' and will thin. This is called **extensional flow** and will often manifest itself at the surface by the formation of crevasses (Figure **8**).

Conversely, a reduction in gradient will cause the glacier to slow down causing it to 'pile up' and become thicker. This is called **compressional flow**. Any crevasses opened up by the earlier extensional flow will now be closed (Figure **8**).

Between the two zones of extensional and compressional flow, the ice moves in a curved or *rotational* manner (Figure **8**). This type of movement is very important in the formation of corries (see 4.8).

In common with a river, the fastest flow of a glacier is towards the centre, away from the effects of friction.

It is possible to envisage a *negative feedback loop* (see 1.1) in operation with glacier movement. All things being equal, thickening of ice (compressional flow) will increase mass and potential erosion. This could lead to a steeper gradient and thus encourage faster extensional flow, a thinning of the ice and a reduction in potential erosion. We therefore have a theoretical negative feedback loop. But, would this happen in reality?

## Controlling factors in ice movement

- *Gravity.* This is the downhill force that encourages ice to move. The steeper the gradient, the greater the pull of gravity.

- *Friction.* If the ice as a whole is to move forward, the friction exerted by the ground on the ice has to be overcome.

- *Mass of the ice.* The heavier the ice, the more potential energy it has to move. However, more force will also be needed to overcome the increased friction caused by the extra weight.

- *Meltwater.* Meltwater lubricates the base of the ice, enabling it to slip downhill. An increase in water pressure beneath the ice may also help to overcome friction by creating a buoyancy effect.

- *Temperature of the ice.* In some environments, such as Greenland and Antarctica, the ice is so cold that it is frozen to the bedrock. Such cold-based glaciers tend to move more slowly than warm-based glaciers where meltwater may be present at the base of the ice.

### Did you know?

Unlike rivers, glaciers can actually flow uphill! This occurs when there is sufficient weight of ice towards the back of a glacier to force the snout up and over a rise in the ground.

▼ **Figure 8** *Extensional, compressional and rotation flow*

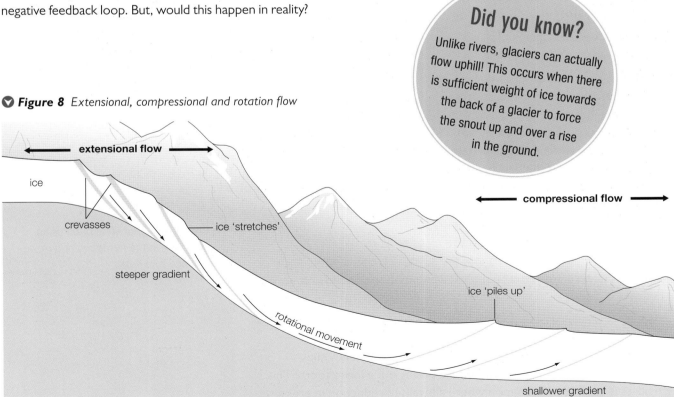

extensional flow

ice

crevasses

ice 'stretches'

steeper gradient

rotational movement

compressional flow

ice 'piles up'

shallower gradient

## EXTENSION

### The relative contribution of internal deformation and basal sliding

The movement of cold-based glaciers will almost exclusively be by internal deformation. This means that the rate of movement will be slow and the action of geomorphological processes will be limited. The rates of erosion and sediment transfer will tend to be low. With warm-based glaciers, internal deformation combines with basal sliding resulting in greater movement and higher rates of erosion and sediment transfer.

Look at Figure **9**. Notice that basal sliding accounts for an equal amount of movement throughout the vertical profile of the glacier (A–A'), which takes it to B–B'. Internal flow and deformation (from B–B' to B'–C), however, varies considerably, with the greatest movement being towards the top and centre of the ice. This is because over time the identity and characteristics of the once discretely individual ice crystals is lost as they combine and become transformed by the intense pressure to become amorphous glacial ice. Closer to the surface, while still behaving as separate ice crystals, internal deformation is more significant. Notice also in Figure **9** how friction with the valley side severely restricts movement of any kind.

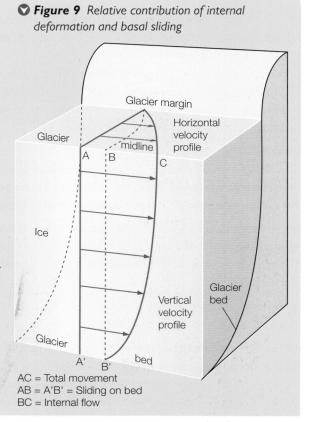

▼ **Figure 9** Relative contribution of internal deformation and basal sliding

AC = Total movement
AB = A'B' = Sliding on bed
BC = Internal flow

## Glacial processes

### Glacial erosion

There are two main forms of glacial erosion – abrasion and plucking.

*Abrasion* is the sandpapering effect of ice as it grinds over and scours a landscape (Figure **10**). It is made possible by the presence of angular, frost-shattered material. Large rocks carried beneath the ice will often scratch the bedrock (these scratches are called *striations*). Striations form useful clues for scientists studying the direction of ice flow in a post-glacial environment. Over time, these rocks become pulverised by the weight of the ice to become fine *rock flour*. This finer material will tend to smooth and polish the underlying bedrock.

*Plucking* or *quarrying* occurs when meltwater freezes part of the underlying bedrock to the base of a glacier. Any loosened rock fragments will be 'plucked' away as the glacier subsequently slips forward. Put very simply, it is rather like pulling out a loose tooth! This process is particularly common where a localised reduction in pressure under the ice has led to *regelation* (refreezing of meltwater).

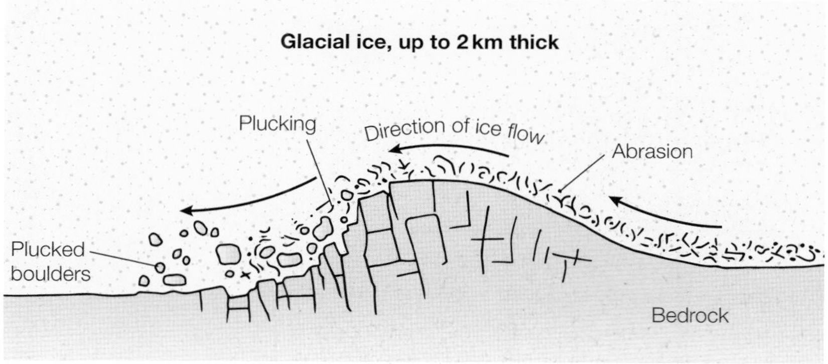

▲ **Figure 10** The processes of abrasion and plucking in the formation of a roche moutonnée (see 4.8).

## Glacial transportation

Glaciers act like giant conveyor belts, transporting material from mountainous areas onto adjoining lowlands. This material can be carried in three ways:

◆ *Supraglacial* – predominantly weathered material carried on top of the ice (Figure **11**).

◆ *Englacial* – once supraglacial material, now buried by snowfall and carried within the ice.

◆ *Subglacial* – material carried beneath the ice, dragged and pulverised by the overlying glacier.

The role of water is not to be underestimated in the transportation of material. In temperate environments, water will flow on top of glaciers, leading to fluvial transport. This water may flow down crevasses or holes in the ice (*moulins*), transporting the material into and beneath the glacier. Warm-based glaciers will have meltwater streams flowing under the ice, carrying material to the glacier's snout and beyond.

**Figure 11** *Supraglacial transport on the Mer de Glace, France*

## Glacial deposition

Deposition of sediment transported by the ice takes place when the ice melts, which is primarily in the ablation zone close to the glacier's snout. Here the sediment on and in the ice simply melts out in much the same way as the items used to decorate a snowman are left on the ground when it finally melts! Unexpected items are sometimes found amongst the sediment (Figure **12**). Water may carry sediment further away from the ice, sometimes over distances of many kilometres.

**Figure 12** *Fragments of an aircraft that were transported by and then melted out of the Gigjökull Glacier, Iceland*

## ACTIVITIES

1 Study Figure **1**. Describe the landscape and suggest the geomorphological processes that are likely to be active.

2 Assess the importance of subaerial processes such as weathering, frost action and nivation in the development of glacial landscapes.

3 Find a photo showing snow patches in a mountain environment. Add annotations to describe the nivation processes at work (Figure **4**). Try to find a photo that shows evidence of two or more of these processes.

4 What evidence would you look for to suggest that a glacier was experiencing extensional or compressional flow? How might these two types of flow be linked by a negative feedback loop?

5 What factors determine the effectiveness of internal deformation and basal sliding?

6 Study Figure **9**. Make a copy of this diagram and add detailed annotations to give a reasoned description of the variations in type and amount of glacial movement.

7 Use Figures **7** and **10** to explain the connection between the basal sliding processes of regelation and basal melting, and the glacial erosional processes of abrasion and plucking. Use simple diagrams to support your answer.

8 With the help of photographs, describe the different forms of glacial transportation.

In this section you will learn about the features and processes associated with periglacial environments

## Features of a periglacial environment

Periglacial environments exist at the edges of glacial/polar environments. While they do not have a permanent covering of ice, they experience extreme cold for much of the year with penetrating frosts and periodic snow cover.

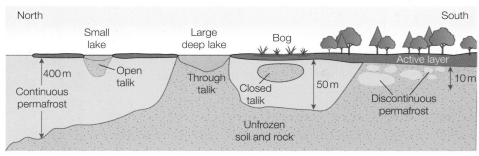

**Figure 1** *Model transect across permafrost in the Arctic*

**Permafrost** is one of the key characteristics of a periglacial environment. It is a condition where a layer of soil, sediment or rock below the ground surface remains frozen for a period greater than a year. During the brief and potentially quite warm summers, lying snow will melt, as will the upper portion of the permafrost – commonly 1–3 m – forming a saturated and potentially mobile *active layer* (Figure **1**).

Permafrost is very widespread, covering some 25 per cent of the Earth's non-glaciated land surface, commonly reaching depths of 400–500 m. In some areas it can reach depths of 1500 m.

Figure **1** shows that there is a transition in the depth and extent of permafrost. Deep, continuous permafrost in the far north gradually becomes less extensive and patchier as conditions become warmer further south. Notice also changes in vegetation and in the depth of the active layer.

**Figure 2** *Thaw lakes near Uniujap, Quebec, Canada*

*Thaw lakes* are common in poorly drained periglacial areas. They form during the summer when any snow melts together with the active layer (Figure **2**). Water retains warmth and its relatively dark surface absorbs radiation from the Sun. This warmth will increase the depth of melting of the underlying permafrost, forming unfrozen zones called *taliks* (Figure **1**).

## Processes in a periglacial environment

Geomorphological processes are mostly associated with frost, ice and snow, together with summer meltwater. You have already learnt about nivation, a common periglacial process in section 4.6. Periglacial processes will impact on a landscape either as a transition to colder glacial conditions or as part of a warming trend.

### Frost action

Frost action is one of the most important processes in a periglacial environment. Not only does it break up a rock surface, making it vulnerable to subsequent erosion, but it also produces rock fragments that can act as tools of erosion.

Frost action is most commonly associated with the weathering process of *frost shattering* (freeze–thaw). This leads to the accumulation of angular scree at the foot of a mountain slope or a boulder-strewn landscape called *blockfield* or *felsenmeer*. Research has indicated that frost-induced fracturing is most effective between temperatures of −4 °C and −15 °C and when rates of cooling are slow.

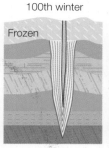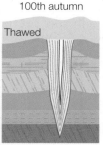

**Figure 3** *The process and formation of ice wedges*

Frost action can also be effective in soils and sediments through the presence of ice in the ground – *ground ice*. Ice can form within pores (*pore ice*) or as *ice needles*, which can force individual soil particles or small stones upwards to the surface – this is called *frost heave* (see 4.6).

On a larger scale, cracks in the ground formed by thermal contraction in winter can become infilled with water in the summer. This water subsequently freezes and enlarges the crack. Over many years an **ice wedge** is formed, which can be up to 3 m wide at the surface and 10 m deep (Figure **3**).

## Mass movement

Mass movement can be defined as the downward movement of material under the influence of gravity. It encompasses a wide range of processes, such as rockfalls, landslides, mudflows and also *solifluction* ('soil flow') (see 3.4). This is when saturated soil (the active layer) slumps downhill during the summer to form solifluction lobes (Figure **4**).

The growth of needle ice in loose sediments can result in the gradual movement of sediment down a slope. This is called *frost creep* (Figure **5**). Particles are raised perpendicular to the ground surface by frost heave, which are then dropped down vertically (due to gravity) on thawing. Repeated cycles cause the soil particles to gradually 'creep' downslope.

**Figure 4** *The process of solifluction*

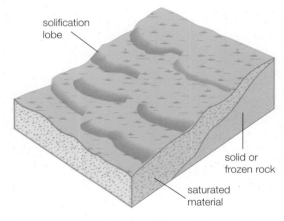

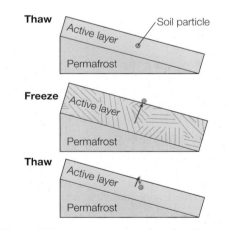

**Figure 5** *Frost creep in periglacial environments*

## ACTIVITIES

1 Study Figure **1**.
   a Describe and account for the variations in depth and extent of permafrost, within the Arctic, from north to south.
   b Explain the differences between the three types of talik.
   c What evidence might scientists look for to support the belief that global warming is causing higher temperatures in the Arctic?

2 Make your own copy of Figure **3**, showing the development of ice wedges. Write detailed labels to describe the processes involved.

3 Use Figures **4** and **5** to create your own labelled diagram to show the processes of solifluction and soil creep operating on a slope under periglacial conditions.

4 Assess the role of ice action in periglacial environments.

In this section you will learn about the origin and development of landscapes of glacial erosion

## What is a glacial erosion landscape?

The photo in Figure **1** was taken from close to the summit of Helvellyn, a peak in the English Lake District. It is looking roughly east/north-east over one of the most iconic glacial landscapes in the UK. The steep ridges, bare rocky outcrops, lakes and wide valleys are typical of a landscape of glacial erosion.

▼ *Figure 1 The Lake District – an iconic landscape of glacial erosion*

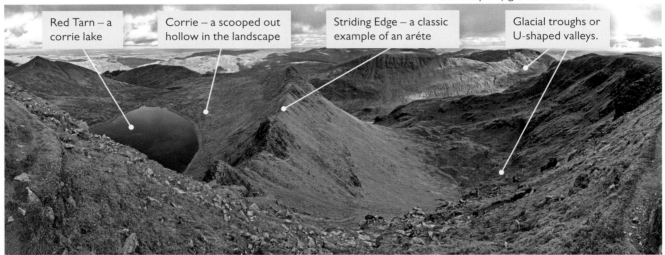

Red Tarn – a corrie lake

Corrie – a scooped out hollow in the landscape

Striding Edge – a classic example of an aréte

Glacial troughs or U-shaped valleys.

## The origin and development of glacial erosional landforms

### Corries

A *corrie* (also known as a *cirque* in France and *cwm* in Wales) is an enlarged, often deep, hollow on a mountainside. Its characteristic features include a steep, cliff-like backwall, often with a pile of scree at its base. At the front of the hollow there is usually a raised rock lip, which can act as a dam trapping water to form a lake called a *tarn* (Figure **1**).

To understand the origins and development of a corrie we need to consider the landscape processes at work before, during and after glaciation (Figure **2**). Remember that most glacial landscapes have experienced a succession of cold glacial periods that account for the scale and complexity of landforms and landscapes.

### Arêtes and pyramidal peaks

When two neighbouring glaciers cut back into a mountainside, the narrow, knife-edge ridge that forms between the two corries is known as an *arête*. They are common in both present-day glacial landscapes such as the Alps and post-glacial landscapes such as the Lake District (Figure **1**).

▼ *Figure 2 How a corrie is formed*

**a)** Periglacial conditions: periglacial processes, particularly those associated with snow (nivation) and frost/ice, slowly increase the size of a hollow or depression on the mountainside.

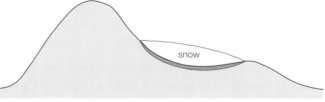

snow

**b)** Glacial conditions: As the climate cools, snow turns to ice in the depression and a corrie glacier develops. Accumulation at the top of the glacier increases its mass and rotational sliding results in a 'scooping out' of the hollow by abrasion.

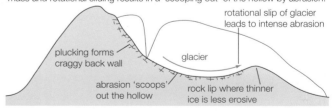

rotational slip of glacier leads to intense abrasion

plucking forms craggy back wall

glacier

abrasion 'scoops' out the hollow

rock lip where thinner ice is less erosive

**c)** Post-glacial conditions: As the climate warms, first periglacial and then temperate processes (including frost and water action) modify the shape of the corrie to create what we see today.

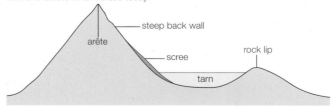

steep back wall

arête

scree

rock lip

tarn

Where three or more corries erode back-to-back, the ridge becomes an isolated peak called *a pyramidal peak* (Figure **3**).

## Glacial troughs

Glaciers are immensely powerful agents of erosion. They can reach thicknesses of several hundred metres and their sheer mass enables them to create spectacular and dramatic landscapes. The Athabasca glacier in Canada is shrinking and, as it does so, more of the impressive valley (*glacial trough*) is revealed (Figure **4**).

Glacial troughs are characteristically steep-sided, broadly flat-bottomed and often several hundred metres in depth. They tend to be largely straight because of the immense power and inflexibility of the glaciers that gouge them out. The main process of erosion is abrasion through basal sliding – plucking does take place but more selectively than abrasion.

There are several factors that affect the extent of erosion by a glacier.

- Mass of the ice (thickness of the glacier) – the thicker the ice, the greater potential energy for erosion.

- Gradient – this can affect the rate of flow of the ice and also its thickness – *extensional* and *compressional* flow (see 4.6).

- Meltwater – the presence of meltwater beneath the glacier enables basal sliding to occur, which is likely to be more erosive than internal deformation (see Figure **8**, 4.6).

- Rock debris – large, angular rocks trapped beneath the ice act as tools that scrape and gouge the underlying bedrock.

- Underlying geology – whether the rocks are strong or weak, massive or thinly layered, jointed or unjointed, can have an important influence on the extent of erosion that takes place.

Glacial troughs may contain deep, narrow lakes called *ribbon lakes*. These result from localised overdeepening due to enhanced erosion caused by:

- weaker bedrock allowing increased vertical erosion

- merging of a tributary glacier which can lead to greater erosion of the valley floor because of the increased mass of ice

- narrowing of the valley and subsequent thicker ice, leading to increased vertical erosion.

## Hanging valleys

A small glacier occupying a tributary valley does not have the same mass as a much larger glacier in a trunk valley; it will be unable to erode down as far as the larger glacier. When the ice melts, the smaller valley will be left 'hanging' high above the main valley – a *hanging valley*.

Look at Figure **5**. This impressive hanging valley in New Zealand is marked by Stirling Falls plunging into Milford Sound, a fjord formed by the sea flooding a huge glacial trough. Notice that the hanging valley itself exhibits the typical characteristics of a glacial trough.

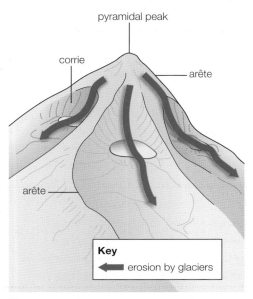

**Figure 3** *How arêtes and pyramidal peaks are formed*

**Figure 4** *The Athabasca glacier, Canada*

**Figure 5** *Stirling Falls, Milford Sound, New Zealand*

## Truncated spurs

In a typical upland river valley, the river will flow around interlocking spurs of rock that jut out into the valley. When the valley is occupied by ice, the rigid and more powerful glacier cuts off or 'truncates' the tips of the rocky spurs as it moves downhill, leaving behind steep cliffs called *truncated spurs* (see Figure **6**).

While abrasion is the dominant process in the formation of truncated spurs, plucking may also take place on the bare rocky surfaces. Subaerial processes, such as frost shattering (freeze–thaw), and rockfalls act upon the exposed rocky cliffs protruding above the ice.

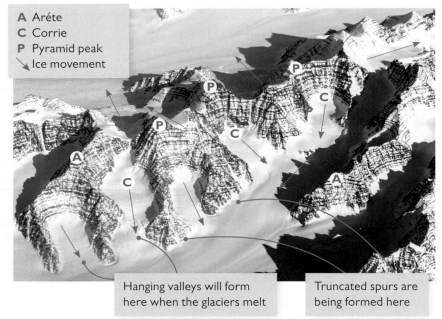

**A** Aréte
**C** Corrie
**P** Pyramid peak
↘ Ice movement

Hanging valleys will form here when the glaciers melt

Truncated spurs are being formed here

⬆ **Figure 6** *Glaciers and glacial landforms in the Watkins Range Mountains, the highest mountain range in Greenland. Enormous glaciers are currently sculpting the landscape, forming several of the landforms previously described. Notice how truncated spurs tend to form between hanging valleys.*

## Roche moutonnées

Bare outcrops of rock on the valley floor that have been sculpted by the moving ice are called *roche moutonées*. Roche moutonnées demonstrate very clearly the two processes of glacial erosion, abrasion and plucking. Look at Figure **7**.

◆ On the upstream side, an increase in pressure caused by the resistance of the rock outcrop to the moving ice, leads to localised pressure melting. This facilitates basal sliding and the process of abrasion as the glacier slides over the rock outcrop. Notice that the abraded upstream side exhibits polishing and striations (scratches).

◆ On the downstream side, the reduction in pressure causes the meltwater to freeze forming a bond between the rocky outcrop and the overlying glacier. As the glacier continues to move forward it plucks away loose rocks leaving behind the jagged surface you can see.

⬇ **Figure 7** *Roche moutonnée, Honister Pass, Lake District*

---

### ACTIVITIES

1 With the aid of diagrams describe the formation of a corrie.

2 What are the distinctive glacial landforms associated with a glacial trough and how are they formed?

**S** 3 Draw a sketch of Figure **7**. Add detailed annotations to describe the physical characteristics and erosional processes responsible for the formation of the roche moutonnée.

## The Cairngorms, Scotland

Figure **8** is an OS map extract showing part of the Cairngorms mountain range in Scotland. The Cairngorms is one of only two National Parks in Scotland. Its landscape is extremely rugged and reflects the action of ice over several glacial periods. Figure **9** is a photo of Loch Einich taken from Sgor Gaoith (902990).

Locate Coire Bhrochain in grid square 9599. This is a corrie. Notice that a semi-circle of bold black cliff symbols marks the backwall of the corrie. This is an arête. The small black dots indicate deposits of scree.

▲ **Figure 9** *Loch Einich from Sgor Gaoith*

⬇ **Figure 8** *1:50 000 OS map extract of the Cairngorms in Scotland*

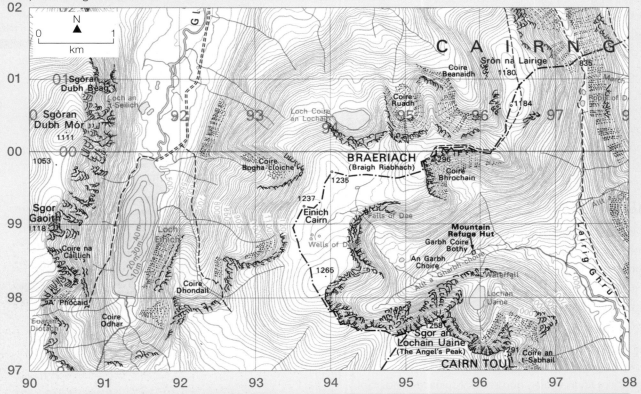

## ACTIVITIES

**(S)**

1 Draw a sketch map of the south-eastern quarter of the map bounded by the 94 and 00 gridlines. Draw and label the landforms of glacial erosion including corries (such as Coire Bhrochain), arêtes and the Lairig Ghru glacial trough. Write the placenames as well as the geographical landforms. Don't forget to include a scale and a north arrow.

2 With the help of Figure **8**, describe the glaciated landscape in the photo (Figure **9**).

3 With reference to evidence from the map and the photo, suggest possible reasons for the formation of Loch Einich, an example of a ribbon lake.

In this section you will learn about the origin and development of landscapes of glacial deposition

## What is a landscape of glacial deposition?

In common with most glaciers in the European Alps – indeed, most glaciers in the world – the Steingletscher glacier (Figure **1**) is shrinking and retreating. As it does so it is revealing some classic landforms of glacial deposition – a debris-strewn valley floor, ridges of sediment and proglacial lakes.

Glaciers act like giant conveyor belts, transporting rock debris from eroded mountains and depositing it on valley floors or lowland plains. Rock debris transported beneath a glacier is pulverised by the sheer weight of the ice above to form fine, splintered *rock flour*. It is this fine-grained sediment that is responsible for turning meltwater streams and proglacial lakes a milky blue colour.

Rock debris carried by the ice and then dumped in situ when the ice melts is called *till*. It is characteristically angular and very poorly sorted due to the lack of water/wind transport. Meltwater streams flowing from the snout of a glacier can carry sediment many kilometres, depositing it as a vast, gravelly, well-sorted *outwash plain*. Figure **2** identifies some of the main depositional landforms associated with a lowland landscape.

## Moraines

*Moraine* is used as a generic term for landforms associated with the deposition of till from within, on and below a glacier and therefore comprises poorly sorted and predominantly angular sediments. Figure **2** shows the distinctive characteristics of moraines.

*Ground moraine* – the sediment transported beneath a glacier that is simply smeared over the underlying bedrock. It can be several metres thick and will, more often than not, form an irregular, hummocky surface topography.

*Terminal moraine* – a ridge of sediment piled up at the furthest extent of an advancing glacier. Terminal moraines commonly appear as a line of hills rather than a continuous ridge due to the erosive action of meltwater streams flowing out of the retreating glacier.

*Recessional moraine* – a retreating glacier often experiences periods of stability during which a secondary ridge can form at its snout – this is a recessional moraine. This has the same characteristics as a terminal moraine but it does not mark the furthest extent of ice flow.

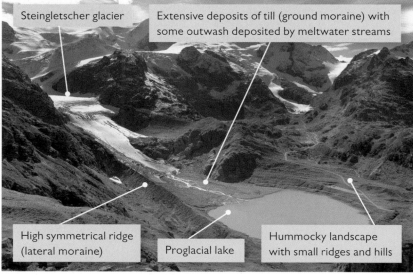

Steingletscher glacier

Extensive deposits of till (ground moraine) with some outwash deposited by meltwater streams

High symmetrical ridge (lateral moraine)

Proglacial lake

Hummocky landscape with small ridges and hills

⬆ **Figure 1** *The Steingletscher glacier in the Swiss Alps*

⬇ **Figure 2** *Landforms of glacial deposition*

**a) Lowland landscape during glaciation**

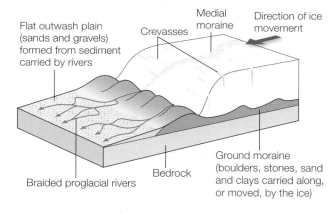

Flat outwash plain (sands and gravels) formed from sediment carried by rivers

Crevasses

Medial moraine

Direction of ice movement

Braided proglacial rivers

Bedrock

Ground moraine (boulders, stones, sand and clays carried along, or moved, by the ice)

**b) The same lowland landscape after glaciation**

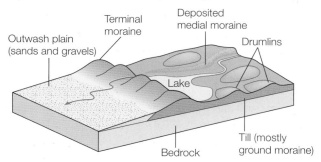

Outwash plain (sands and gravels)

Terminal moraine

Deposited medial moraine

Drumlins

Lake

Bedrock

Till (mostly ground moraine)

*Lateral moraine* – Figure **1** shows a superb example of a lateral moraine – a high, almost symmetrical ridge formed alongside a glacier primarily from the build-up of scree slopes. Lateral moraines can attain heights of several metres.

*Medial moraine* – when two glaciers merge, lateral moraines at the edges of the two glaciers join to form a medial moraine. This line of debris travels down the centre of a glacier towards the snout. When the glacier melts, the medial moraine is deposited to form a low ridge.

## Till plain

A *till plain* is an extensive plain resulting from the melting of a large sheet of ice that became detached from a glacier. The till effectively levels out the topography to create a mostly flat landscape (Figure **3**).

## Erratics

An *erratic* is most commonly a boulder or a smaller rock fragment that has been deposited in a location that is foreign to its origin. You can see that the erratic in Figure **4** is clearly a different rock type than the surrounding landscape. Erratics give important clues in establishing the direction of ice movement and in recreating paleo-environments (fossil or past environments). Find the source of the erratic and you know where the ice has come from!

## Drumlins

A typical *drumlin* is an oval- or egg-shaped hill and is essentially composed of glacial till and aligned in the direction of ice flow (Figure **5**). Their size varies, but are most commonly 30–50 m in height and 500–1000 m in length. They usually occur in clusters or 'swarms' on flat valley floors or lowland plains. Drumlins are common in parts of northern England, Scotland, Ireland, Sweden, Finland and Canada – all previously glaciated regions.

While some drumlins have a rocky core, with sediment moulded around it, most do not. Some drumlins consist, at least in part, of fluvial sediments as well as glacial till, suggesting an important role for meltwater. It may well be that these characteristic landforms of glacial deposition owe their formation to a combination of several processes. Given their formative environment – beneath hundreds of metres of ice – no-one has actually observed drumlin formation, hence the various theories and controversy (see following page).

⬥ **Figure 3** *Rich agricultural soils developed on a till plain in Idaho, USA*

⬥ **Figure 4** *An erratic boulder deposited on limestone in the Yorkshire Dales*

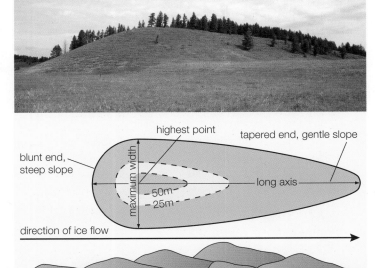

⬥ **Figure 5** *Drumlin in Alberta, Canada*

## EXTENSION

## The drumlin controversy: a study by University of Sheffield

There have been many hypotheses and theories that attempt to explain their formation. Put simply, drumlins may have formed by a successive build-up of sediment to create the hill (i.e. deposition or accretion) or pre-existing sediments may have been depleted in places leaving residual hills (i.e. erosion), or possibly a process that blurs these distinctions. Hypotheses have been proposed for all these cases but most common have been those involving some form of sediment accretion (deposition). These, however, have difficulty explaining drumlins with cores of pre-existing fluvially sorted sediments.

### The deforming bed model

Observations of the nature of the bed of contemporary ice sheets have revealed that the forward motion of ice, can, in part, be accomplished by deformation of the soft sedimentary bed. This has led to the deforming bed model of glacier flow, which has become the most widely accepted (but still unproven) mechanism for drumlin formation. If the sediments of the bed are weak, they may deform as a result of the sheer stress imparted by the overlying ice. If parts of this deforming till layer vary in relative strength, then the stronger, stiffer portions will deform less and remain static, while the intervening weaker portions will deform more readily and become mobile. So a till layer with spatially variable strength will have static or slow-moving strong patches, around which the weaker, more deformable till will flow. This can explain the cores of drumlins (strong patches; rock-cored, coarse-grained or with preserved fluvially sorted sediments) surrounded by more easily deformed till that is responsible for the streamlining. It also explains the occurrence of folds and thrusts commonly observed in drumlins.

### The cavity-fill meltwater model

An alternative to the above is a model developed by Shaw et al. (1989). It views drumlins and other subglacial bedforms to be the result of meltwater erosion and deposition as a consequence of large floods beneath the ice. In this meltwater model, regional scale outburst floods from the central regions of the ice sheet produce sheet flows of water, tens to hundreds of kilometres wide, and deep enough to separate the ice from its bed. Turbulent water during the flood stage erodes giant drumlin-shaped scours in the base of the ice, which are then infilled with sediment as the flood wanes and as the ice presses down onto its bed. This is the cavity-fill drumlin and explains how fluvially derived sediments may appear in drumlins.

The difficulty in evaluating these theories arises from the fact that the deforming bed and meltwater models are each so comprehensive as to be able to predict the wide variety of observed drumlin characteristics. This makes it hard to use geomorphological observations to test between them. Also, both are still at the stage of qualitative theories rather than physically-based models. Deforming beds have been observed to exist and so have subglacial floods: the question remains as to which are capable of producing drumlin forms and over the widespread patterns for which they are observed.

(University of Sheffield www.sheffield.ac.uk/drumlins/drumlins)

## ACTIVITY

(S) Outline the two theories of drumlin formation. Use simple sketches and annotate to illustrate the processes involved. What further evidence would be required to establish which (if either) theory is most likely to be correct? What do you think?

## Till fabric analysis

Till fabric analysis involves the study of the orientation and 'plunge' of rock fragments within a till deposit. The orientation of the long axis can suggest the direction of ice flow at the time when deposition occurred. Studies involving 50 or more rock fragments conducted at a number of localities can provide scientists with evidence of ice flow patterns, especially when used in association with other evidence, such as erratics or drumlin orientation. The angle of 'plunge' can suggest thrust motions within the till as the glacier deforms the underlying till.

Data is usually recorded in categories (classes), say every 10° or 20° (as in Figure **6**), and then presented in the form of a rose diagram (Figure **7**). Preferred orientations can be identified and compared with other evidence. In Figure **7** there is a clearly preferred orientation in a NW–SE direction.

| Class (degrees) | Midpoint (degrees) | Number of particles |
|---|---|---|
| 350–009 | 000 | 2 |
| 010–029 | 020 | 7 |
| 030–049 | 040 | 6 |
| 050–069 | 060 | 4 |
| 070–089 | 080 | 3 |
| 090–109 | 100 | – |
| 110–129 | 120 | – |
| 130–149 | 140 | 2 |
| 150–169 | 160 | 3 |
| 170–189 | 180 | 10 |
| 190–209 | 200 | 9 |
| 210–229 | 220 | 2 |
| 230–249 | 240 | – |
| 250–269 | 260 | – |
| 270–289 | 280 | – |
| 290–309 | 300 | – |
| 310–329 | 320 | 1 |
| 330–349 | 340 | 1 |

 **Figure 6** *Rock particle orientation from a till sample (Briggs, 1977)*

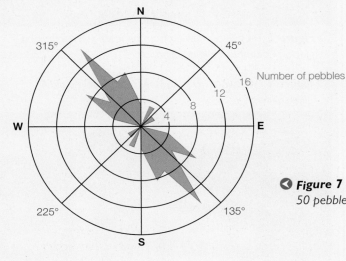

 **Figure 7** *Till fabric analysis showing pebble orientation of 50 pebbles from the Endon till, Endon Valley, Leek, Staffordshire*

## ACTIVITIES

**S**

**1** Use the photos in this section, and others sourced from the internet, to describe the typical landscape of glacial deposition. Annotation on copies of the photos may help with your answer.

**2** Draw a simple annotated diagram to describe the location and formation of the different types of moraine.

**3** Describe how scientists make use of landforms of glacial deposition to suggest the extent and movement of ice in the past.

**S**

**4** Figure **6** is a table of data showing pebble orientation for a sample of rock particles taken from a till sample.

**a** Construct a rose diagram using either the degree sectors or (as with Figure **7**) the midpoint values.

**b** Suggest a preferred orientation and critically evaluate the level of confidence in your conclusion.

**c** What additional evidence would you seek from a field locality to increase confidence in drawing conclusions about the direction of past ice flow?

In this section you will learn about fluvioglacial processes and landforms of erosion and deposition

## What are fluvioglacial landscapes?

Fluvioglacial landscapes are associated with flowing water – essentially meltwater – in glacial or periglacial environments. Meltwater is seasonally very abundant in temperate glacial and periglacial environments and is often seen at a glacial snout flowing out from under the ice (Figure **1**). It is much less common in the world's coldest environments, characterised by cold-based glaciers.

Figure **2** shows a typical fluvioglacial landscape, with a wide, multi-channelled (braided) river flowing over a vast area of sediment. It is possible to identify two types of sediment in this environment:

◆ poorly sorted and angular glacial till

◆ well sorted and more rounded outwash, typically gravelly or sandy in nature.

A fluvioglacial landscape is very dynamic, with river channels constantly migrating and changing course. This makes it a very hazardous environment for hiking and camping – deposits of dangerous quicksand are also common!

## What are fluvioglacial processes?

As with rivers, meltwater erodes, transports and deposits sediment and forms many of the typical river features, such as meandering channels, levees and deltas. Occasionally huge quantities of meltwater are trapped either beneath ice or as surface lakes. When these eventually burst – a glacial outburst – the surging meltwater has the power to carve deep channels or gorges in the landscape.

Meltwater plays a crucial role in several glacial and periglacial processes and can directly contribute to the formation of distinctive glaciated landscapes.

◆ In the process of nivation, meltwater assists in the removal of broken rock material during the summer when the outer edges of the snow patch melts. It is also vital for the process of freeze–thaw (see 4.6).

◆ In warm-based glaciers, meltwater lubricates the base of a glacier, enabling basal sliding to occur (see 4.1 and 4.6).

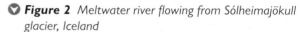

**Figure 1** *Meltwater flowing from beneath the Athabasca glacier, Columbia icefield, Canada*

**Figure 2** *Meltwater river flowing from Sólheimajökull glacier, Iceland*

◆ Meltwater flowing beneath a glacier facilitates abrasion by providing rocks that are used as 'tools' for erosion. Meltwater also freezes the base of the ice to rock fragments, enabling rocks to be 'plucked'.

◆ Meltwater erodes meltwater channels and forms distinctive depositional features both beneath the ice and in front of it.

## What are the distinctive fluvioglacial landforms?

A **meltwater channel** usually takes the form of a steep-sided valley carved into the landscape – it may not necessarily be occupied by a river. There are many possible scenarios leading to the formation of a meltwater channel. The most common of these is overspill from a lake alongside or in front of a glacier – vast quantities of highly erosive water surge across a landscape and carve a deep valley.

One of the best examples of a meltwater channel in the UK is Newtondale in North Yorkshire (Figures **3** and **4**). During the last ice advance (70 000–10 000 years ago), the North York Moors remained largely unglaciated, forming an ice-free 'island' surrounded by a vast ice sheet. At the edges of the ice, meltwater formed lakes in the lower valleys of the North York Moors. As water levels rose in these lakes they spilled out into adjacent valleys, eroding deep meltwater channels. Newtondale was formed when water overflowed from Lake Wheeldale in a southerly direction towards Lake Pickering. When the ice finally retreated, these lakes emptied. Today, Newtondale forms a narrow wooded gorge some 80 m deep and 5 km in length.

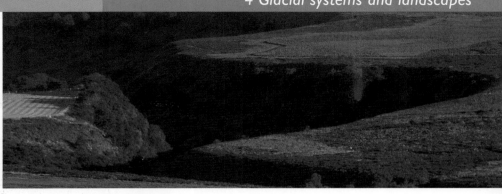

**Figure 3** *Newtondale, North York Moors*

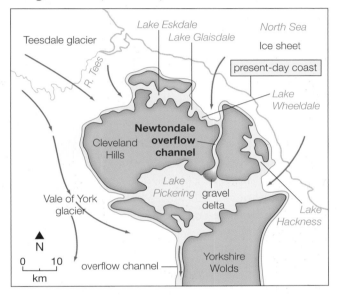

**Figure 4** *The formation of Newtondale meltwater channel*

An **outwash plain** is an extensive, gently sloping area of sands and gravels formed in front of a glacier (see 4.9). As the name implies, it results from the 'outwash' of material carried by meltwater streams and rivers. During a glacial period, meltwater and the deposition of outwash will be very seasonal, being mainly restricted to the summer months. However, at the end of a glacial period, huge quantities of material will be spread out over the outwash plain by great torrents of meltwater. Today, some of the most extensive outwash plains can be seen in Iceland and Alaska, where large *braided rivers* choked with sediment meander their way across vast floodplains (Figure **5**).

**Figure 5** *An outwash plain in Canada with a braided river system*

187

## Dahlen Esker, North Dakota, USA

One of the best examples of an esker in the USA is the Dahlen Esker, located in north-east North Dakota (Figure **6**). The esker is about 7 km in length, 120 m wide and up to 25 m high. It comprises a mixture of fluvial sands and gravels together with till and some boulders that would have fallen into the submarine river from the melting of the overlying ice.

In places the esker has a distinctive native prairie vegetation, due to its sandy soil, whereas the flat and fertile till plains on either side are farmed intensively.

Scientists believe that the Dahlen Esker was deposited by a meltwater stream flowing in an ice-walled channel, or possibly through a tunnel in the ice, near the edge of the glacier. The stream flowed mainly southward, towards the margin of the glacier.

▶ **Figure 6** *Dahlen Esker, North Dakota, USA*

Both **eskers** and **kames** are ice-contact depositional landforms. Meltwater plays an important role in the formation of both these landforms (Figure **7**).

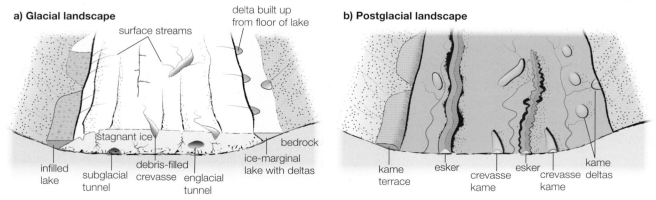

**a) Glacial landscape**

surface streams
delta built up from floor of lake
stagnant ice
bedrock
infilled lake
subglacial tunnel
debris-filled crevasse
englacial tunnel
ice-marginal lake with deltas

**b) Postglacial landscape**

kame terrace
esker
crevasse kame
esker
crevasse kame
kame deltas

▲ **Figure 7** *How eskers and kames are formed*

Eskers are long, sinuous (winding) ridges made of sand and gravel that can be up to 30 m high and stretch for several kilometres (Figures **6** and **7**). They usually take the form of meandering hills running roughly parallel to the valley sides. Their meandering shape suggests that they were formed by subglacial river deposition during the final stages of a glacial period, as the ice was melting and no longer moving forward. Today, eskers are often discontinuous hills, having been eroded away in places by glacial meltwater and post-glacial rivers.

As with eskers, kames are also largely made of sand and gravel, which has been deposited on the surface of the ice by streams in the final stages of a glacial period.

There are different types of kame (Figure **7b**).

◆ *Kame terrace* – this is the most extensive type of kame. It results from the infilling of a marginal glacial lake. When the ice melts, the kame terrace is abandoned as a ridge on the valley side.

◆ *Kame delta* – a smaller feature that forms when a stream deposits material on entering a marginal lake. They form small, mound-like hills on the valley floor and can be identified by their deltaic sedimentation characteristics.

◆ *Crevasse kame* – some kames result from the fluvial deposition of sediments in surface crevasses. When the ice melts they are deposited on the valley floor to form small hummocks.

## ACTIVITIES

1  What are the characteristics of a fluvioglacial landscape?

2  Outline the role of meltwater in glacial environments.

3  On a visit to Newtondale on a field trip, what evidence would you look for to suggest that this was a meltwater channel rather than a 'normal' river valley? Find Newtondale on OS map OL27 or on digimapforschools. edina.ac.uk.

4  Use the internet to find a photo of an outwash plain. Write annotations to describe and account for the main characteristics associated with this environment.

5  What are the distinctive characteristics of an esker?

6  Explain how you would attempt to identify each of the three types of kame in the field.

7  Study Figure **8**. The orientation of a pebble is measured using a compass to examine the orientation of its long axis. This is done in situ. Flowing water will orientate pebbles in the direction of their long axis so, by studying glacial sediments, it is possible to suggest flow directions of fluvioglacial rivers. Outwash plains are characterised by multi-directional flow patterns, so a sediment sample often has more than one preferred orientation.

  a  The main orientation trend of the pebbles studied is NE–SW. How can you tell this from the graph?

  b  What does this suggest about the dominant direction of river flow during the deposition of the sediment?

  c  Describe and suggest reasons for the secondary trend shown on the graph.

  d  Suggest how and why a rose diagram for a sample of till would differ from Figure **8** (see 4.9).

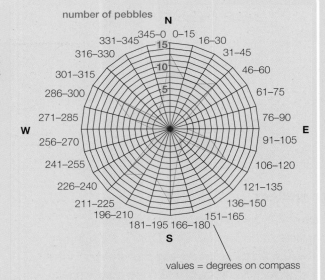

▲ *Figure 8* A rose diagram showing the orientation of pebbles collected from fluvioglacial sediments in the Nant y Llyn Valley in North Wales

In this section you will learn about periglacial landforms

## What is a periglacial landscape?

Periglacial landscapes, such as in Alaska and parts of Northern Europe, Russia and Canada, are typically wide expanses of largely featureless plains, either strewn with rocks (blockfields) or clothed with low-growing, marshy vegetation (Figure **1**). Lakes or streams are common in the summer when the snow has melted and some of the permafrost has thawed.

## What are the distinctive periglacial landforms?

Permafrost – soil, rock or sediment that has been frozen for at least two consecutive years – is one of the most characteristic features of a periglacial environment (see 4.7). Many periglacial landforms owe their formation, at least in part, to the presence of permafrost and the action of ice. Figure **2** identifies the main periglacial landforms.

⬆ **Figure 1** *Periglacial environments, Northwest Territories, Canada*

⬇ **Figure 2** *Periglacial landforms*

## Ice wedges

In extremely low temperatures, the ground contracts and cracks develop. During the summer, meltwater fills these cracks and then freezes in the winter to form *ice wedges,* which increase in size through repeated cycles of freezing and thawing (Figure **3**, 4.7). They also affect the ground surface as *frost heave* (see 4.7). leads to the formation of narrow surface ridges, between which ponds may form in the summer (Figure **3**).

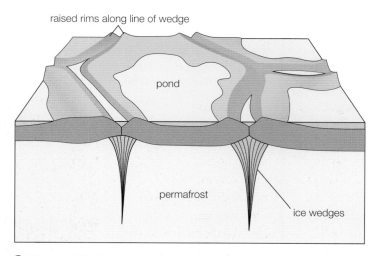

⬆ **Figure 3** *Surface features associated with ice wedges*

## Patterned ground

As ice wedges become more extensive, a polygonal pattern may be formed on the ground, with the ice wedges marking the sides of the polygons – **patterned ground** (Figure **4**). Scientists believe that patterned ground may result from a number of different processes associated with ice action.

*Stone polygons* are a feature of patterned ground and tend to form on flat ground (Figures **4** and **5**) and are directly associated with ice wedges. Frost heave causes expansion of the ground and lifts soil particles upwards. Smaller particles may be removed by wind or meltwater, leaving a concentration of larger stones lying on top of the ice wedges marking out the polygonal pattern.

Sloping ground can distort the polygons as stones gradually slide or roll downslope, leading to the formation of *stone stripes* rather than polygons (Figure **4**).

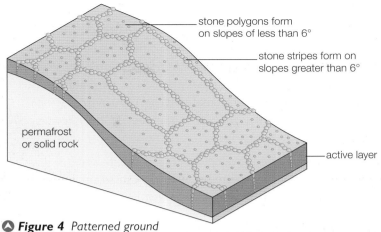

stone polygons form on slopes of less than 6°

stone stripes form on slopes greater than 6°

permafrost or solid rock

active layer

▲ **Figure 4** *Patterned ground*

### EXTENSION

**The patterned ground mystery**

A single process cannot explain the variety of symmetrical, geometric shapes that are known as patterned ground. Many of these features appear to be caused by freeze–thaw action selectively moving coarse particles to the edge of the shape or to its surface. Some polygon forms seem to be caused by the same thermal processes that create ice wedges. Yet, the formation of some types of patterned ground remains unexplained.

## Blockfields

Periglacial landscapes, particularly in mountainous regions, can be characterised by extensive frost-shattered bedrock consisting of broken up angular fragments of rock (Figure **6**). Such areas, called *blockfields* or *felsenmeer*, are subject to repeated cycles of freezing and thawing. This indicates that this landform is more likely to occur in moderate rather than extreme periglacial conditions.

▲ **Figure 5** *An aerial photograph of ice wedge polygons, Hudson Bay Lowlands, Manitoba, Canada. They occur in permafrost peatlands which are composed mainly of dry sphagnum moss. The brown polygons mark the location of massive ice wedges that are 2 to 3 m deep.*

▼ **Figure 6** *Blockfield or felsenmeer developed on granite bedrock, Minatoba, Canada*

## Pingos

Look at Figure **7**. It shows a rounded hill feature called a **pingo**. Reaching heights of up to 90 m with some stretching 800 m across, pingos are one of the most spectacular periglacial landforms found in the Arctic and sub-Arctic. In the Mackenzie Delta region of Northern Canada there are estimated to be 1400 pingos. While they are often green and vegetated on the outside, their core is solid ice.

The formation of pingos is a subject of much debate and there may be several different formative scenarios. Canadian pingos are thought to form on the site of a lake that has gradually been infilled with sediment – a 'closed' pingo (Figure **8**).

In the summer, part of the ice core melts and the centre may collapse. The resulting central hollow or depression resembles a volcanic crater, which may subsequently fill with water.

⊘ **Figure 7** *Pingo in the Mackenzie Delta, Canada*

### Formation of a 'closed' pingo

**Stage 1**

- Lake infills with sediment, insulating the ground beneath.
- Liquid water is trapped in the unfrozen ground (talik) between the lake sediment and permafrost.

**Stage 2**

- Water freezes as climate cools to form ice core.
- Ice expands due to increased hydrostatic pressure and the talik is squeezed.
- Lake sediment pushed up to form a pingo. In summer, part of the ice core may collapse and fill with water.

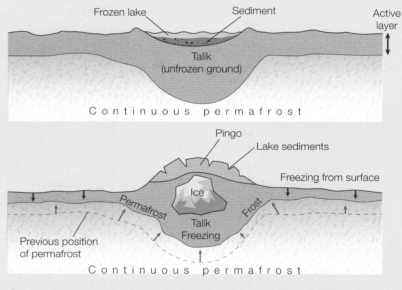

⊘ **Figure 8** *Diagram showing the formation of a 'closed' pingo stages 1 and 2*

## Solifluction lobes and terracettes

The process of *solifluction* (see 4.7) involves the downslope movement of rock and soil material in response to gravity in a periglacial environment. It is most likely to occur in areas experiencing significant summer melting, where there is a reasonably thick and saturated active layer.

One of the characteristic features of solifluction is the formation of small-scale tongue-like landforms called *solifluction lobes* (Figure **9** and 4.7). In the summer, when there is a lot of surface water trapped above the permafrost, the saturated active layer will simply slip

⊘ **Figure 9** *Solifluction lobes, Skookum Pass, Alaska, USA*

slowly downhill on gradients as low as 2 degrees. The formation of lobate features suggests variations in the degree and depth of saturation and possible micro-variations in gradient.

The saturated active layer may well be affected by a combination of solifluction and frost heave, particularly in spring and autumn. Frost heave causes soil particles to move perpendicular towards the surface due to expansion caused by freezing. On thawing, the particles fall back vertically. On a slope, this displacement causes individual particles to move gradually downhill with every cycle of freezing and thawing. On some slopes, distinct 'steps' can form – these are called **terracettes** (Figure **10**).

## Thermokarst

The term *karst* describes a landscape formed by the dissolving action of water. It is usually associated with limestone landscapes, which are characterised by hollows and depressions, rocky outcrops, gorges and sparse vegetation. In periglacial environments, a similar micro-topography called **thermokarst** can result when ground ice melts and settles unevenly to form a landscape of marshy hollows and hummocks (Figure **11**). This type of landscape is most commonly associated with flat lowland plains in the Arctic, particularly in the far north of Russia, Canada and Alaska.

The development of thermokarst varies enormously depending on the characteristics of vegetation, climate, topography, soils and the nature of the permafrost itself. The extent and location of melting is also considered to be highly variable and can be linked to heat sources within the ground and trapped water bodies, as well as warmth from above. Many scientists are very concerned about the release of high concentrations of stored carbon and methane trapped within the permafrost and their contribution to climate change.

▲ *Figure 10* Terracettes at Maiden Hill, Dorset

▼ *Figure 11* Thermokarst lakes, Taymir, Russia

## ACTIVITIES

1 Describe the typical characteristics of a periglacial landscape. Consider over what timescale and in what ways this landscape is likely to change in the future.

2 Using simple sketches, explain in your own words the formation of patterned ground.

3 Study Figure **7**. Make a careful sketch of the pingo in the photograph. Write annotations to describe its characteristics and likely mode of formation.

4 How and why do ice wedges change over time? What effect do they have on the surface micro-topography?

5 Outline the processes responsible for the formation of solifluction lobes and terracettes. What factors are likely to determine which of the two landforms is present in a particular place?

6 Suggest how and why thermokarst features are likely to vary from place to place.

In this section you will learn about the environmental fragility of cold environments

## What is environmental fragility?

If a product is described as being 'fragile' it usually means that it needs to be handled with care because it may be easily damaged or broken, with an implication that it cannot easily be mended or repaired. In essence, environmental fragility is much the same. It describes a sensitive environment, that is on the edge of survival, where even the slightest change can have significant effects (Figure **1**).

▲ **Figure 1** *Changes in the fragile Arctic environment can have serious implications for wildlife such as the polar bear*

### Environmental fragility: a definition

Some ecosystems can cope with wide variations in climatic conditions and changes in patterns of land use, whereas others are much more sensitive to any environmental change. The effects of small shifts in rainfall patterns or ambient temperatures can often do great harm to fragile environments and these effects can act as indicators of imminent threats elsewhere.

Natural events can precipitate sudden changes. Increasingly, however, the anthropogenic effects of human activity — intensive agriculture, deforestation, urbanisation, etc. — are causing specialised habitats to change, shrink and become fragmented to the extent that they may no longer be self-sustainable. In addition, the accidental or deliberate introduction of invasive non-native species can also severely affect communities of indigenous species.

(Partnership for European Environmental Research) www.peer.eu/research/fragile-environments

## Why are cold environments fragile?

There are several reasons why cold environments are fragile:

◆ Slow ecosystem development and highly specialised habitats – plants and animals have had to adapt to the lack of daylight during the winter and harsh climatic conditions of strong, drying winds, lack of rainfall, the presence of permafrost and the short growing seasons. Ecosystems are surprisingly diverse
but they take a very long time to become established.

◆ Sensitive to change – plants and animals that have adapted to particular environmental conditions are very sensitive to change. Many scientists are extremely concerned about the possible consequences of climate change in Arctic and sub-Arctic regions.

◆ Once damaged, an ecosystem can take a very long time to recover or it might never recover. It is said that just treading on tundra vegetation can result in the footprint remaining for a decade.

▲ **Figure 2** *The Arctic's oil and gas reserves*

# Human impacts on fragile cold environments?

### Oil spills in Siberia, Russia

The Arctic holds extensive and highly valuable reserves of oil and natural gas (Figure **2**). The United States Geological Survey estimates that over 87 per cent of the Arctic's oil and natural gas resource (about 360 billion barrels oil equivalent) is located in seven huge Arctic basins, two of which are in Siberia, Russia.

According to Greenpeace, the Russian oil industry spills more than 30 million barrels on land each year, and every 18 months more than four million barrels spews into the Arctic Ocean, poisoning the water, killing wildlife and destroying the livelihoods of local indigenous people.

Oil contaminates the soil, kills all plants that grow in it and destroys habitats for mammals and birds. Figure **3** shows the effects of a recent small-scale oil spill near the Russian town of Usinsk, close to the Arctic Circle. The cause of the spill was a decommissioned oil well, the rusty valves of which oozed thick, toxic, inflammable crude oil, which formed lakes on the tundra and polluted rivers. Rusty pipelines and old wells are a common source of spills.

⬆ **Figure 3** *Russia's black gold causing ecological catastrophe*

### Tourism in alpine environments

Tourism in Europe's mountains has increased significantly in recent years due to higher average levels of wealth and increased mobility and leisure time for many. According to the World Wide Fund for Nature, 120 million people visit the Alps each year, for example.

It can be argued that the most destructive human impacts are associated with the ski industry. Figure **4** shows illegal ski developments in the Pirin Mountains in Bulgaria, widely acknowledged to be one of Europe's last wilderness areas. The mountains, with some of the last remaining old-growth forests in Europe, are home to two-thirds of the continent's populations of brown bears, wolves and lynx. The highly lucrative ski industry has led to some illegal developments with scant regard for environmental conservation. Forests have been stripped to make way for ski developments and infrastructure, resulting in habitat loss or fragmentation.

Modern adventure sports (mountain-biking, canyoning, or paragliding) and some motor-based leisure activities are now entering areas previously untouched by tourism. They are is causing a major disturbance to wildlife and pose a very direct threat to biodiversity

⬆ **Figure 4** *Illegal ski developments at Bansko, Pirin, Bulgaria*

## ACTIVITIES

1  What are the key characteristics associated with environmental fragility in cold environments?

2  With the aid of additional research, suggest how and why Siberian oil spills are so common and have such devastating environmental impacts.

3  Select one of the oil producing regions in Figure **2**. Use the internet to investigate the environmental impacts of oil production in your chosen region.

4  Consider the environmental impacts of the ski industry in Europe.

## STRETCH YOURSELF

Evaluate how the Arctic ground squirrel has adapted to Arctic conditions. Examine other adaptations of flora and fauna such as bears, lemmings and marmots. You may find this website useful

www.scilogs.com/frontier_scientists/ squirrels-impact-climate-carbon

Also search for 'umassk tundra adaptations' and look at the pdf at www.umassk12.net

In this section you will learn about the impacts of climate change on cold environments

Look at Figure **1**. The Arctic is red, indicating that the trend over this 50-year period is an increase in air temperature of more than 2 °C, which is greater than for any other part of the globe. This trend is supported by the graph of temperature increase relating to latitude. Alpine environments are also witnessing increases in temperature, leading to reductions in snowfall, retreating glaciers and melting permafrost.

▼ **Figure 1** *Map showing trends in mean surface air temperature (1960–2011)*

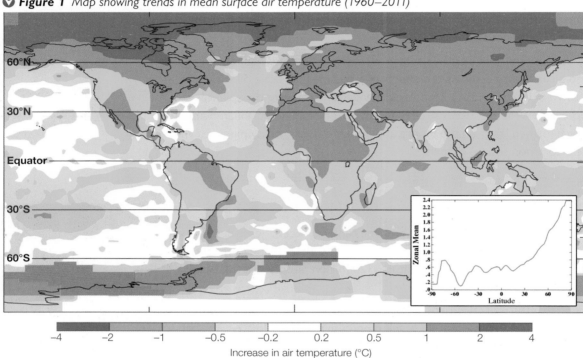

## How is climate change affecting fragile cold environments?

One of the key geographic definitions of 'fragile' is sensitivity to change. For this reason, climate change in cold environments – particularly in the Arctic – threatens to have a very significant impact on natural ecosystems and habitats. There are already signs that climate change is having an impact and this is likely to become more severe and wide-ranging in the future, if current trends continue.

### Impacts of climate change in the Swiss Alps

#### Flooding and landslides

In Switzerland almost all glaciers are retreating. In the summer, significant melting occurs and water can be ponded-up behind debris dams to be released suddenly as powerful torrents of water. Flooding and mudslides have hit the village of Saas Balen three times in the past 50 years.

#### Melting permafrost

Permafrost covers about 5 per cent of Switzerland. Predicted thawing has increased the threat of rock avalanches and mudslides. Several mountain communities are under threat, and so are the cable cars and chair lifts, the supports of which are anchored in permafrost rather than in solid rock.

The masts of the ski lifts on the Titlis Glacier have to be repositioned three or four times every year because of melting permafrost (Figure **2**).

#### Ecosystems

Alpine ecosystems are predicted to experience significant change as the climate warms – a rise in temperature of 1 °C pushes the treeline up by 100 m and some of the specialist high-mountain flora and fauna would become extinct. Habitats would change, and some species such as marmot would be threatened.

## Economic costs

A study carried out by the Swiss National Research Programme has estimated that a rise in temperature of 2 °C will cost Switzerland about CHF 3 billion (£2 billion) a year, mostly affecting winter tourism. The study also said that expensive measures would have to be implemented to deal with the increase in flooding and other natural disasters.

## Ski industry

Research by the Swiss National Science Foundation suggests that by 2050 only those resorts at 1500 m and above (62 per cent of all resorts) will be able to offer skiing for at least a hundred days a year. This figure is currently 1200 m and above (85 per cent of all resorts). Several low-altitude resorts are already suffering from declining numbers of tourists, with local businesses being forced to close down.

▲ **Figure 2** *Ski lift on the Titlis Glacier, Switzerland*

## EXTENSION

### What is 'Arctic amplification'?

The Arctic acts like a refrigerator by giving off more heat than it absorbs. Scientists are concerned that this process may be affected by three significant positive feedback loops that may develop as a result of climate change. Once initiated, these loops may be irreversible.

1 As the extent of sea ice is reduced, a smaller fraction of the Sun's radiation is reflected directly back to space. More energy is absorbed by the ice-free water, raising temperatures which, in turn, further reduces the ice cover (see 1.13).

2 Melting permafrost (often evidenced by thermokarst – see 4.11) is releasing trapped carbon dioxide and methane into the atmosphere increasing the concentration of greenhouse gases and contributing to global warming. This in turn leads to an increase in surface temperatures, which leads to more melting of the permafrost (Figure 3).

3 Reduction in snow cover means that an increasing area of bare rock is exposed. This increases heat absorption from the Sun, leading to higher temperatures and increased melting of adjacent snow.

Together, these three positive feedback loops will exacerbate the problem, hence the term *Arctic amplification*.

Climate change may, however, result in negative feedback loops. For example, if warm temperatures expose more land that was previously covered by ice or snow, more plants can survive and, with longer growing seasons, will sequestrate more carbon from the air (see 1.13). However, scientists suspect that the positive feedback effects will outweigh the negative effects.

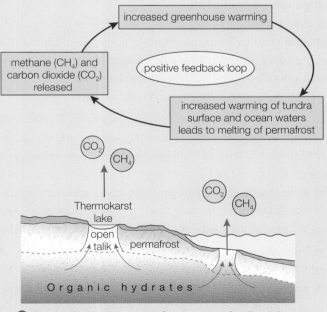

▲ **Figure 3** *Melting permafrost positive feedback loop*

## Impact of climate change in the Arctic

'The Arctic is warming faster than the rest of the globe and is experiencing some of the most severe climate impacts on Earth.' (*Greenpeace*)

The warming of the Arctic has resulted in a decline in the thickness and extent of the Arctic sea ice. Satellite data show that over the past 30 years, Arctic sea ice cover has declined by 30 per cent in September, the month that marks the end of the summer melt season (Figure **4**). Satellite data also show that snow cover over land in the Arctic has decreased, and glaciers in Greenland and northern Canada are retreating. In addition, frozen ground (permafrost) in parts of the Arctic has started to thaw out (see 1.10 and 1.13).

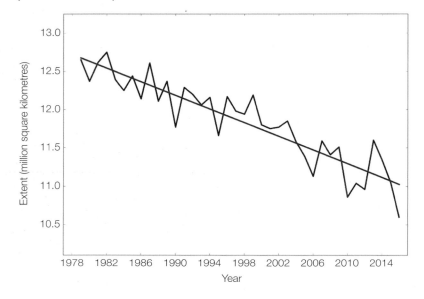

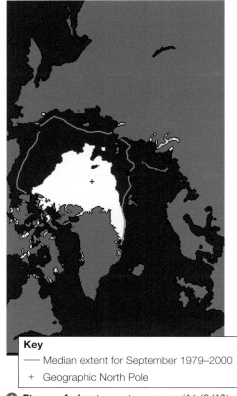

**Key**
— Median extent for September 1979–2000
+ Geographic North Pole

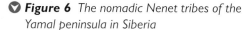
**Figure 4** *Arctic sea ice extent (16/9/12). This was the lowest minimum extent of sea ice since records began in 1979.*

**Figure 5** *Average June Arctic sea ice extent (1979–2016)*

## Indigenous population

For over a thousand years the nomadic Nenet tribes have migrated along the remote Yamal peninsula in north-west Siberia (Figure **6**). In summer they head north, taking their reindeer with them, across a landscape of boggy ponds, rhododendron-like shrubs and wind-blasted birch trees. In winter they return southwards.

This region is now threatened by global warming. Traditionally, the Nenets travel across the frozen Ob River in November and set up camp in the southern forests around Nadym. Warmer temperatures are delaying the freezing of the river, preventing tribes from crossing until December. This delay results in the reindeer going hungry due to the lack of pasture. Additionally, the early melting of snow prevents the reindeer pulling sledges, an essential element of the nomadic lifestyle of the Nenets.

In the high Arctic the reduction in sea ice, together with rising sea levels, is making coastal communities more vulnerable to flooding from high waves and storm surges. The ice-based animals, an important source of food, have become scarcer and less accessible. One positive impact is that harbours are ice-free for longer periods, enabling more opportunities for fishing.

**Figure 6** *The nomadic Nenet tribes of the Yamal peninsula in Siberia*

## Sea ice ecosystems

Polar bears are dependent on sea ice for their entire lifecycle – from hunting seals, their main prey, to raising their cubs. Researchers are reporting an increasing number of polar bears drowning because they have to swim longer distances between ice floes. The lack of available food has resulted in polar bears fasting on land or even cannibalising each other.

Ice-dependent seals, such as the ringed seal and the spotted seal, are particularly vulnerable to the current and prospective reductions in Arctic sea ice. They give birth and nurse their pups on the ice and use it as a resting platform. They also forage near the ice edge and under the ice (Figure **7**). Many scientists believe it is very unlikely that these species could adapt to life on land in the absence of summer sea ice.

▲ *Figure 7* *Spotted seal foraging on the edge of an ice floe*

## Land ecosystems

Climate change – in particular a warming trend – could have significant impacts on ecosystems.

- ◆ Arctic plants have adapted to cope with seasonal patterns of freezing and thawing with limited access to nutrients. Rapid environmental change associated with global warming, such as earlier springs, extended growing seasons and changes to nutrient availability, may occur quicker than the ability of the plants to adapt.

- ◆ Some animals, such as musk oxen and reindeer, depend on snow cover to insulate plants (Figure **8**). With reduced snow, plants may be less able to survive the winter, which impacts on the animals' food supply.

- ◆ Foreign species of insects from the warmer south could suppress the indigenous insects that have adapted to the Arctic conditions.

- ◆ Insect pollination in the Arctic has evolved over a long period of time. If plants start to flower earlier but changes to insect life cycles do not happen at the same rate, pollination will be less successful.

▲ *Figure 8* *Musk ox depend on snow that insulates plants*

## ACTIVITIES

1 Study Figure **1**. What is the evidence that the Arctic region is at the greatest risk from global warming?

2 Describe how positive feedback loops may exacerbate the rate and consequences of climate change.

3 Write a detailed report on the prospective impacts of climate change on the Swiss Alps. Consider tourism (specifically the ski industry), flooding and landslides, melting permafrost, ecosystems and economic costs. The following websites will support your research.

www.geocases2.co.uk/swissalps1.htm

http://proclimweb.scnat.ch/portal/ressources/794.pdf

4 Suggest how the lives of the Nenet tribes are likely to be affected by climate change.

5 Discuss the consequences of a reduction in the Arctic sea ice on people and wildlife.

Ⓢ 6 Study Figure **5**.

 a Describe the trend of the average monthly Arctic sea ice extent for June (1979–2016).

 b Use the blue trend line to calculate the percentage change per decade.

 c Redraw the graph and extrapolate the trend line into the future to suggest when there will be no sea ice in June, if current trends continue.

 d Do you think June is a fair and appropriate month to make comparative measurements?

In this section you will learn about the management and future planning of cold environments

*Case study*

## What are the principles of environmental management?

At the heart of all environmental management is the need to act sustainably, implementing policies and taking actions that will preserve and conserve cold environments and the people who depend upon them. Cold environments are fragile and extremely sensitive to change and are easily damaged by human actions. Impacts on indigenous people, local economies and wildlife are already being felt across a wide range of cold environments (see 4.13). Management of these environments needs to address the current issues and consider alternative possible futures as determined by economic, political and environmental drivers.

It is possible to identify three management approaches:

1 *Prevention* – attempting to prevent a harmful event occurring (e.g. an oil spill or damage caused by deforestation).

2 *Reaction* – responding to an event once it has occurred (e.g. trying to clean up an oil spill or restore a damaged ski slope).

3 *Adaptation* – learning to live with change (e.g. changes to the world's climate).

### Oil exploitation in Alaska, USA

Northern Alaska has extensive reserves of oil and gas (Figure **1**). Drilling for oil began in a piecemeal fashion in the early 1900s but it was not until the discovery of the Prudhoe Bay oilfield in 1967 that oil exploitation really took off.

The Prudhoe Bay oilfield is the largest oilfield in North America and it has produced a significant proportion of the USA's energy needs for four decades. Currently, the USA supplies about three-quarters of its own petroleum needs. In a world of increasing energy insecurity, the Alaskan reserves are extremely important to the US both politically and economically.

### The Trans-Alaska pipeline – successful management?

The most ambitious project in the region has been the Trans-Alaska oil pipeline (1.2 m in diameter and 1287 km long). The pipeline snakes its way through the rugged Alaskan countryside from Prudhoe Bay in the north to the ice-free port of Valdez in the south. In order to satisfy both economic and environmental demands, the pipeline is insulated and raised on stilts (Figure **2**), not only to prevent the warm oil melting the underlying permafrost but also to allow the migration of caribou. Interestingly, if climate change results in the Arctic Ocean becoming ice-free, oil tankers may be able to access Prudhoe Bay directly, thereby rendering the Trans-Alaska pipeline redundant.

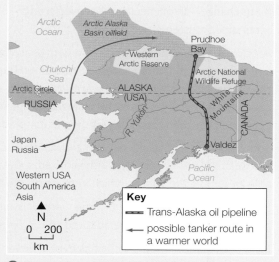

**Figure 1** *Oilfields and protected areas in Alaska, USA*

▶ **Figure 2** *Trans-Alaska pipeline*

### Recent developments in oil exploration

As the Prudhoe Bay oilfield reserves have begun to decline, oil companies have considered alternative locations. Both are considered to be fragile cold environments.

◆ The Western Arctic Reserve, with its extensive wetlands and large population of threatened species, is home to groups of Native Americans who depend on the wildlife for their food, clothing and shelter. Almost 500 000 caribou – the largest herd in Alaska – live and migrate through the reserve.

◆ The Chukchi Sea is considered to be a pristine and highly sensitive marine environment.

In recent years, oil prices have fluctuated (Figure **3**). Following the global economic crash of 2008, oil prices rose steadily to plateau at $100–120 per barrel between 2011 and 2014. Throughout this time of high oil prices, there was a strong economic case in the US for increasing oil production.

In 2008, Royal Dutch Shell started to explore for oil in the Chukchi Sea off Alaska's north coast (Figure **1**). The US Bureau of Ocean Energy Management estimated that the oilfield could hold nearly 30 billion barrels of oil. However, the project faced challenges from the start.

◆ In 2010, BP's catastrophic oil spill off the Gulf of Mexico sparked a government crackdown on offshore drilling, halting Shell's drilling programme

◆ On Shell's voyage to the Arctic in 2012, it wasn't able to drill to depths where oil could be found because of the failure of its spill response system during testing.

◆ At the end of the season, one of its drilling rigs ran aground as it made its way back to Seattle. The crew had to be helicoptered to safety and the rig was scrapped.

Shell had hoped that its best prospect, a drilling area called Burger J, would prove the region's huge potential. Instead, the field became one of the industry's most expensive *dry holes* (a well with no significant reserves). In September 2015 the $7 billion plan was abandoned. This was welcome news for environmentalists, who were concerned about the impact of exploitation in these pristine Arctic waters. However, the decision was largely driven by economics, with the price of oil having plummeted to $50 per barrel.

'Oil companies have looked longingly at the Arctic for years, but its often icebound seas and treacherous weather make exploring expensive and dangerous. Shell's decision could spell the end of Arctic drilling for some time, although low oil prices and geopolitics – not environmental concerns – are the main reason.'

(*Wall Street Journal*, 29 Sept 2015)

◉ **Figure 3** *Fluctuating global oil prices (2000–2015)*

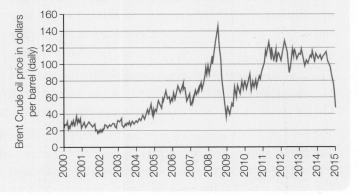

### Planning for alternative possible futures

Oil exploration and extraction is an extremely costly and time-consuming business, often taking many years to establish drilling sites, construct rigs and assemble the necessary infrastructure. Since 2014, a surplus in the production of oil has resulted in low prices, which analysts expect to be maintained for some time. Issues of energy insecurity are likely to continue due to the political tensions in the Middle East.

To further complicate matters, there is the recent global accord (Paris, 2015) to set carbon emissions targets, which implies a future reduction in the use of fossil fuels.

So, given the varied and complex economic, political and environmental drivers, there are several alternative possible energy futures, making planning for the years ahead extremely challenging.

## Adapting to climate change in the European Alps

A report funded by the European Union, *Climate Change and Its Impacts on Tourism in the Alps* (2011), has identified opportunities and threats associated with climate change and painted a picture of alternative futures for the region.

'There is a widespread consensus that Alpine tourism needs to be rethought and both public institutions and private stakeholders have to meet the challenge of a new idea of tourism which goes beyond the traditional vision of winter sports.' (Balbi and Giupponi and Bonzanigo, Climate Change and Its Impacts on Tourism in the Alps, 2011)

▲ **Figure 4** *Snow cannon and snow groomer at Riederalp, Valais, Switzerland*

◆ Summer tourism could benefit from climate change. Hotter summers (as in 2003) would bring more people to the mountains and the tourism season could be extended.

◆ Droughts may become more frequent in the summer.

◆ Winter tourism faces serious challenges due to the expected decrease in snow and ice cover. Already, 57 of the main ski resorts in the European Alps are considered not to be snow-reliable. Many ski resorts are using snow cannons to create artificial snow (Figure **4**), which use huge amounts of water and energy – up to 20 per cent of some ski areas are covered by artificial snow.. A chemical additive used to raise the temperature at which snow forms is sometimes used in the creation of artificial snow. It is unclear whether this chemical may have a detrimental impact on ecosystems.

▼ **Figure 5** *Misurina, Italy*

◆ Climate change creates an opportunity for those resorts that are snow-reliable, as they will face less competition in the future.

### The Marzon valley, Veneto region, Italy

The Marzon valley is an important winter tourism destination in north-east Italy, but its low altitude means that it faces an uncertain future. After a preliminary consultation with the local public administration, a plan was devised to develop winter tourism over the next 40 years, in the context of climate change (warming effect on snow availability) and changing market demand (an ageing population).

The plan involves the construction of ski lifts to connect with the high-altitude ski area of Misurina 2 km away (Figure **5**). Misurina has two ski-lifts to five high-altitude pistes in the Col de Varda. It also offers 17 km of cross-country ski loops.

The area as a whole is well-suited to the further development of winter tourism, with several alternative activity centres such as the Somadida Forest, offering cross-country skiing, the Marmarole sled-dog centre and an ice-kart circuit located in Palus. The 1956 Winter Olympic Games were held in the area, at Cortina d'Ampezzo, which has a modern system of ski lifts and the capacity to create artificial snow. It remains one of Europe's most popular winter destinations.

| | 2009/10 | 2010/11 | 2011/12 | 2012/13 | 2013/14 | 2014/15 | 2015/16 |
|-----|---------|---------|---------|---------|---------|---------|---------|
| **Nov** | 3/34 | 19/58 | 0/21 | 2/19 | 1/7 | – | 2/5 |
| **Dec** | 25/99 | 39/132 | 18/37 | 19/103 | 24/82 | 5/110 | 15/28 |
| **Jan** | 91/230 | 43/135 | 29/140 | 32/126 | 102/195 | 18/107 | 17/45 |
| **Feb** | 80/312 | 45/140 | 22/87 | 45/165 | 167/341 | 35/143 | – |
| **Mar** | 80/261 | 42/137 | 23/75 | 43/179 | 115/307 | 19/119 | – |
| **Apr** | 138/279 | 25/119 | 15/75 | 14/133 | 29/106 | 3/54 | – |

◀ **Figure 6** *Snow slope depth record for Cortina D'Ambrezzo, Italy (2009/10–2015/16)*

Lower slopes snow depth/upper snow slopes depth (in cm)

**The Giandains protection dam, Pontresina, Switzerland**

High-altitude permafrost in the Swiss Alps is melting as a result of climate change. This has increased the risk of rock avalanches and landslides, and several local communities have had to respond to the threat. The village of Pontresina, the first village in the Alps to tackle the problem, has constructed a huge dam to protect itself from both avalanches and the possible consequences of the thawing permafrost. The Giandains Protection Dam (Figure **7**) was completed in 2003 at a cost of almost £1.3 billion.

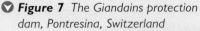 **Figure 7** *The Giandains protection dam, Pontresina, Switzerland*

## ACTIVITIES

 1 Describe the trend in oil price since 2000. Find out the current price of oil and update Figure **3**. To what extent has the changing price of oil determined the growth and development of the oil industry in Alaska?

2 Using the internet for additional research, evaluate how the main features of the Trans-Alaska pipeline have addressed environmental concerns. Do you consider the pipeline to be a success?

3 Outline the alternative futures for the oil industry in Alaska.

4 What strategies can be adopted by the alpine tourist industry to adapt to future climate change?

5 Carry out your own research to examine some more of the issues associated with melting permafrost.

 6 Study Figure **6**. Represent the data using a graphical technique of your choice to compare the lower and upper slope snow depths during the winter months for the seven seasons shown. Use the internet to update the figures (search for 'Cortina historic snow depth').

7 To what extent is there evidence to suggest that Cortina D'Ambrezzo is experiencing variable (unreliable) snowfall, with particularly low falls on the lower slopes? Include some statistical comparisons (e.g. mean, range, percentage difference, etc.) to support your answer.

# Challenges and opportunities for development – Svalbard, Norway

In this section you will learn about the challenges and opportunities for development in Svalbard, Norway

## Where is Svalbard?

Svalbard is a Norwegian territory located in the Arctic Ocean some 1000 km north of Norway (Figure **1**). It is the most northerly permanently inhabited archipelago in the world, extending for some 700 km from Rossoya in the north at 81°N to Bjornoya in the south at 74°N. Much of Svalbard has a dry, steppe-like climate, with average annual rainfall below 200 mm in places. Average temperatures are 5–6 °C in summer (July–August) and −16 °C in winter (February), falling as low as −40 °C during severe winters.

Svalbard's land area is roughly equivalent to that of the Republic of Ireland. About 60 per cent is currently covered with ice, 30 per cent is barren ground (rock, scree, moraines, fluvial deposits) and 10 per cent vegetated. Permafrost exists almost everywhere, to a depth of 100–150 m in the valleys and up to 400 m in the mountains. Almost all of Svalbard's population of about 2700 live in the town of Longyearbyen on Spitzbergen, the largest and most densely populated island.

## What are the opportunities and challenges for development?

Svalbard is considered to be the largest wilderness area in Europe, yet despite the challenges associated with its extreme climate, its mountainous terrain and its sheer remoteness, it offers many opportunities for development.

### Coal mining

Until the end of the nineteenth century, Svalbard was of no particular interest to European powers. Its only economic activities were associated with whaling and hunting for ivory, oil, furs and seal products. The Industrial Revolution in Europe sparked an interest in Svalbard's known reserves of coal, gold, marble, gypsum, asbestos and copper (Figure **2**). Despite the construction of exploratory mines and quarries, there has been very limited mineral extraction apart from coal.

Coal mining in Svalbard presents many challenges due to the extremely cold working conditions, the long hours of winter darkness, challenging sea conditions affecting transportation and the remoteness of the mines. It was not until 1899 that the first coal reached Norway, but production soared to reach almost 500 000 tonnes in 1948.

▲ **Figure 1** *The location of Svalbard*

▼ **Figure 2** *Mineral resources in Svalbard*

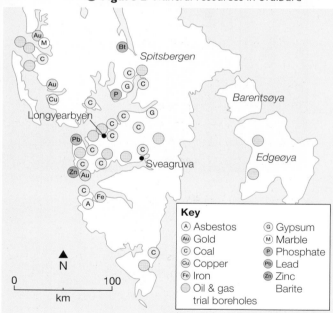

Key
- Ⓐ Asbestos
- Ⓐ Gold
- Ⓒ Coal
- Ⓒ Copper
- Ⓕ Iron
- ◯ Oil & gas trial boreholes
- Ⓖ Gypsum
- Ⓜ Marble
- Ⓟ Phosphate
- Ⓟ Lead
- Ⓩ Zinc
- Barite

▶ **Figure 3** *Coal-fired power station at Longyearbyen*

Originally, the centre of the coal mining industry was the Longyear Valley and a mining settlement was established at Longyearbyen in 1906. In the decades that followed, as coal prices fell, production declined and mines closed. Today, only Mine 7 remains operational in the Longyear Valley. It feeds the power station in Longyearbyen (Figure **3**), which supplies all of Svalbard's energy but is now inefficient and in need of upgrading or replacing.

The bulk of the present-day coal mining takes place at Sveagruva (Figure **4**), some 50 km south-east of Longyearbyen. The state-owned mining company, Store Norske, employs around a third of all workers on Svalbard, and extracts high-quality coal from reserves that are expected to last until about 2030. However, the company is now in economic and political difficulties, with job losses and calls from environmentalists to bring an end to mining on Svalbard, which would be disastrous for the community.

**Figure 4** *The Lunckefjell coal mine in Sveagruva was opened in 2014 and needed a new access road to be built through the mountain and across the Marthabreen glacier. It is now threatened with closure.*

## Fishing

The cold waters of the Barents Sea to the south of Svalbard are some of the richest fishing grounds in the world, with an estimated 150 species of fish, including cod, herring and haddock. However, storms and sea ice make fishing here extremely challenging and hazardous. Norway and Russia jointly control and monitor fishing practices to ensure that they are sustainable and that the ecosystem is protected – these waters are extremely important breeding and nursery grounds for fish stocks and must be protected from pollution and unsustainable exploitation if fish stocks are to be maintained.

**Figure 5** *SVALSAT – Svalbard satellite station, near Longyearbyen, Svalbard*

### Polar research

Svalbard has a long history of polar research, involving studies of marine ecosystems, geology and meteorology. Norway, Russia and Poland all run permanent research stations on Svalbard. Currently, a great deal of research is focused on the atmosphere, analysing changes that might be linked to climate change. Amongst these changes is the forecasted significant increase in Arctic temperature and the subsequent impacts on ecosystems and physical systems, such as glaciers.

Close to Longyearbyen is the SVALSAT receiving station, where huge antennae-studded 'golf balls' collect data from satellites encircling the Earth (Figure **5**). The ground station at Svalbard is unique. Because of its latitude, data from all 14 daily passes of polar orbiting satellites can be obtained, allowing for a more continuous download of information than can be obtained at lower latitudes. NASA, NOAA and the ESA have all invested heavily in this facility www.aerospace-technology.com/projects/svalbard.

However, research activities are faced with fewer developmental regulations and restrictions than, say, tourism or mining, and actually result in environmental damage themselves because of the associated infrastructure – construction of access roads and research stations. Research vessels now account for much of the shipping traffic in the more remote and highly sensitive environments on Svalbard, where oil spills or waste discharges could have a significant impact.

Case study

## Tourism

As early as the 1820s, people were visiting the archipelago on cruise ships or for hunting. Since the opening of the airport at Longyearbyen in 1975, tourism has grown significantly. In 2015, 130 000 people visited Longyearbyen, of whom 38 000 were cruise ship passengers. About 90 per cent of the land-based visitors travel from Norway.

Most people visit Svalbard to explore the natural environment – the glaciers, fjords and the wildlife (polar bears, seals, walrus) – or to study the historical development of the islands. Adventure tourism is becoming increasingly popular, with opportunities for hiking, snow mobile safaris and kayaking (Figure **6**). In the winter, tourists visit to see the Northern Lights.

Currently, some 480 local people benefit directly from being employed in the tourist sector, and the town of Longyearbyen has witnessed significant growth in tourist facilities such as hotels, shops, restaurants and tour operators.

Tourism on land is limited to Longyearbyen and its immediate vicinity. This is due to the high cost of road construction and maintenance. Frost action, for example, can cause damage to roads, resulting in uneven surfaces and cracking. Construction on permafrost presents engineering challenges too, in that surface melting can lead to structural instability.

## Human responses to living and working in Svalbard

People can respond to the challenges of living and working in extremely cold conditions by demonstrating **resilience**, **mitigation** and **adaptation**.

### Resilience

The long hours of darkness for half the year, extremely low winter temperatures, strong winds and, at times, heavy snowfall, require considerable resilience. This trait applies particularly to those working in the coal mines or conducting research in remote environments. On leaving the settlement of Longyearbyen there is almost total wilderness – no roads, settlements, shops or services.

### Mitigation

People need to mitigate the hostile and potentially dangerous environmental conditions by wearing appropriate clothing and footwear (Figure **6**). The use of layering and ensuring protection against the strong winds are essential principles, along with wearing gloves and warm headgear. Houses are well insulated to protect against the cold.

### Adaptation

Providing health and social care is an issue in such a remote location. Although there is a hospital and dental surgery, people with long-term needs or require extensive hospital attention are encouraged to fly to Norway for treatment. Elderly people with special health requirements cannot be catered for and may

**Figure 6** *Sea kayaking at Spitsbergen, Svalbard archipelago*

### Key to map symbols

|  |  |
|---|---|
| · | House or cabin, possibly unserviceable |
| ⊞ | Church |
| ⬭ | Cemetery |
| ♦ | School |
| ⊞ | Hospital |
| ✖ | Mine in operation |
| ✖ | Mine out of operation |
| ✈ | Airfield |
| —▵— | Aerial cableway, conveyor band |
| —+— | Power line |
| —— | Nature conservation area |
| —— | Road |
| ------- | Path or track |
| ▵ *294* | GPS-measured bench mark |
| ▵ *1050* | Trig. point |
| · *968* | Photogrammetric point |
| ⚙ | Pingo |
| ⬭ | Depression contour |
| ✦ | Beacon |
| | Quay |
| | Built-up area |
| | Ocean |
| | Lake |
| | Glacier, elevation contours |
| | Braided river |
| | Foreshore flat |
| | Moraine |
| | Contour lines, interval 100m |

**Figure 7** *Key for Svalbard map (Figure 8)*

Adaptation and mitigation are required to cope with varying periods of light and dark during the year that can have significant psychological effects if sleep patterns are affected. The use of blackout curtains in the summer and, where possible, artificial 'natural' light in the winter helps to mitigate the impact of these problems. Energy security is vital during the long winters and this is provided by Svalbard's own coal-fired power station in Longyearbyen.

The rocky terrain and permafrost present challenges in ensuring the supply of essential services such as water, electricity and sanitation to people's homes and places of work. In most cases, insulated pipes run above ground because underground pipes run the risk of rupturing during the summer as the permafrost thaws, or freezing during the winter. Maintenance is also easier.

▼ **Figure 8** *Extract from Blad 1 Svalbard Sorvest (1:250 000) Norsk Polarinstitutt (2010)*

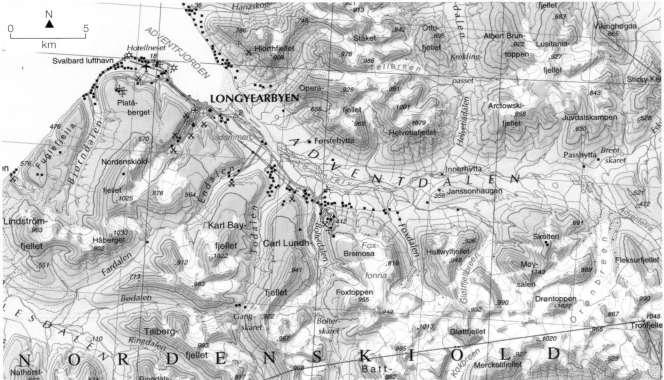

## ACTIVITIES

1 Use the internet to find out more about the physical challenges for human occupation and development on Svalbard. Consider the climate, the physical geography and the natural ecosystem (polar bears are a real hazard, for example).

2 Construct a summary table to consider the opportunities and challenges facing the following activities in Svalbard: mining, fishing, polar research and tourism.

3 Suggest how the four human activities in activity **2** may change in relative importance in the future. Consider the main drivers of change.

4 Use the internet to discover what opportunities are available for visitors travelling to Svalbard. To what extent is Svalbard an 'expensive and exclusive destination'?

5 Explain how people living on Svalbard have shown resilience, mitigation and adaptation.

Ⓢ Study Figures **7** and **8**, the key and map extract of Svalbard and answer the following synoptic questions. You must use evidence from the map to support your answers. Also consider using simple sketch maps.

6 What is the evidence that this is a glaciated landscape?

7 What is the evidence that this area has been developed by people? Suggest some of the physical challenges that have had to be addressed.

The following are sample practice questions for the Glacial systems and landscapes chapter. They have been written to reflect the assessment objectives of the Component 1: Physical geography Section B of your A Level.

These questions will test your ability to:

◆ demonstrate knowledge and understanding of places, environments, concepts, processes, interactions and change, at a variety of scales [AO1]

◆ apply knowledge and understanding in different contexts to interpret, analyse and evaluate geographical information and issues [AO2]

◆ use a variety of quantitative, qualitative and fieldwork skills [AO3].

---

**1** Outline the concept of the glacial system. (4)

**2** Study Figure **1** below and Figure **2** on page 160. Compare the differences between the extent of ice during the last glacial maximum and the current distribution of environments affected by ice. (6)

◉ **Figure 1** *Present-day global distribution of cold environments*

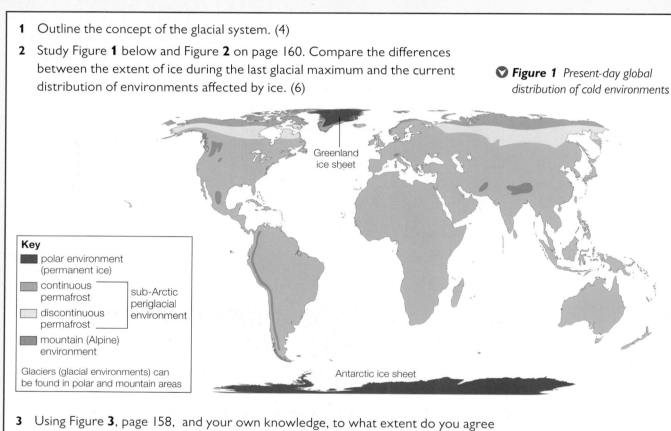

**Key**
- polar environment (permanent ice)
- continuous permafrost — sub-Arctic periglacial environment
- discontinuous permafrost — sub-Arctic periglacial environment
- mountain (Alpine) environment

Glaciers (glacial environments) can be found in polar and mountain areas

**3** Using Figure **3**, page 158, and your own knowledge, to what extent do you agree that a glacial system best typifies the concept of dynamic equilibrium? (6)

**4** 'The glacial system is as much shaped by its links with other physical systems as it is by flows and transfers within itself.'
To what extent do you agree with this statement? (20)

---

**5** Outline the differences between warm- and cold-based glaciers. (4)

**6** Study Figure **1**, page 166. Analyse the data shown. (6)

**7** Using Figure **5**, page 183, and your own knowledge, evaluate the different theories proposed regarding the process of drumlin formation. (6)

**8** Assess the view that the study of past glaciated landscapes can help inform planning for, and adaptation to, future climate change. (20)

**Tip**

**Evaluate** is a command word asking you to consider different options, ideas or arguments before coming to your own conclusion/ opinion. Your conclusions must be supported by evidence.

**9** Examine contrasting approaches associated with managing cold environments. (4)

**10** Study Figure **6**, page 202 and Figure **2** below. Average winter snow depths from 2009/10 to 2015/16 in the Marzon Valley have been plotted on dispersion diagrams. (*The dispersion diagrams show how data are dispersed or clustered within a range – with the range describing the spread.*) Analyse the data presented. (6)

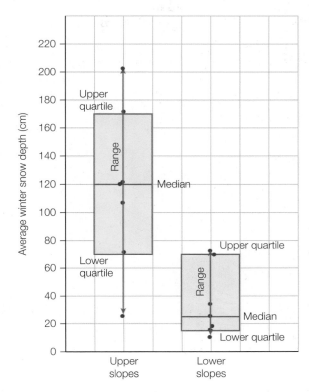

▲ **Figure 2** *Average winter snow depths from 2009/10 to 2015/16 in the Marzon Valley, north-east Italy*

**11** Using Figure **7**, page 206 and Figure **8**, page 207, and your own knowledge, explain the distribution of human habitation in this landscape and other glacial landscapes. (6)

**12.** Assess the extent to which cold environments are more fragile today than at any other time in human history. (20)

# 5 Hazards

Hazards like the earthquake in Nepal, on 25 April 2015, shook the nation to its very core — what do you think life is like there today?

## Your exam

**AL** Hazards is an optional topic. You must answer one question in Section C of Paper 1: Physical geography, from either Hazards or Ecosystems under stress. Paper 1 makes up 40% of your A Level.

**AS** You must answer one question in Section B of Paper 1, Physical geography and people and the environment, from a choice of two: Hazards or Contemporary urban environments. Paper 1 makes up 50% of your AS Level.

## Your key skills in this chapter

**S** In order to become a good geographer you need to develop key geographical skills. In this chapter, whenever you see the skills icon you will practise a range of quantitative and relevant qualitative skills, within the 'hazards' theme. Examples of these skills are

♦ Interpreting photographs and remotely-sensed images 5.7, 5.16, 5.18

♦ Using atlases and other map sources 5.9, 5.10, 5.14

♦ Drawing and annotating sketch maps 5.13

♦ Measurement and geospatial mapping 5.7, 5.13

♦ Presenting field data and interpreting graphs 5.1, 5.14

♦ Analysing data, including applying statistical skills to field measurements 5.10

♦ Drawing and annotating diagrams 5.3, 5.16,

## Fieldwork opportunities

In the UK, rivers and coasts provide many fieldwork opportunities to explore flooding as a natural hazard and the extent to which we can reduce its effects on people and property. Secondary sources of data support fieldwork. Did you know that you can follow your local river on social media to get regular flood risk updates?

*1. Flood risk investigation*

An investigation of flood risk requires a calculation of both the likelihood of a flood and its potential impact, in other words, its severity. If flood risk is the product of these two factors, it is a calculation not unlike those you'll need to do before a field trip with regard to potential health and safety risks. At the coast, the likelihood of flooding may be scored according to the relative height of a sea defence above the height of high tide or a recent storm tide. Secondary data such as photographs of storm events will allow you to agree on an accurate storm tide height before you start. Severity (in terms of damage to property) can be calculated and compared for different areas using a sliding scale for different types of land use (you'll need to survey this) where high-value retail and office buildings rank 5, and open space, car parks and derelict land scoring 1 or 2. Alternatively, property values for houses or businesses located on the seafront can be easily sourced online.

*2. Analysis of cost of flood defence*

A cost-benefit analysis of flood risk protection may be resourced using data provided by local councils about how much they have spent on flood defences. For example, data about historic spend on flood defences along the Holderness coast is freely available from East Riding of Yorkshire Council. The benefit of those same defences and the value of coastal protection for the local community may prove harder to quantify, justifying a qualitative approach. How do you think local and regional policy-makers take informed decisions about how to spend their budgets? Do they value all opinions voiced by the community equally? And do aesthetics and cultural heritage fit in? Find out what sort of public consultations have already taken place.

In this section you will learn about hazards in a geographical context

## What is a hazard?

'Loss of property from natural hazards is rising in most regions of the earth, and loss of life is continuing or increasing among many of the poor nations of this world.'
(Burton, Kates and White, 1978)

This was written almost 40 years ago, yet is still relevant today. Even though the overall death rate has fallen, more people have been affected by natural hazards. This is because of the increase in population and the wide-reaching effects of events such as earthquakes, volcanic eruptions, tropical storms, floods, wildfires and droughts (Figure **1**).

A hazard is the threat of substantial loss of life, substantial impact upon life or damage to property that can be caused by an event. These events can be caused by human actions (explosions, chemical release into the atmosphere, nuclear incidents) or are mainly natural (earthquakes, storms, volcanoes, wildfires), although natural events can be a consequence of human actions (wildfires can be ignited by human carelessness, floods can be brought about by poor land-use management).

A disaster occurs as a result of a hazard. For example, living on or near a fault line is a hazard, whereas an earthquake on the fault line that has enormous impacts on people and property is a disaster.

## What are the potential impacts of natural hazards?

The impacts of natural hazards depend on a number of general factors, such as the location of the hazard relative to areas of population and the magnitude and extent of the hazard. Each type of hazard has its own determining factors that affect the impacts, for example, the type and explosivity of a volcano, the nature of the continental shelf and shoreline for tsunamis or the availability of vegetation to fuel wildfires.

There are three main types of hazard, although each hazard has different driving forces, some of which overlap. For example, volcanoes can have impacts on the atmosphere, which affects weather patterns.

◆ Geophysical – driven by the Earth's own internal energy sources, for example, plate tectonics, volcanoes, seismic activity.

◆ Atmospheric – driven by processes at work in the atmosphere, for example, tropical storms, droughts.

◆ Hydrological – driven by water bodies, mainly the oceans, for example, floods, storm surges, tsunamis.

Impacts can be primary and secondary. Primary impacts are those that have an immediate effect on the affected area, such as destruction of infrastructure and buildings. Secondary impacts happen after the disaster has occurred, such as disease, economic recession and contamination of water supplies

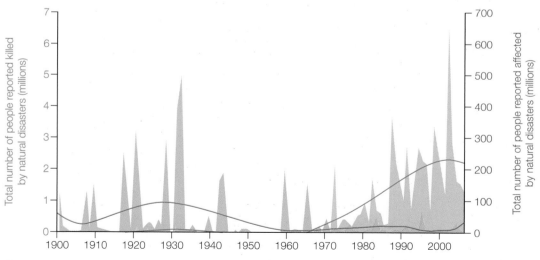

▲ **Figure 1** *Impacts of natural disasters 1900–2006; the line graph shows the smoothed trends for number of fatalities and number affected*

## Perception of hazards

How we perceive a hazard is determined by the effect that it may have on our lives. This increases if people have direct experience of a particular hazard and also how long term the impact of this experience has been.

It is only by the presence of people that a natural event becomes a hazard. The pressure of an increasing population and subsequent demand for land has resulted in building in areas that are at increased risk. Population expansion itself can increase the threat of a hazard, for example, increasing population at the peripheries of large urban areas may increase the risk of wildfires (see 5.16).

The advantages of living with the threat of hazards sometimes outweigh the risk. Making use of fertile soils on flood plains or in the vicinity of a volcano can be considered a risk worth taking and living with the threat is accepted as a part of everyday life (Figure **2**).

▼ **Figure 2** *Flooded agricultural land, Thailand*

A natural disaster can have catastrophic effects on an economy, not just in the countries that are directly affected, but also globally. In highly developed economies (HDEs) these effects tend to do little long-term damage to the economy – there is enough wealth and potential for redevelopment to be able to rebuild infrastructure and support those that are directly affected. Less developed economies (LDEs) are much more reliant on support and aid, both in the immediate aftermath of an event and also in the long term as they try to repair the damage physically, socially and economically.

Despite living in what we perceive as an obviously hazardous area, many still underestimate the risk of hazards. In 1971, Robert Kates found that of those people who had experiences of storm damage to their property on the east coast of USA, most of them did not expect such damage to occur again. Age, social status and religious beliefs can be determining factors when it comes to leaving behind in an evacuation all that has been worked for in a lifetime.

## Human responses

The natural human response to a hazard is to reduce risk to life and equity. At a local level this involves saving possessions and safeguarding property; globally this means coordinating rescue and humanitarian aid. The intensity and magnitude of the event as well as the original state of the infrastructure (and how badly it has been damaged) affects the speed of the international response (see 5.15).

Response times have been reduced by the development of the Automatic Disaster Analysis and Mapping system (ADAM), a database that pools information from the US Geological Survey, World Bank and World Food Programme. This allows almost immediate access to such information as the scale of the disaster, what supplies are available locally and local infrastructure. Previously a manual search of several databases took hours, rather than minutes.

## Fatalism

Doing nothing can be seen as a defeatist attitude to take but it is an acceptance that hazards are natural events that we can do little to control and losses have to be accepted. In fact, interference with the natural processes can have a detrimental effect on ecosystems.

'Command-and-control attitudes towards fire have become pervasive, to the detriment of ecological communities.'
(School of Ecosystem and Forest Sciences, University of Melbourne)

The point being made here is that while fires can be hazardous to human activity, they are also a natural regenerative process within forest ecosystems and should be allowed – in certain circumstances – to take their course.

## Prediction

As technology increases, the methods of predicting hazardous events becomes more sophisticated. Remote sensing and seismic monitoring give clues to activity that may lead to a disaster and need to be acted upon (Figure **3**). Advances in communications mean that information from all parts of the world can be shared and analysed quickly. Warnings can be communicated promptly and reach a greater number of those at risk.

## Adaptation

Once we accept that natural events are inevitable, we can adapt our behaviour accordingly so that losses can be kept to a minimum. This is the most realistic option for many people and often proves to be effective and cost-effective for governments (see 5.15).

## The hazard management cycle

For areas at risk there is a cycle that manages both the pre- and post-event situations (Figure **4**):

◆ *Preparedness* – large-scale events can rarely be prevented from happening, but education and raising public awareness can reduce the human causes and adjust behaviour to minimise the likely impact of the hazard. Knowing what to do in the immediate aftermath of an event can speed up the recovery process. In areas of high risk, the level of preparedness will be greater than in areas where such events are rare.

◆ *Response* – the speed of response will depend on the effectiveness of the emergency plan that has been put in place. Immediate responses focus on saving lives and coordinating medical assistance. Damage assessment helps plan for recovery.

◆ *Recovery* – restoring the affected area to something approaching normality. In the short term this will be restoration of services so that longer-term planning and reconstruction to the pre-event levels can begin.

◆ *Mitigation* – actions aimed at reducing the severity of an event and lessening its impacts. This can involve direct intervention, such as building design that can withstand earthquakes or hurricanes (see 5.10), or preparing barriers or defensible zones that may slow down or even halt the advance of wildfires (see 5.16). Most desirable is the long-term protection of natural barriers such as coral reefs, which protect the shore against storm surges (see 6.10). Support after a disaster in the form of aid and insurance can reduce the long-term impacts. However, insurance may not be available at all in high-risk areas, even in HDEs, and is something that may not be available at all in LDEs, which are often those that need it most.

▲ **Figure 3** *Volcano monitoring station at Mount Etna, Sicily*

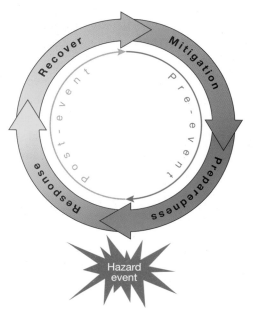

▲ **Figure 4** *The hazard management cycle*

## The Park model of human response to hazards

A hazard event causes disruption to everyday life. The type of disruption depends on factors such as the type of hazard, the intensity or magnitude, the immediate environment and infrastructure. The Park model describes three phases following a hazard event – relief, rehabilitation and reconstruction (Figure **5**).

*Relief.* The immediate local and possibly global response in the form of aid, expertise and search and rescue.

*Rehabilitation.* A longer phase lasting weeks or months, when infrastructure and services are restored, albeit possibly temporarily, to allow the reconstruction phase to begin as soon as possible.

*Reconstruction.* Restoring to the same, or better, quality of life as before the event took place. This is likely to include measures to mitigate against a similar level of disruption if the event occurs again.

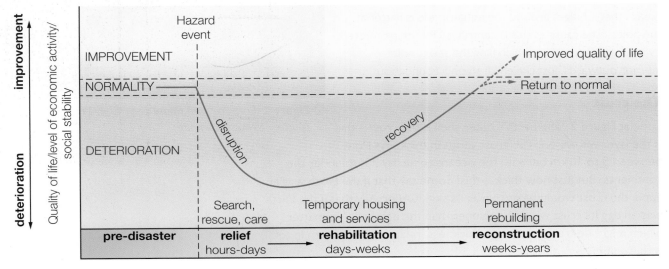

▲ **Figure 5** *The Park model of human response to hazards. The steepness of the downward curve during disruption depends on the nature of the event. A tsunami or earthquake would have immediate disruptive impacts. A volcano may give weeks of warning, during which time preparations can be made to mitigate the impacts, so would have a less steep curve. The depth of the curve is a factor of the scale of the disaster, which is dependent on the magnitude of the event and the nature of the locality.*

## ACTIVITIES

**(S)**

**1** Study Figure **1**. Describe the trends of numbers of fatalities and those affected and explain the reasons behind these trends.

**2** Explain the meaning of fatalism. Discuss the reasons why some people think it is best to do nothing and let nature run its course.

**(S)**

**3** Study Figure **5**. Make a copy of the axes and sketch a graph that you think fits the response to a) a powerful earthquake in a LDE with high population density b) flooding after prolonged rain in a HDE.

## STRETCH YOURSELF

Look at the website http://earthobservatory.nasa.gov/ NaturalHazards, which shows hazards that have happened in the last three months, along with associated images. Choose one of the events and assess the causes and impacts. You may need to source further information on the internet.

In this section you will learn about the Earth's structure and internal energy sources

## Earth structure and internal energy sources

Over 2000 years ago the Greek philosopher Plato was considering the structure of the Earth. But it wasn't until 1692 that Edmond Halley (after whom the famous comet was named) first proposed a theory to describe the Earth's structure. He suggested that it was made up of hollow spheres – rather like Russian nesting dolls. Halley considered that each sphere was actually habitable.

Look at Figure **1**. While the Earth appears to be a perfect sphere when seen from space, it is in fact a geoid. This means that it bulges around the equator and is flatter at the poles. The cause of this is centrifugal forces, generated by the Earth's rotation, which fling the semi-molten interior outwards, just like children on a roundabout!

### The crust

Look at Figure **2**. The Earth's outer shell is the crust – this is the layer we live on. The crust varies in thickness from between 5 to 10 km beneath the oceans to nearly 70 km under the continents. But just how thick is this? Some say that if the Earth was an apple the crust would be as thin as its skin. Others say that if the Earth was an egg its crust would be thinner than the eggshell! No matter which might be true, its average thickness, relative to the Earth in total, is thin – very, very thin.

There are two types of crust:

◆ Oceanic – an occasionally broken layer of basaltic rocks known as *sima* (because they are made up of **si**lica and **ma**gnesium).

◆ Continental – bodies of mainly granitic rocks known as *sial* (because they are made up of **si**lica and **al**uminium).

Sial is the upper layer of the Earth's crust and forms the continental land masses. Sima is the lower layer of the Earth's crust and is found beneath the oceans as well as grading into the lower part of the sial beneath the continents. Sial is much thicker than the oceanic sima, but it is less dense.

### The lithosphere

Look again at Figure **2**. Together, the crust and the upper mantle are known as the lithosphere. It is in this zone that the tectonic plates are formed.

**Figure 1** *Earth as seen from Apollo 17 in 1972 (The Blue Marble). What lies within?*

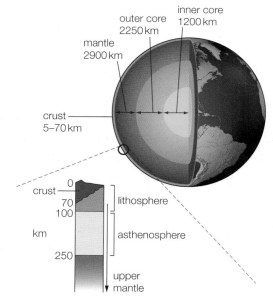

**Figure 2** *The internal structure of the Earth*

## The mantle

The mantle is the widest section of the Earth – it is 2900 km thick. Due to the great heat and pressure within this zone, the mainly silicate rocks are in a thick, liquid state, which become denser with depth.

The rocks in the upper mantle are solid and sit on top of the *asthenosphere*, a layer of softer, almost plastic-like rock. The asthenosphere can move very slowly, carrying the lithosphere on top. Densities within the mantle increase as you go down into the lower mantle.

## The core

The core is the centre and hottest part of the Earth – temperatures can reach 5000 °C. It is mostly made of iron and nickel and is four times as dense as the crust.

The core is actually made up of two parts. The outer core is semi-liquid and is mainly iron; the inner core is solid and is made up of an iron-nickel alloy. It is not known for certain but it is thought that as the Earth rotates, the liquid outer core spins, which creates the Earth's magnetic field.

The core's internal heat is the major cause of the Earth's tectonic activity. Some part of this heat may be primeval – retained from the ball of dust and gas from which the Earth evolved. But we now know that by far the greatest source of heat energy within the Earth is derived directly from radioactivity. Natural radioactive decay of uranium, thorium, potassium and other elements provides a continuous but slowly diminishing heat supply. The Earth is, in effect, a vast nuclear power station and, without this internal energy source, would be a completely dead and inert planet.

The phenomenal heat at the core generates convection currents within the mantle above. These currents spread very slowly within the asthenosphere – they are important but not solely responsible for the movement of the tectonic plates (see 5.3).

### Did you know?

The Earth is 4.6 billion years old. But just how old is this? Compress it into 24 hours and humans have only occupied the last few seconds!

### Associated key terms

**Magma**: molten rock, gases and liquids from the mantle accumulating in vast chambers at great pressures deep within the lithosphere. On reaching the ground surface magma is known as *lava*.

**Igneous rocks**: rocks formed by the cooling of molten magma, either underground (intrusive) or on the ground surface (extrusive).

**Intrusive**: magma that cools, crystallises and solidifies slowly below the surface is intrusive. It forms coarse-grained igneous rocks, such as granite and dolerite. Vertical dykes and horizontal or inclined sills may only become part of the landscape once erosion removes the overlying rocks.

**Extrusive**: lava that is in contact with the air or sea. It cools, crystallises and solidifies far quicker than magma that is still underground. The resulting igneous rocks, such as basalt, tend to be fine-grained with small crystals.

### Think about

Learning about the inner structure of the Earth is tricky. The depth and hot temperatures mean that we cannot drill into or physically see most of the Earth – the deepest mines and boreholes are effectively only pinpricks in the crust. Instead, scientists determine and map the interior composition and structure by a combination of heat-flow measurement, astronomical observation, satellite remote sensing and, most importantly, monitoring how seismic waves, either from earthquakes or human-made blasts, travel through the various layers.

### ACTIVITIES

In no more than 100 words, describe the structure of the Earth.

In this section you will learn about:
- plate tectonic theory of crustal evolution and sea-floor spreading
- tectonic plates and plate movement

## Plate tectonic theory of crustal evolution and sea-floor spreading

Look at Figure **1**. This artist's impression shows the world's ocean floors as monitored and mapped from nuclear submarines since the 1950s. The underwater mountain chains (mid-ocean ridges) and deep ocean trenches proved to be seismically active. These discoveries added compelling evidence in support of Alfred Wegener's controversial 1912 theory of continental drift, updated in 1962 by American geologist Harry Hess.

Hess studied the age of the rocks on the floor of the Atlantic Ocean, finding that the youngest rocks were in the middle and the oldest nearest the USA and the Caribbean. With the newest rocks still being formed in Iceland, this was compelling evidence that the Atlantic sea floor was spreading outwards from the centre, a concept known as *sea-floor spreading*. The rate of spreading was estimated at about 5 cm a year, which was confirmed by studies of the Earth's magnetic field in rocks – *palaeomagnetism*. Every 400 000 years or so, the Earth's magnetic field switches polarity, causing the magnetic north and south poles to swap. Magnetite (iron oxide) in lava erupted onto an ocean floor records the Earth's magnetic orientation of that time. Sea-floor spreading from mid-ocean ridges is shown by mirror-imaged patterns of 'switches' or reversals (Figure **2**).

Sea-floor spreading and palaeomagnetism would suggest that the Earth would be getting bigger, were it not for the discovery of deep ocean trenches where the ocean floor was being pulled downwards (subducted) and destroyed. With all this mounting evidence, plate tectonic theory has evolved and refined from Wegener's radical origins, and is now universally accepted.

## Tectonic plates and plate movement

Look at Figure **1** in section 5.10. The Earth's surface is made up of seven major and several minor tectonic plates. Each plate is an irregularly shaped 'raft' of lithosphere effectively floating on the 'plastic' asthenosphere beneath. There are two types of plate material: continental (mainly sial) over 1500 million years old; oceanic (mainly sima) less than 200 million years old.

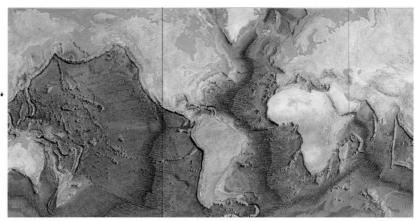

**Figure 1** *The Earth without oceans; this map outlines clearly many of the world's tectonic plates*

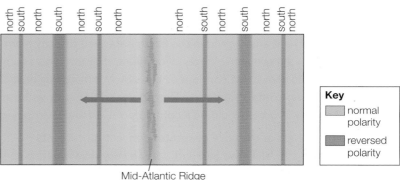

north south north south north south north    north south north south north south north

**Key**
- normal polarity
- reversed polarity

Mid-Atlantic Ridge

**Figure 2** *Magnetometers towed behind ships in the Atlantic Ocean first showed the symmetrical pattern of reversals in the Earth's magnetic field as it moves away from the Mid-Atlantic Ridge. The resulting fossil compasses can be likened to a mirrored magnetic supermarket bar code.*

### Did you know?
Tectonic plate movements vary in speed from the rate our finger nails grow to the rate our hair grows.

Continental plates are permanent and may extend far beyond the margins of current land masses. They will not sink into the asthenosphere because of their relatively low density. In contrast, denser oceanic plates are continually being formed at mid-ocean ridges and destroyed at deep ocean trenches – hence their relatively young age.

The resulting jigsaw of continental and oceanic plates is always changing, albeit very slowly. The plates move relative to each other at varying rates from 2 cm to 16 cm a year.

As already discussed, radioactive decay within the core of the Earth generates exceptional temperatures. *Hot spots* around the core heat the lower mantle, creating *convection currents*, which rise towards the surface before spreading in the asthenosphere, then cooling and sinking again (Figure **3**).

Tectonic plates can move sideways, towards each other or away from each other. They cannot overlap at their boundaries, so must either push past each other, be pushed upwards on impact, or be forced downwards into the asthenosphere and destroyed by melting.

No gaps can occur between plates, so if they are moving apart, new oceanic plate must be formed. In fact, because the Earth is neither expanding nor shrinking, any new oceanic plate that is formed must be compensated for by the subduction and destruction of an older plate elsewhere. As a result, zones of earthquake, volcanic and fold mountain activity are located along these great faults and it is here that that most of the world's major landforms occur (Figure **4**).

### Gravitational sliding, ridge push and slab pull

Look again at Figure **3**. Conventional explanations suggest that the tectonic plates are driven solely by these vast, slow-moving convection currents. But convection cells thousands of kilometres wide, yet sinking to such relatively shallow depths, seem implausible, especially now that we know of multiple convection cells beneath the Pacific plate, for example, moving in different directions.

**Gravitational sliding**, therefore, may be more significant in the explanation of tectonic plate movement. The lithosphere thickens with distance (and time) away from the mid-ocean ridge. This is because it cools with distance and the boundary between the solid lithosphere and plastic asthenosphere becomes deeper. The result of this thickening with distance from the ridge is that the lithosphere/asthenosphere boundary slopes away from the ridge. Gravity acting on the weight of the lithosphere near the ridge 'pushes' the older part of the plate in front (**ridge push**). Furthermore, following subduction, the lithosphere sinks into the mantle under its own weight, helping to 'pull' the rest of the plate with it (**slab pull**).

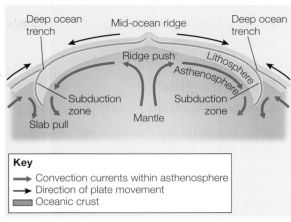

▲ **Figure 3** *Factors explaining tectonic plate movement*

**Key**
→ Convection currents within asthenosphere
→ Direction of plate movement
▬ Oceanic crust

▲ **Figure 4** *The Himalayas – fold mountains marking the collision between the Indo-Australian and Eurasian plates*

### ACTIVITIES

1  Draw a series of simple sketches to illustrate the concept of palaeomagnetism. Explain how this forms compelling evidence of sea-floor spreading and so helps to prove plate tectonic theory?

### STRETCH YOURSELF

Alfred Wegener died in 1930. During his lifetime, and for many years after, he had been ridiculed as a 'crank'. Research the evidence that proved this not to be the case and prepare an argument defending his theory.

In this section you will learn about plate margins, magma plumes and landforms associated with plate tectonics

## Plate margins

### Constructive (divergent) plate margins

When two plates separate (diverge) they form a *constructive margin*. There are two types of divergence:

◆ in oceanic areas, sea-floor spreading occurs on either side of mid-ocean ridges (e.g. the Mid-Atlantic Ridge)

◆ in continental areas, stretching and collapsing of the crust creates rift valleys (e.g. the Great African Rift Valley).

It is at constructive plate margins that some of the youngest rocks on the Earth's surface are to be found. This is because new crust is being formed as the gap created by the spreading plates is filled by magma rising from the asthenosphere. As the magma cools, it solidifies to form dense new basaltic rock.

### Mid-ocean ridges

Oceanic divergence forms chains of submarine mountain ridges that extend for thousands of kilometres across the ocean floor. If we could drain water from the oceans, the ridges would look like giant, bending spinal cords snaking along the constructive plate margins! Regular breaks called **transform faults** cut across the ridges – similar to the way our discs break up our spines.

These faults occur at right angles to the plate boundary, separating sections of the ridge. They may widen at different rates, which leads to frictional stresses building up, with shallow-focus earthquakes (those that occur at a depth of less than 70 km) releasing the tension (Figure **1**).

Mid-ocean ridges can rise up to 4000 m above the ocean floor. The middle of the ridges is marked by deep rift valleys in all but the most rapidly separating plate margins, which are found in the east Pacific. Over centuries, the rift valleys are widened by magma rising from the asthenosphere, which cools and solidifies to form new crust.

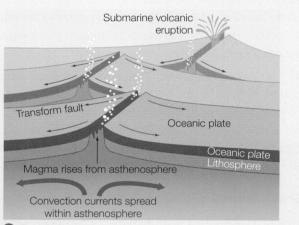

⊙ **Figure 1** *Transform faults along a mid-ocean ridge*

Volcanic eruptions along the ridges can build **submarine volcanoes**. Over time these may grow to rise above sea level, creating volcanic islands such Ascension Island, Tristan da Cunha and Surtsey in the mid-Atlantic Ocean.

### Rift valleys

Continental divergence forms massive rift valleys. These valleys are formed when the lithosphere stretches, causing it to fracture into sets of parallel faults. The land between these faults then collapses into deep, wide valleys that are separated by upright blocks of land called *horsts*. The Great African Rift Valley is especially interesting because it may eventually mark the formation of a new ocean as eastern Africa splits away from the rest of the continent.

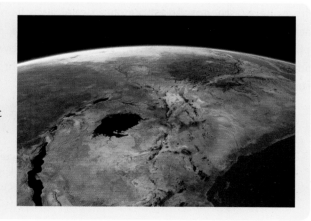

⊙ **Figure 2** *East Africa's Great Rift Valley extending south from the Red Sea*

## Destructive (convergent) plate margins

When two plates collide (converge) they form a destructive plate margin. Three types of convergence are possible:

◆ oceanic plate meeting continental plate, such as along the Pacific coast of South America

◆ oceanic plate meeting oceanic plate, such as along the Mariana Trench in the western Pacific

◆ continental plate meeting continental plate, such as the Himalayas.

### Oceanic plate meets continental plate

The colliding of oceanic plate with continental plate is associated with **subduction**. This involves one plate diving beneath the other and being destroyed by melting.

Oceanic plate is denser than the lighter continental plate and so subducts underneath it. The exact point of collision is marked by bending of the oceanic plate to form a *deep ocean trench* such as the Peru–Chile trench along the Pacific coast of South America. As the two plates converge, the continental land mass is uplifted, compressed, buckled and folded into chains of *fold mountains* (Figure **3**) such as the Andes. As compression continues, simple folding can become asymmetrical, then overfolded (making a *recumbent* fold). Increasing the compression yet further would make the middle section so thin that it might break, creating a *nappe*.

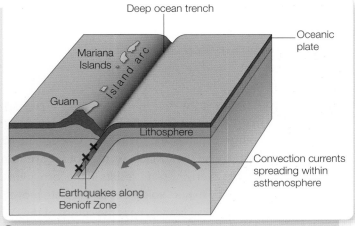

◯ **Figure 3** *Oceanic plate meets continental plate*

The descending oceanic plate starts to melt at depths beyond 100 km and is completely destroyed by 700 km. This zone of melting is called the *Benioff Zone*. Melting is caused by both increasing heat at depth and also friction. This friction may also lead to tension (stresses) building up, which may suddenly be released as intermediate (70–300 km deep) or deep-focus (300–700 km deep) earthquakes. The melted oceanic plate creates magma, which is less dense than the surrounding asthenosphere and, as a result, rises in great plumes. Passing through cracks (faults) in the buckled continental plate, the magma may eventually reach the surface to form explosive volcanic eruptions.

### Oceanic plate meets oceanic plate

When two oceanic plates collide, one plate (the faster or denser) subducts beneath the other. This leads to the formation of a deep ocean trench and melting, as described previously. The resulting rising magma from the Benioff Zone forms crescents of submarine volcanoes along the plate margins which may grow to form *island arcs* (see 5.7). The Mariana Trench and the Mariana Islands in the western Pacific illustrate this particularly well.

Here the Pacific plate is being subducted beneath the smaller Philippine plate.

◯ **Figure 4** *Oceanic plate meets oceanic plate*

**Continental plate meets continental plate**

Continental plates are of lower density than the asthenosphere beneath them. This means that subduction does not occur. The colliding plates, and any sediments deposited between them, simply become uplifted and buckle to form high fold mountains such as the Himalayas (Figure **5**). Volcanic activity does not occur at these margins because there is no subduction, but shallow-focus earthquakes can be triggered. Young fold mountains, such as the Himalayas, are continually compressing and growing higher. Everest is growing 5 mm every year, so each climber to reach the summit really has climbed higher than any other person before!

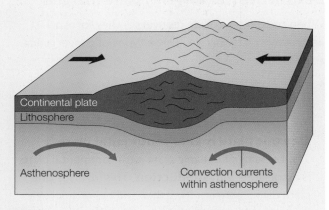

⬤ **Figure 5** *Continental plate meets continental plate*

## Conservative plate margins

When two plates slide past each other they form a conservative plate margin (Figure **6**). Along these margins crust is not being destroyed by subduction. There is no melting of rock and, therefore, no volcanic activity or formation of new crust. Despite the absence of volcanic activity, these margins are extremely active and are associated with powerful earthquakes.

Friction between the two moving plates leads to stresses building up whenever any 'sticking' occurs. These stresses may eventually be released suddenly as powerful shallow-focus earthquakes, such as in Los Angeles (1994) and San Francisco (1906 and 1989). These earthquakes occurred along California's infamous San Andreas fault system (Figure **7**, also Haiti, 5.12).

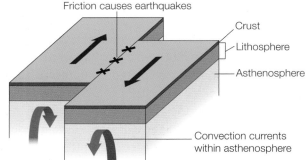

⬤ **Figure 6** *Conservative plate margin*

## Magma plumes

Look at Figure **1**, in section 5.10. Earth shaking (**seismicity**) and volcanic activity (**vulcanicity**) are strongly associated with plate tectonics. In fact 95 per cent of the world's earthquakes and most volcanoes are located along plate margins.

Even seemingly notable exceptions, such as the volcanic Hawaiian Islands, are explained in part by plate movements. As already described, radioactive decay within the Earth's core generates very hot temperatures. If the decay is concentrated, hot spots will form around the core. These hot spots heat the lower mantle creating localised thermal currents where **magma plumes** rise vertically. Although usually found close to plate margins, such as beneath Iceland, these plumes occasionally rise within the centre of plates and then 'burn' through the lithosphere to create volcanic activity on the surface. As the hot spot remains stationary, the movement of the overlying plate results in the formation of a chain of active and subsequently extinct volcanoes as the plate moves away from the hot spot. The Hawaiian Islands, near the centre of the Pacific Plate, are a classic example of this (see Figure **2**, page 279).

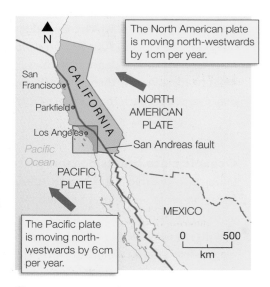

⬤ **Figure 7** *The San Andreas fault system*

**Figure 8** *The relationship between seismicity, vulcanicity and plate tectonics*

| Type of plate margin | Processes | Geographical features and examples |
|---|---|---|
| **Constructive margins** | Plates diverge – move away from each other | • New oceanic crust is formed by basaltic magma rising from the asthenosphere<br>• New basaltic rocks<br>• Mid-ocean ridges broken up by transform faults (Mid-Atlantic Ridge)<br>• Shallow-focus earthquakes<br>• Basic volcanoes (Eyjafjallajökull, see 5.6)<br>• Volcanic islands (Ascension Island, Tristan da Cunha and Surtsey)<br>• Continental rift valleys (The Great African Rift Valley) |
| **Destructive margins**<br><br>• Subduction zones<br><br><br><br><br><br><br><br><br><br><br>• Collision zones | Plates converge – move towards each other | *Oceanic v. oceanic*<br>• Oceanic crust is destroyed by subduction and melting at depth<br>• Deep ocean trenches (Mariana Trench)<br>• Island arcs (West Indies, Aleutian and Mariana Islands)<br>• Shallow, intermediate (70–300 km deep) and deep-focus (300–700 km deep) earthquakes<br>• Explosive, acid volcanoes (Krakatoa, Merapi, see 5.6 and Montserrat, see 5.7).<br><br>*Oceanic v. continental*<br>• Oceanic crust is destroyed by subduction and melting at depth<br>• Deep ocean trenches (Peru–Chile Trench)<br>• Continental land mass is uplifted, compressed and buckled into fold mountains (Andes)<br>• Intermediate and deep-focus earthquakes<br>• Explosive, acid volcanoes (Cotopaxi)<br><br>*Continental v. continental*<br>• Colliding plates, and any sediments between them, uplift and concertina into particularly high fold mountains (Himalayas)<br>• Shallow-focus earthquakes<br>• Continued compression and overfolding can result in fracture creating a thrust fault and nappe |
| **Conservative margins** | Plates move sideways past each other | • Shallow-focus earthquakes (Haiti, see 5.12 and the San Andreas fault system, Figure **7**) |
| **Exceptions to plate margins** | Hot spots near the centre of a plate | • Basaltic volcanoes (Mauna Loa, Kilaue (Hawaii) and Yellowstone) |

## ACTIVITIES

1 Study the information about the different types of plate margins.

   **a** Explain why submarine volcanoes are formed at constructive plate margins.

   **b** What is meant by subduction and why is it an important process at destructive plate margins?

   **c** Why is there no volcanic activity at a continental collision margin?

   **(S) d** Describe and explain the pattern of earthquakes at the different destructive margins. Use simple sketches to support your answer.

2 **a** Label Figure **9** with the names of specific tectonic plates. Use Figure **1** in section 5.10 and your own knowledge, Begin by locating New Zealand and the Andes on the map.

   **b** Finish your diagram by adding annotations to the suggested boxes.

3 Explain the relationship between earthquakes, volcanic activity and plate margins. Be sure to use examples in your answer.

**Figure 9** *Block diagram of tectonic plate margins*

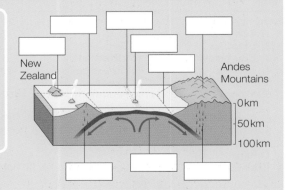

New Zealand

Andes Mountains

0km
50km
100km

In this section you will learn about the spatial distribution of volcanoes and their magnitude, frequency, regularity and predictability

### Spatial distribution of volcanoes

Look at Figure **1**. The relationship between volcanoes and tectonic plate margins is clear. For example, the so-called 40 000 km 'Pacific Ring of Fire' shows particularly high densities of volcanoes stretching from the Aleutian Islands, through Japan, the Philippines and across to New Zealand. But closer examination in conjunction with a tectonic plates map showing types of margin (Figure **1**, 5.10) demonstrates that although volcanic activity is common at constructive (divergent) and destructive (convergent) margins, it is absent at conservative (transform) margins. Also, some volcanoes occur within the centres of plates, such as the Hawaiian hot spot (see Figure **2**, page 279), and along rift valleys, such as the Great African Rift Valley. The type and magnitude of eruption vary according to location, which has an effect on the type of magma – for example, basaltic lava is erupted at constructive plate boundaries; andesitic or rhyolitic lava at destructive plate boundaries.

⊽ **Figure 1** *The spatial distribution of volcanoes and their relationship to plate tectonics*

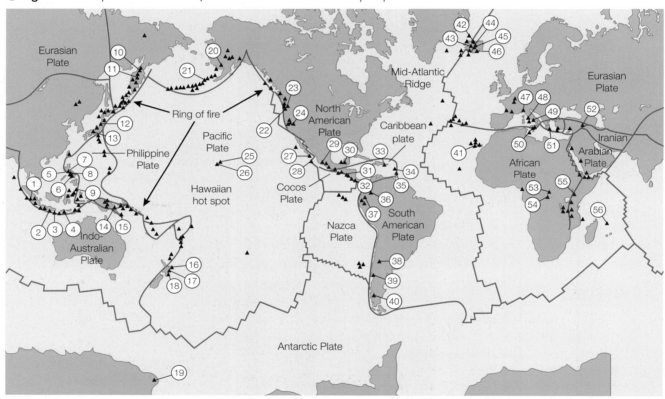

| | | | | |
|---|---|---|---|---|
| 1 Krakatoa | 13 Unzen | 25 Mauna Loa | 37 Cotopaxi | 49 Etna |
| 2 Mt Merapi | 14 Barn | 26 Kilauea | 38 Aconcagua | 50 Vulcano |
| 3 Galunggung | 15 Lamington | 27 Paricutin | 39 Villarrica | 51 Santorini |
| 4 Tambora | 16 Ngauruhoe | 28 Jorullo | 40 Hudson | 52 Ararat |
| 5 Pinatubo | 17 Tarawera | 29 Popocatépetl | 41 Pico de Teide | 53 Nyamuragira |
| 6 Mayon | 18 Ruapehu | 30 El Chichón | 42 Laki | 54 Nyiragongo |
| 7 Taal | 19 Erebus | 31 Acatenango | 43 Hekla | 55 Kilimanjaro |
| 8 Hibok-hibok | 20 Katmai | 32 Cosequina | 44 Helgatell | 56 Piton de la Fournaise |
| 9 Awu | 21 Iliamna | 33 Mt Pelée | 45 Surtsey | |
| 10 Bezymianny | 22 Mt St. Helens | 34 Soufrière Hills | 46 Eyjafjallajökull | |
| 11 Klyuchevskoy | 23 Mt Rainier | 35 La Soufrière | 47 Vesuvius | |
| 12 Mt Fuji | 24 Lasson Peak | 36 Nevado del Ruiz | 48 Stromboli | |

## Magnitude

Since 1982 the magnitude of volcanic eruptions has been measured using a logarithmic scale from 0 to 8, called the Volcano Explosivity Index (VEI). The adjectives used to describe each classification leave nothing to the imagination. For example, 1 is 'gentle', 4 'cataclysmic' and 8 'mega-colossal'. Typical Hawaiian eruptions are rated 1, while Mount St Helens in 1980 was a 4 (Figure **2**). When, rather than if, a **supervolcano** such as under Yellowstone National Park erupts again, the VEI might exceed 8 and change global climates for years afterwards!

Volcanic classifications based upon the violence of the eruption may include a VEI rating, but are more useful if readily related to tectonics by including details of the type of magma and even frequency (Figure **3**).

**Figure 2** *Eruption of Mount St Helens, Washington, USA*

| Name | Type of magma | Characteristics of eruption |
| --- | --- | --- |
| Icelandic | basaltic | Lava flows gently from fissures. |
| Hawaiian | basaltic | Lava flows gently from a central vent. |
| Strombolian | (thicker) basaltic | Frequent, explosive eruptions of tephra and steam. Occasional, short lava flows. |
| Vulcanian | (thicker) basaltic andesitic and rhyolitic | Less frequent, but more violent eruptions of gases, ash and tephra (including lapilli). |
| Vesuvian | (thicker) basaltic, andesitic and rhyolitic | Following long periods of inactivity, very violent gas explosions blast ash high into the sky. |
| Peléean | andesitic and rhyolitic | Very violent eruptions of nuées ardentes. |
| Plinian | rhyolitic | Exceptionally violent eruptions of gases, ash and pumice. Torrential rainstorms cause devastating lahars. |

**Figure 3** *A classification of volcanic eruptions*

225

## Frequency and regularity

However, frequency and regularity of eruptions are rarely, if ever, measurable to any degree of predictable accuracy. Recent floods in the Lake District, for example, give a painful reminder that 1 in 200 years events can occur repeatedly in the space of a decade. Volcanic periodicity, too, can seem to be a random concept.

The uncomfortable reality is that volcanoes such as Mount St Helens and Chances Peak, Montserrat (see 5.7) have proved to be dormant rather than extinct. Even for those volcanoes known to be dormant, relying on average cycles of activity can only alert volcanologists of the necessity for heightened observation. For example, Mount Vesuvius in Italy (Figure **4**) is one of the world's most carefully monitored volcanoes. Most famous for its explosive 'Plinian' eruption in AD79, it last erupted in 1944. Seventy-year 'cycles' may be statistically interesting but these calculations are, in reality, of questionable use. Vesuvius erupted six times in the eighteenth century, eight times in the nineteenth century (most notably in 1872), and three times in the twentieth century. But 3 million people live within 15 km of it and so *contingency planning* has to be thorough. The current plan assumes that a minimum of two weeks' warning will be given of any eruption and foresees a week-long emergency evacuation of 600 000 people from a 'red zone' already identified as at greatest risk from *pyroclastic flows* or *nuée ardente* ('glowing cloud').

## Predicting volcanic eruptions

Unlike earthquakes, where there is little or no warning, volcanic eruptions tend to follow weeks of seismic activity and other warning signs. As long as active and dormant volcanoes are monitored, using equipment such as seismometers and seismographs (Figure **5** and **6**), warnings of imminent eruptions can be issued to governments and civil authorities. Prediction is generally very successful and can be aided by evidence from previous eruptions – such as **lahar** and pyroclastic deposits following river valleys. As part of the contingency planning, hazard maps can be produced to identify those areas most at risk, which are therefore prioritised for evacuation to safe zones.

**Figure 4** *Mount Vesuvius overlooking the Port of Naples*

### Did you know?

The difference between an active volcano and a dormant one is simply one of time-scale. Active volcanoes have erupted in living memory, whereas dormant ones have erupted within historical record.

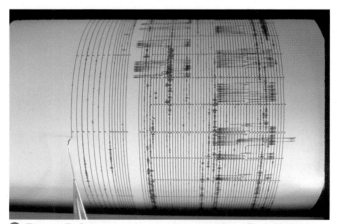

**Figure 5** *Seismograph recording activity at Taal, Philippines*

| What is being monitored? | What does it indicate? |
|---|---|
| Seismic activity is measured using seismometers and recorded using a seismograph | Microquakes indicate rising magma fracturing and cracking the overlying rocks |
| Ground deformation is measured using tiltometers and laser-based electronic distance measurement | Bulging (inflation) of the ground is caused by rising magma. Both slope angles and the increasing distance between set points can be measured very accurately |
| Upward movement of iron-rich magma is measured using magnetometers | Changing magnetism within the volcano is a common geophysical indication of rising magma |
| Rising groundwater temperature and/or gas content is measured using hydrological instrumentation | Rising magma will both heat groundwater and corrupt it with gases such as sulphur (increasing its acidity) |
| Warning signs such as small eruptions, emissions of gases, landslides and rockfalls can be recorded in real time using remote sensing equipment | Remote solar-powered digital camera surveillance is a particularly powerful and safe tool to record physical changes in and around the main crater. Thermal imaging and gas sampling of (poisonous) emissions such as chlorine can also be included in remote sensing |

**Figure 6** *Methods and instruments used in the monitoring of active and dormant volcanoes*

## Mudflows and lahars

Despite the simplistic name, mudflows possess extraordinary destructive power. In Indonesia all types of mudflows are called lahars (see 4.2), but the term is now applied more specifically to flows of ash, cinder, soil and rock that have been changed to clay by acids in volcanic gases and hot-spring waters. Over the centuries, lahars have destroyed more property than any other volcanic action – and killed thousands of people.

They may be hot or cold depending upon their origin. Common causes include:

- eruptions ejecting water directly from a crater lake, or through the broken crater walls

- rapid melting of ice or snow on the volcano's slopes

- eruption-induced heavy rainfall mixing with loose material on the volcano's slopes

- pyroclastic flows (nuées ardentes) entering streams.

The Colombian Armero tragedy of 1985 illustrated the destructiveness of lahars. Rising magma and gas and steam emissions started melting the summit glacier of Nevado del Ruiz, which had not erupted since 1595. Melting continued until the full eruption of 13 November that melted the remaining snow and ice. The resulting meltwater (combined with torrential rain and raging flood waters from the Lagunillas River) mixed with ash from previous eruptions to send a 30-metre-high lahar travelling at over 80 km per hour, towards the sleeping town of Armero. 21 000 inhabitants perished in the tragedy (Figure **7**).

**Figure 7** *The morning after; a layer of mud, up to 8 m deep, covers Armero and the surrounding area*

### Did you know?

Steam is produced during volcanic eruptions, which may also contain nuclei of sulphur. Heat from the eruption encourages convection currents to rise. High above the volcano, the steam condenses into localised cumulus clouds resulting in bursts of (sulphuric) acid rain.

### ACTIVITIES

**S**

1 Study Figure **1**. Describe and comment on the global distribution of active and dormant volcanoes.

2 What is a hot spot and why is it associated with volcanic activity?

3 Outline the purpose, methods and value of remote sensing techniques adopted in the monitoring of active and dormant volcanoes.

In this section you will learn about the impacts of volcanic activity

## The impacts of volcanic activity

Look again at Figure **3** in section 5.5. The impacts of volcanic activity very much depend upon the type of volcano.

◆ Fissure eruptions of basic lava, such as blocky (*aa*) or smooth and ropy (*pahoehoe*), can create extensive lava plateaus. Hollows in the existing landscape are filled to create flat, featureless basalt plains such as the Deccan Traps in central India, where there are multiple layers of basalt 2000 m thick and covering 500 000 km². The impact of fissure eruptions cannot be underestimated. They represent the largest contributors to global climate change and large-scale landscaping.

◆ Basic shield volcanoes are shallow-sided and broad. They are formed by relatively pure basalt that cools as it runs down from the summit crater. Typical of constructive plate margins, rift valleys and hot spots, their impact is most obvious by the vast scale of the resulting volcanic cones. Eruptions, however, are gentle enough to become tourist attractions, such as in Hawaii and Iceland.

◆ Acid dome volcanoes are steep-sided convex cones associated with thick (viscous), silica-rich gaseous lava that solidifies before running too far down slope. Associated with destructive plate margins, their explosive eruptions of pyroclastic flows have a deadly impact. For example, in 1902, Mount Pelée on the Caribbean island of Martinique caused the worst volcanic disaster of the twentieth century when pyroclastic flows killed 30 000 people in St Pierre just minutes after erupting.

◆ Composite cones (strato-volcanoes) are formed from alternating eruptions of ash, **tephra** and lava, which builds up the volcano in layers. The layering produces weaknesses that can be exploited by the magma. Consequently, the classic conical shapes of Mounts Fuji (Figure **1**) and Merapi are relatively rare compared to the irregular-shaped composite volcanoes covered by numerous secondary (parasitic) cones and fissures, such as Mount Etna.

◆ Calderas result from violent eruptions that blow off the volcano's summit. This empties the magma chamber, causing the sides of the volcano to collapse inwards. The resulting vast pit crater can be many kilometres in diameter and is left to be flooded by the sea or fill as a lake. One famous example is Santorini in the Mediterranean, where the sea has broken through the walls of the caldera and produced an impressive ring of small islands (Figure **2**).

▲ *Figure 1*  *An aerial view of the classic cone of Mount Fuji, Japan*

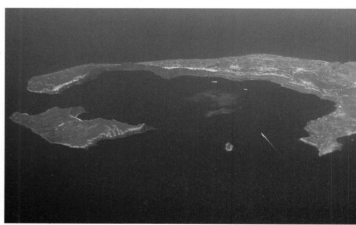

▲ *Figure 2*  *Santorini caldera from the air*

### Eyjafjallajökull, Iceland, April 2010

Eyjafjallajökull is one of a number of volcanoes in Iceland covered by great thicknesses of ice. Subglacial eruptions cause the ice to melt, which frequently leads to considerable flooding. Following preliminary seismic activity and small fissure eruptions in March 2010, Eyjafjallajökull's eruption caused torrents of meltwater (called jökulhlaups or glacial bursts) to wash away part of Iceland's main perimeter road. No deaths or injuries were reported, but 800 people were evacuated. Thick deposits of ash made farming impossible and contaminated water sources with fluoride.

Winds carried the ash cloud south and east towards Europe (Figure **3**), causing the progressive closure of airspace over the space of a week. A total of 100000 flights were cancelled, over 10 million people were left stranded and airlines lost US$1.7 billion in revenue. Cargo and freight traffic were also disrupted, affecting the movement of fresh produce. For example, Kenyan farm workers were laid off because the fresh vegetables and flowers they produced for European markets could not be transported. The travel and tourism industry faced massive payouts in compensation and even schools faced disruption as pupils and teachers were stranded abroad following Easter holidays.

⬆ **Figure 3** *Eyjafjallajökull's ash plume spread across Europe*

## Mount Merapi, Java, October 2010

Mount Merapi, on the island of Java, is the most active volcano in Indonesia. Erupting frequently and violently, the volcano represents a constant threat to the tens of thousands of people who live and farm on its fertile flanks. In October 2010, after more than 500 earthquakes were registered beneath the volcano, the evacuation of 20000 villagers living within a 20 km radius was advised. Within days a series of powerful eruptions began, which over subsequent weeks blasted pyroclastic flows down the strato-volcano's flanks (Figure **4**). Fires, burns, respiratory failure and blast injuries caused 350 fatalities, most commonly to those who had refused to evacuate or had returned to their homes between eruptions. In total, up to 350 000 people were displaced because their homes were destroyed and farmland

⬆ **Figure 4** *Villagers attempt to flee Mount Merapi's pyroclastic flows, October 2010*

was smothered in thick deposits of ash and lava bombs. As with Eyjafjallajökull, the resulting ash cloud also caused considerable disruption to aviation across Java.

## ACTIVITIES

1  Study Figures **1** and **3** (5.4) and Figure **1** (5.5). Compare and contrast the 2010 Eyjafjallajökull and Mount Merapi eruptions. (You should comment on location, tectonics, levels of development, preparedness, and primary and secondary impacts.)

2  Study Figure **5**. In pairs, research the eruptions of Eyjafjallajökull and Mount Merapi in greater detail. Choose one each.

  **a**  Answer the questions in Figure **5**.

  **b**  Rate the risk of disaster on a sliding scale. For example, 1 = highly likely to 5 = highly unlikely.

  **c**  Share and justify your results with your partner.

  **d**  Discuss and write a paragraph explaining which volcano represents the greatest risk of disaster.

**Natural hazard**
- When did it last occur?
- What is the probability that it will occur again?
- How quickly did it start?
- How big was the event?
- How long did it last?
- How large was the affected area?

**Risk of disaster**

**Vulnerability**
- What is the population?
- How developed is the economy?
- What is the land used for?
- Are emergency services prepared?
- Are buildings, roads and wider infrastructures hazard proof?
- How are vulnerable (e.g., elderly) people and buildings (e.g., historic) protected?

Risk = Hazard × Vulnerability

⬆ **Figure 5** *The risk of disaster – when vulnerability and a natural hazard collide*

In this section you will learn about short- and long-term responses to a volcanic eruption

## Short- and long-term responses to volcanic eruptions

We have already established that volcanic prediction can be remarkably successful (see 5.5). Furthermore, continual reviewing of contingency planning allows preparedness and eventual responses to be as efficient as possible for any monitored volcano. Even though volcanoes do not always behave predictably, and so still pose a major threat, volcanic prediction and contingency planning (including hazard mapping) have undoubtedly saved many thousands of lives. We increase our knowledge with each eruption that is studied, and so preparedness, short-term (immediate) and long-term responses can continue to improve.

**Key**
| | |
|---|---|
| ☐ | Safe zone |
| ▨ | Area evacuated by July 1996 |
| ▩ | Area evacuated by April 1996 |

⟩ *Figure 1* *Montserrat in the Lesser Antilles island arc*

### Montserrat, 1995–present

The British Overseas Territory of Montserrat is situated within the northern part of the Lesser Antilles (Figure **1**). This is an island arc formed where the South American tectonic plate subducts beneath the Caribbean tectonic plate (see 5.4). Most of the islands are composite volcanoes, formed as a result of repeated, violent eruptions. Throughout Montserrat's history, lava domes have been created as a result of viscous, silica-rich lava building up at the top of the volcanoes. When the lava eventually becomes too heavy, the domes collapse, resulting in andesitic lava and pyroclastic flows.

Eruptions of ash and dust from the southern Chances Peak volcano in the Soufrière Hills in July 1995 prompted an immediate response. Scientists assembled to monitor seismic activity, gases and changes in the volcano's shape. They discovered that it was only dormant, and not extinct as previously thought. One month later, the evacuation of the south began with residents being moved to churches and halls in the north of the island. In April 1996, the entire population was forced to leave the capital, Plymouth.

On 25 June 1997, Chances Peak catastrophically erupted. The dome of the volcano collapsed, sending 5 million cubic metres of hot rocks and gases down its sides. The south of the island, including Plymouth, was covered by these pyroclastic flows of hot ash, rocks, boulders and lahars (Figure **2**).

Fires associated with the pyroclastic flows killed 19 people. There were also a number of burn and inhalation injuries, and two-thirds of all houses were either buried by ash or flattened by rocks.

⌃ *Figure 2* *Adventure tourism now exploits the ash-covered ruins of Plymouth, the former capital*

The island's airport was in the direct path of the main pyroclastic flow and was completely destroyed. Farmland, vegetation and three-quarters of all other infrastructure were also destroyed and more than half of the population of 11 000 people was evacuated to Antigua, the USA and the UK.

Further eruptions and associated evacuations occurred in December 2006, July 2008 and February 2010. This most recent activity was particularly explosive – a **vulcanian** eruption ejecting a 15 km column of ash into the atmosphere and propelling pyroclastic flows in many directions (Figure **3**).

⬆ **Figure 3** *Pyroclastic flows on Montserrat in 2010*

| Short-term | Long-term |
|---|---|
| The Montserrat Volcano Observatory was set up in 1995 and successfully predicted its eruption on 25 June 1997. | A three-year redevelopment programme for houses, schools, medical services, infrastructure and agriculture was funded by the UK. |
| Exclusion zones were defined in the south of the island and visits severely restricted. | In 1998 the people of Montserrat were granted full residency rights in the United Kingdom, allowing them to migrate if they chose. British citizenship was granted in 2002. A top-heavy population structure resulted as many younger people did not see an economic future on the island. |
| NGOs, such as the Red Cross, set up temporary schools and provided medical support and food. | By 2005 many people had moved back, but the south including Plymouth, remained an exclusion zone. The 2015 population of 5250 is less than half that of 1995 and is now administered from Brades near the port of Little Bay. |
| Warning systems were set up to alert inhabitants – sirens, speaker systems and via the media. | Vegetation is slowly regrowing as the ash, lava and lahar deposits break down. Fertile soil means that land will again be used for cash crops such as cotton. But two-thirds of the island remains an exclusion zone and warning sirens are tested daily at 12.00 noon. |
| Troops from the USA and the British Navy came to aid the evacuation process. | Rebuilding tourism with the volcano itself as an attraction. Montserrat was selected by National Geographic Adventure Magazine as one of the 'Top 25 New Trips of 2010'. A new airport, hotel and dive shop have been built as the island carves a niche in adventure tourism. There is potential to exploit geothermal energy and the export of volcanic sand for construction. |
| £17 million in UK aid paid for temporary buildings and water purification systems. | UK financial aid since 1995 has exceeded £420 million. |

⬆ **Figure 4** *Summary of short- and long-term responses and risk management*

## ACTIVITIES

**1** Study Figures **1** and **5**.

   **a** Identify the human-made feature marked **X** on the satellite image.

   **b** List evidence to suggest that the volcano is active.

   **c** Approximately what area of the island is considered 'safe'?

   **d** Explain why the 'safe zone' is several kilometres from the site of the volcano.

**2** Evaluate Montserrat's responses to the Chances Peak eruptions since 1995.

Red areas are vegetation. Grey/white areas are areas of new flow deposits

⬆ **Figure 5** *Satellite image of Montserrat following the 2010 eruption*

# Mount Etna, Sicily

**In this section you will learn about impacts and human responses to a recent volcanic event**

Out of an estimated 1500 active volcanoes globally, 50 or so erupt every year. In 2015, global eruptions were widespread and included Mount Etna in Sicily, Italy, Cotopaxi in Ecuador, Villarrica in Chile, Piton de la Fournaise on Réunion Island and Kilauea on Hawaii.

## Mount Etna, Sicily, Italy

Rising to an altitude of 3350 m and covering an area of 1250 km$^2$, Mount Etna is Europe's most active volcano (Figure **1**). It is a strato-volcano with a classic, elegant conical shape that can only be best appreciated from a distance. Close up, Etna is a cartographer's nightmare – a complex, dynamic geology of summit craters, and numerous secondary (parasitic) cones and fissures. In short, Etna has the longest documented record of eruptions of any volcano in the world. Its fertile slopes are a potentially hazardous home to over 900 000 people, and rarely a decade passes without some spectacular activity involving a wide variety of eruptive styles (Figure **3**).

Etna's geological and eruptive complexity results from challenging tectonics. The volcano is a result of the collision between the African and Eurasian tectonic plates, but there is no agreement on the exact cause of each eruption. Most theories suggest they are linked to *rifting*, normally associated with constructive plate margins. As seen in the Great African Rift Valley, this pulling apart of the Earth's crust is thought to be separating eastern Sicily from the rest of Italy.

▲ **Figure 1** *Location of Mount Etna, Sicily, Italy*

Volcanologists classify the eruptions of Etna as mostly effusive (low-viscosity lava flows) and occasionally mild *strombolian*. The strombolian-type eruptions, with short lava flows, tend to occur on the summit (Figure **2**). Less frequent, but more effusive, are the eruptions from fissures on the sides of Etna.

At the summit, volcanic cones mark the locations of the active Northeast and Southeast Craters. The remaining summit craters, the Voragine and the Bocca Nuova, actually lie within the pre-existing 250 metre-wide Central Crater.

▲ **Figure 2** *Voragine Crater eruption, Mount Etna, December 2015*

Secondary volcanic cones and jagged fissures combine to create a rich and varied landscape on the slopes of Etna. A catastrophic collapse of Etna's eastern slopes resulted in the impressive 7 km-long horseshoe-shaped Valle del Bove (Valley of the Oxen). This natural funnel is layered with ancient and recent lava flows. By drilling through these layers, scientists have been able to uncover a detailed geological record of past eruptions. In contrast, ridges such as the active Northeast Rift, stand upright and may be identified by eruptive cones separated by deep fissures.

⯆ **Figure 3** *Mount Etna eruptions 1991–2018*

| Date | Nature of eruption | Impacts and responses |
| --- | --- | --- |
| Dec 1991–Mar 1993 | Effusive lava flows and Hawaiian fountaining from a fissure on the eastern side of Southeast Crater. The eruption lasts for 473 days. | Town of Zafferana Etnea was threatened by the largest volume of lava in hundreds of years. So:<br>• an earth dam was built to temporarily hold the lava flow<br>• rock and concrete blocks were dropped to 'plug' the lava channel<br>• explosives used to divert the lava flow into a new human-made channel. |
| July–Aug 2001 | Seven fissures erupt effusively on the south and north-east flanks. | Significant damage to tourist facilities. The town of Nicolosi is threatened. Catania airport is intermittently forced to close as a result of the ash fall. |
| Oct 2002 | Strombolian, Hawaiian fountaining and *phreatomagmatic* (magma is mixed with water) flank eruptions. | The most explosive flank eruption in the last 150 years. Lava flows threaten the mountain village of Rifugio Sapienza. Catania airport is closed again. |
| Nov 2009 | Strombolian eruptions in the Southeast Crater. | 4.4 magnitude earthquake beneath the southwest flank. |
| Apr 2010 | Summit ash eruption from the lower east flank of the Southeast Crater. | The eruption increases the crater width from 10 m to 50 m. |
| Jan–Oct 2011 | Hawaiian fountaining and lava flows from the Southeast Crater. The ash column reaches several kilometres high. | A lava flow descends the western slope of the Valle del Bove. It is successfully diverted from the ski resort at Sapienza Refuge (the main tourist hub on the volcano). |
| Oct 2013 | Renewed eruptions at New Southeast Crater and Northeast Crater. | |
| Jan–Jun 2014 | Strombolian activity at New Southeast Crater. | Lava flows travel towards the Valle del Bove and also north-east in the direction of Monte Simone.<br>Catania Airport is temporarily closed. |
| Dec 2015 | Strombolian eruptions from the Voragine Crater. Lava fountaining 1 km in height and the ash column rose to 7 km above the summit. | Sulphur dioxide plumes drift to Tunisia, Libya, Egypt, Iran and Turkmenistan.<br>Catania Airport is closed for a few hours. |
| May 2016 | Strombolian eruptions from Northeast and New Southeast crater. | Lava flow extends into Valle del Bove. |
| Mar 2017 | Phreatic eruptions eject lava and tephra from New Southeast Crater. | Ten reported injuries from steam and tephra include a BBC film crew reporting the events. |
| Dec 2018 | Lava effusion and ash fall from fissure eruptions on the upper south-eastern flank. | Lava flow extends into Valle del Bove. Temporary closure of airspace over Sicily. |

## ACTIVITIES

1 Access the Osservatorio Etneo Istituto Nazionale di Geofisica e Vulcanologia website at www.ct.ingv.it/en/real-time-seismic-signal.html. Here you can see a live seismic feed and also access an Etna webcam. Explore the website to see what other information is available about monitoring Mount Etna. Comment on the management issues that arise from your observations.

Ⓢ 2 You will find a map of recent earthquakes in the Mount Etna area and the Aeolian Islands at www.ct.ingv.it/ufs/analisti/maps.php. Comment on the current seismic activity and locate the epicentres of recent earthquakes.

In this section you will learn about the nature and underlying causes of seismic hazards

## Seismic hazards – earthquakes and tsunamis

The Earth rumbles, twitches, jolts and shakes thousands of times a day – an inevitable product of a dynamic planet. As tectonic plates move over, under and against each other, the stresses generated through frictional drag build to breaking point, resulting in earthquakes ranging from unnoticeable jiggles to apocalyptic tsunamis. But not all earthquakes coincide with plate margins – they can happen anywhere. The UK, hundreds of kilometres from the nearest plate margin, experiences more than 300 every year, but most are too small to notice.

### A shaky story...

Siberian legend says that the Earth sits on a gigantic sled that is driven around the cosmos by the great god Tuli. The flea-infested dogs that smoothly pull the sled along occasionally have to stop to scratch, shaking the Earth in the process.

### Causes of seismicity

Earth shaking (seismicity) can be caused by human activities such as mining, fracking or reservoir construction. But it is most strongly associated with plate tectonics. Plate movements produce energy of extraordinary proportions, although these movements are not smooth. Friction along plate margins builds stresses in the *lithosphere*. When the strength of the rocks under stress is suddenly overcome, they fracture along cracks called faults, sending a series of seismic *shockwaves* to the surface. The breaking point is called the *focus* (hypocentre) of the earthquake. The *epicentre* is the point on the surface directly above the focus. It commonly experiences the most intense ground shaking (Figure **1**). The shaking then becomes progressively less severe the further from the epicentre, like ripples spreading outwards in a pond. Earthquake tremors usually last for less than a minute and are followed by several weeks of aftershocks as the crust settles.

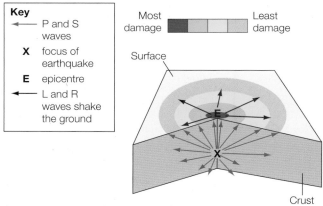

**Key**
←  P and S waves
**X**  focus of earthquake
**E**  epicentre
←  L and R waves shake the ground

Most damage ▮▮▮▮ Least damage

⊙ **Figure 1**  *The focus (hypocentre) and epicentre of an earthquake*

### Seismic shockwaves

Scientists have identified several different types of seismic waves (Figures **1** and **2**).

- *Primary* or *pressure (P)* waves are the fastest and reach the surface first. P waves are like sound waves – high-frequency and pushing like balls in a line. They travel through both the mantle and core to the opposite side of the Earth.

- *Secondary* or *shear (S)* waves are half as fast and reach the surface next. Like P waves they are high-frequency but shake like a skipping rope. They can travel through the mantle, but not the core, so cannot be measured at a point opposite the focus or epicentre.

- Surface *Love (L)* waves are the slowest waves and cause most of the damage.

- *Rayleigh (R)* waves radiate from the epicentre in complicated low-frequency rolling motions.

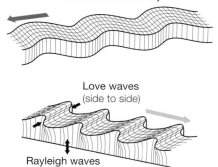

Primary waves 'push' through crust, mantle and core

compressions

Secondary waves 'shake' through crust and mantle only

Love waves (side to side)

Rayleigh waves (up and down)

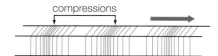

**Key**
➡ from focus   ➡ from epicentre

⊙ **Figure 2**  *Primary, Secondary, Love and Rayleigh waves*

## Tsunamis

From the Japanese *tsu* (meaning harbour) and *nami* (meaning wave), *tsunamis* are sometimes called *tidal waves*. This is misleading as they have nothing to do with tides, or even the wind like normal ocean waves. Tsunamis are usually generated by seismic activity such as ocean floor earthquakes or submarine volcanic eruptions (Figure **3**). Massive landslides into the sea, submarine debris slides, and even meteor or asteroid strikes can also cause them.

Tsunamis differ from normal ocean waves in many respects.

◆ The wave height is very low (less than 1 m), but on reaching the shore they can rise to over 25 m.

◆ The wavelength (distance between crests) is very long. In open water the wavelength can be anything from 100 to 1000 km.

◆ They travel very quickly, reaching speeds between 640 and 960 km per hour.

◆ They usually consist of a series of waves, with the first not always being the biggest.

◆ A long time between each wave (the wave period) – between 10 and 60 minutes.

◆ On approaching the coast, especially if funnelled into an inlet, they will slow down and pile up as a massive wall of water before breaking.

> **Did you know?**
> Around 95 per cent of the world's earthquakes occur along or near tectonic plate margins.

Up to 90 per cent of tsunamis are associated with seismicity along the 'Pacific Ring of Fire'. The effects of tsunamis will vary according to the factors above but are also influenced by population density, the coastal relief and the land-use of the coastal region.

Effective warning systems do exist, for example, the Pacific Tsunami Warning System based in Hawaii, which give many hours warning of approaching waves following 'important', 'major' and 'serious' seismic events. But with no such warning, the first sign – the apparent draining away of the sea in front of the tsunami (known as a drawdown) – will be too late. The tsunamis will sweep away its victims, uproot vegetation, destroy wooden buildings and bridges, and wash boats hundreds of metres inland. The destructive power of tsunamis is legendary, especially if caused by ocean floor earthquakes.

◀ **Figure 3** *Characteristics of a tsunami*

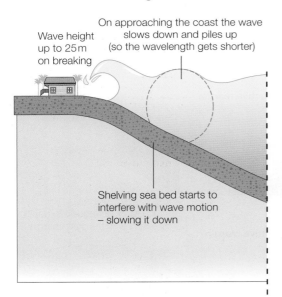
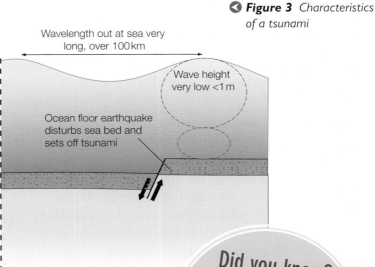

Wave height up to 25 m on breaking

On approaching the coast the wave slows down and piles up (so the wavelength gets shorter)

Shelving sea bed starts to interfere with wave motion – slowing it down

Wavelength out at sea very long, over 100 km

Wave height very low <1 m

Ocean floor earthquake disturbs sea bed and sets off tsunami

## ACTIVITIES

Access the website of the USGS National Earthquake Information Centre (http://earthquake.usgs.gov). They publish up-to-date maps of recent seismic patterns for the last few days or weeks. Find out what seismic activity is happening now.

> **Did you know?**
> Tsunamis generated by the eruption of Krakatoa in 1883 killed 36 000 people. The waves caused catastrophic flooding and were estimated to be over 40 m high.

In this section you will learn about the spatial distribution of earthquakes and their magnitude, frequency, regularity and predictability

## Spatial distribution of earthquakes

The relationship between volcanoes and tectonic plate margins has already been established (see 5.4). Earthquake clustering along plate margins would, in consequence, be expected (Figure **1**). But closer examination of these linear belts reveals further information about the spatial distribution of earthquakes:

◆ Some plate margins are more seismically active than others.

◆ Some plate margins show a wider extent.

Furthermore, both the intensity and depth of earthquakes varies according to the type of plate margin – whether constructive, destructive or conservative (Figure **2**).

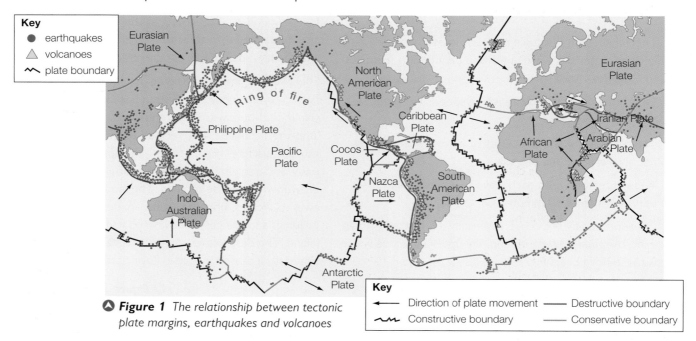

**Figure 1** *The relationship between tectonic plate margins, earthquakes and volcanoes*

**Key**
→ Direction of plate movement    — Destructive boundary
〰 Constructive boundary    — Conservative boundary

## Magnitude, frequency, regularity and predictability

The magnitude of an earthquake has long been measured by the Richter scale. This logarithmic scale applied a simple mathematical formula to interpret the distance moved by the vibrating pen on a seismograph. The scale starts at 0 and each number is ten times the magnitude of the one before it, so a slight increase in value equates to an enormous effect on the ground. Destructive earthquakes tend to have a value in excess of 6 – they rarely exceed 9 (9.5 in Chile in 1960 is the highest recorded). However, values above Richter 8 would send a seismograph into a frenzy. So, since 1977, an adaptation of the scale has been used that uses complex mathematics to calculate the total energy released in an earthquake – the Moment Magnitude Scale.

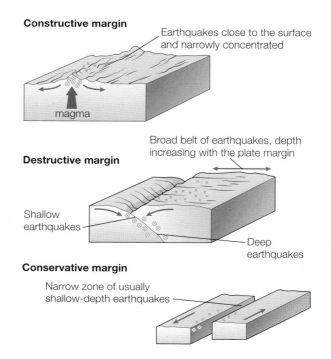

**Figure 2** *The extent and depth of earthquakes associated with constructive, destructive and conservative plate margins*

Just as with volcanic eruptions, the frequency and regularity of earthquakes show no predictability. While an understanding of plate tectonics allows seismologists to know which areas are most at risk, it is impossible to predict either exactly when or where a seismic disaster will strike. For example, even well-documented evidence of migrating sequences of 'stress-triggered' earthquakes along Turkey's North Anatolian Fault, and the correct assumption that Izmit would, at some point, be hit (as it was in 1999), gave no certainty that Istanbul is next in line. It can only be calculated that there is a 62 per cent probability that Istanbul will be struck by an earthquake of magnitude 8+ in the next 30 years.

## Mitigation – the best defence?

Scientists have identified a number of events that can occur before an earthquake strikes. These include:

◆ microquakes before the main tremor

◆ bulging of the ground

◆ decreasing radon gas concentrations in groundwater

◆ raised groundwater levels

◆ electrical and magnetic changes within local rocks

◆ increased argon gas content in the soil

◆ curious animal behaviour.

Even in areas where earthquakes are expected, issuing warnings based on such flimsy evidence could never be risked as panic and chaos may result. But risk assessment, contingency planning and earthquake engineering is possible.

New buildings, bridges, roads, pipelines and power lines can be designed to withstand ground shaking (Figure **3**). Geographic Information Systems (GIS) are used to prepare hazard maps that show the areas at greatest risk and aid planning of urban growth and development. Public education may be as simple as providing earthquake preparation checklists and practising evacuation drills in schools, offices and public buildings. In Japan, these are as common as practice fire drills.

## Modified Mercalli Intensity (MMI) scale

The actual intensity of damage caused is measured on the Modified Mercalli Intensity scale. This scale uses observations on the ground of the actual impact of the earthquake. The twelve-point scale ranges in single units from I (imperceptible) to XII (catastrophic). For example:

◆ I – 'imperceptible' is measured only by seismometers.

◆ IV – 'moderate' is likely to rattle doors and windows.

◆ VIII – 'destructive' will cause considerable damage. Chimneys and monuments will fall.

◆ XII – 'catastrophic' results in complete destruction. Waves would even be seen on the ground.

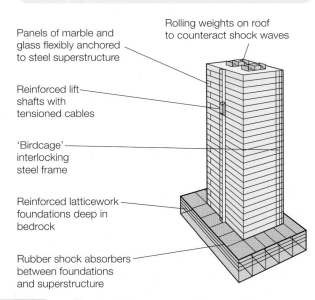

Panels of marble and glass flexibly anchored to steel superstructure

Rolling weights on roof to counteract shock waves

Reinforced lift shafts with tensioned cables

'Birdcage' interlocking steel frame

Reinforced latticework foundations deep in bedrock

Rubber shock absorbers between foundations and superstructure

▲ *Figure 3* *A modern earthquake-proof building*

## ACTIVITIES

1 Study Figures **1** and **2**.

   **a** Describe and comment on the global distribution of earthquakes.

   **b** Compare the number, horizontal spread and depth of earthquakes along constructive, destructive and conservative plate margins.

2 Study Figure **3**.

   **a** Describe how buildings can be designed to withstand earthquakes.

   **b** 'The best approach to dealing with earthquakes is to learn to live with rather than try to predict them.' How far do you agree with this statement? Explain your answer.

## STRETCH YOURSELF

To what degree can the level of earthquake prediction available today help governments who are faced with potential future earthquakes? Explain your answer.

In this section you will learn about the primary and secondary impacts of the Indian Ocean tsunami in 2004

## The impacts of seismic activity

Earthquakes are major natural hazards that can cause death and destruction, particularly in urban areas. Collapsing buildings bury people, with many more dying in the fires, chaos and confusion that follows, and in the smaller, crust-settling aftershocks that destroy weakened structures. The consequences of an earthquake will depend on:

- the magnitude and depth of the earthquake
- geological conditions
- the distance from the epicentre
- population density, preparedness and education
- the design and strength of buildings
- the time of day
- the impact of indirect hazards such as fires, landslides and tsunamis.

Earthquakes and tsunamis have immediate and often well-publicised consequences. The long-term impacts are less well reported and may cost an economy billions (US$) to put right (Figure **1**).

**Did you know?**
The earthquake that triggered the 2004 Indian Ocean tsunami released energy equivalent to 23 000 Hiroshima-type atomic bombs! The tsunami waves travelled at the speed of a Boeing 747.

⏷ **Figure 1** *Potential environmental, social, economic and political impacts of earthquakes*

| Primary impacts – immediate impact | Secondary impacts – impacts resulting as a direct consequence | Long-term impacts |
|---|---|---|
| Ground shaking causes:<br>• buildings and bridges to collapse<br>• windows to shatter<br>• power lines to collapse<br>• road and railway damage<br>• water/gas mains and sewers to fracture. | Fires caused by broken gas pipes and power lines are difficult to put out<br><br>Emergency services are hindered<br><br>Diseases spread from contaminated water | Higher unemployment as not all businesses recover<br><br>Repair and reconstruction of buildings and infrastructure may take months or years<br><br>Immediate suffering leading to longer-term illness and/or reduced life expectancy |
| Schools, colleges and universities destroyed | Education suspended for immediate future | A 'lost generation' affecting the ability to develop local/regional economy |
| Immediate deaths and injuries from crushing, falling glass, fire and transport accidents | Bodies not buried or cremated spread diseases such as cholera<br><br>Injuries may result in long-term disability or death if not treated promptly | Trauma and grief may take months or years from which to recover<br><br>Disability and reduced life expectancy |
| Shocked, hungry people forced to sleep outside | Non-governmental organisations (NGOs) provide tents, water and food | Emergency prefabricated homes may become permanent fixtures |
| Slope failures set off landslides and avalanches | Further deaths and injuries<br><br>Flooding from blocked rivers creating 'quake lakes' | Loss of farmland and food production<br><br>Permanent disruption to natural drainage patterns |
| **Liquefaction** of saturated soils | Building foundations subside, resulting in more collapses, deaths and injuries | Repair to buildings are difficult and reconstruction expensive |
| Damage to power stations | Power cuts affect emergency services, restricting immediate medical care | Reconstruction of power stations is very expensive |
| Panic, fear and hunger | Civil disorder, looting and direct intervention by civil authorities, such as police and army | Problem restoring trust in neighbours and civil authorities |

# The Indian Ocean tsunami, December 2004

On 26 December 2004, an earthquake with a magnitude of between 9.0 and 9.3 (the second biggest in history) ruptured the ocean floor 240 km north-west of Sumatra, Indonesia (see 3.8). The 15 m vertical displacement of 1600 km of sea bed (see Figure **3**, 5.9), in two phases over a period of two to three minutes along the subduction zone where the Indian Plate slides under the Eurasian Plate, caused massive tsunamis that travelled thousands of kilometres across the Indian Ocean. This cost the lives of approximately 300 000 people across two continents (Figure **2**).

There was no effective tsunami warning system in the Indian Ocean – radio and television alerts were not broadcast in Thailand until nearly an hour after the first waves hit! Only 15 minutes after the initial earthquake, waves of up to 20 m began to crash into Sumatra – the first of many coastlines to be devastated (Figure **3**).

▲ **Figure 2** *The tsunami wall of water hitting Ao Nang, Thailand*

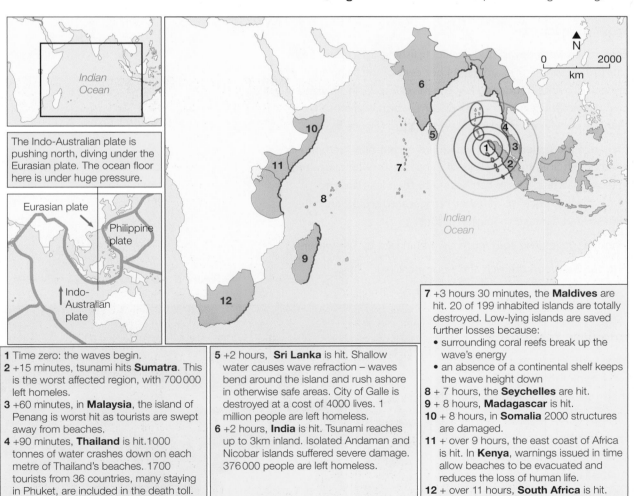

Indian Ocean

The Indo-Australian plate is pushing north, diving under the Eurasian plate. The ocean floor here is under huge pressure.

Eurasian plate

Philippine plate

Indo-Australian plate

Indian Ocean

**1** Time zero: the waves begin.

**2** +15 minutes, tsunami hits **Sumatra**. This is the worst affected region, with 700 000 left homeles.

**3** +60 minutes, in **Malaysia**, the island of Penang is worst hit as tourists are swept away from beaches.

**4** +90 minutes, **Thailand** is hit. 1000 tonnes of water crashes down on each metre of Thailand's beaches. 1700 tourists from 36 countries, many staying in Phuket, are included in the death toll.

**5** +2 hours, **Sri Lanka** is hit. Shallow water causes wave refraction – waves bend around the island and rush ashore in otherwise safe areas. City of Galle is destroyed at a cost of 4000 lives. 1 million people are left homeless.

**6** +2 hours, **India** is hit. Tsunami reaches up to 3km inland. Isolated Andaman and Nicobar islands suffered severe damage. 376 000 people are left homeless.

**7** +3 hours 30 minutes, the **Maldives** are hit. 20 of 199 inhabited islands are totally destroyed. Low-lying islands are saved further losses because:
- surrounding coral reefs break up the wave's energy
- an absence of a continental shelf keeps the wave height down

**8** + 7 hours, the **Seychelles** are hit.

**9** + 8 hours, **Madagascar** is hit.

**10** + 8 hours, in **Somalia** 2000 structures are damaged.

**11** + over 9 hours, the east coast of Africa is hit. In **Kenya**, warnings issued in time allow beaches to be evacuated and reduces the loss of human life.

**12** + over 11 hours, **South Africa** is hit.

▲ **Figure 3** *Countries affected by the 2004 Indian Ocean tsunami*

## Primary impacts

◆ Up to 300 000 people died – thousands of bodies have never been recovered.

◆ Vegetation and top soil were removed up to 800 m inland.

◆ Infrastructure was destroyed. For example, the Andaman and Nicobar Islands were all but cut off as jetties were washed away.

◆ Coastal settlements were devastated. For example, the city of Banda Aceh in Sumatra was obliterated.

**Figure 4** *Debris carried by the powerful waves along the southern Indian coastline*

## Secondary impacts

◆ Widespread homelessness. For example, 500 000 people were forced into refugee camps in the worst hit region of Aceh Province, Indonesia.

◆ Economies were devastated, including fishing, agriculture and tourism sectors. For example, 44 per cent of the population living in Aceh Province, Indonesia lost their livelihoods. In Thailand the cost to the fishing industry was £226m, which was much more than the widely reported impact on tourism (Figure **5**).

◆ Negative multiplier effects weakened economies further. For example, in Sri Lanka the loss of deep-sea trawlers resulted in fewer catches. These catches were then less likely to reach market because of fewer refrigerated trucks and the widespread damage to infrastructure.

◆ Water supplies and soils were contaminated by salt water.

◆ The gap between the rich and poor increased.

**Figure 5** *Boat stranded in Bandah Aceh, Indonesia*

## Immediate responses

◆ Massive international relief efforts were established that involved more than 160 aid organisations and UN agencies (Figure **6**).

◆ Foreign military troops provided assistance. For example, the Australian Air Force improved air traffic control at Banda Aceh airport, Indonesia.

## Longer-term responses and their effectiveness

- Large-scale programmes of reconstruction were implemented, but there were still many thousands left in tents one year on (Figure **6**).

- Political barriers slowed aid distribution. For example, in Sri Lanka aid was delayed to areas held by rebel Tamil Tigers.

- Inequalities were highlighted. For example, in India, the *dalits*, an underclass, did not receive as much help from the government.

- Tourist resorts were quickly rebuilt, such as Phuket, Thailand. But elsewhere some native coastal communities were forced out by new developments, such as Andhra Pradesh, India.

- A United Nations group set up a tsunami warning system for the Indian Ocean. However, as individual governments are responsible for sending out their own alerts there are huge contrasts in their possible effectiveness. For example, there are sirens on tourist beaches in Thailand, yet isolated rural communities receive their warnings by radio.

- Education on tsunami awareness began in schools.

- Practice drills and evacuation plans were established.

- Coastal zones were hazard-mapped to identify areas most at risk.

▲ **Figure 6** *UN refugee camp in Aceh Province, Indonesia*

▼ **Figure 7** *2004 Indian Ocean tsunami; number of dead and missing*

| Country affected | Dead and missing |
| --- | --- |
| Indonesia | 236 169 |
| Sri Lanka | 31 147 |
| India | 16 513 |
| Thailand | 5395 |
| Somalia | 150 |
| Myanmar | 61 |
| Maldives | 82 |
| Malaysia | 68 |
| Tanzania | 10 |
| Seychelles | 3 |
| Bangladesh | 2 |
| Kenya | 1 |
| **Total** | **289 601** |

### ACTIVITIES

1 'An earthquake is as much a human as a natural disaster'. Discuss this statement.

2 Study Figure **3**.

  a Describe the cause of the 2004 Indian Ocean tsunami with reference to plate tectonics.

  b Describe and explain the relationship between distance and the impact on the named countries affected.

**S** 3 Study Figure **7**. Use this data and also Figure **3** to assess which countries suffered the most. Include reference to any specific link between the dead and missing toll and the time taken for the waves to reach individual countries.

In this section you will learn about short- and long-term responses to seismic hazards

## Short- and long-term responses to seismic hazards

The generally chaotic nature of seismic hazards makes it impossible to predict both when and exactly where a disaster will strike. But our understanding of plate tectonics does allow us to focus on preparedness in those areas identified as most at risk. It is preparedness that is our best defence for mitigating against the impacts of these hazards by:

◆ identifying areas of high risk, generally along certain sections of active plate margins

◆ limiting development wherever possible to **life-safe** earthquake-proof designs

◆ building tsunami defence walls wherever possible

◆ contingency planning to cope with the aftermath of a seismic event.

As with volcanic eruptions, each new seismic event studied teaches us more, and so preparedness, short-term (immediate) and long-term responses generally improve.

### Port-au-Prince earthquake, Haiti, January 2010

Haiti is a multi-hazard environment that is not only at risk of being hit by tropical storms (see 5.15), but is also a seismically active LDE. It is located where the Caribbean and North American plates slide past one another in an east-west direction (Figure **1**). This complex strike-slip fault follows a conservative plate margin and is not unlike California's San Andreas fault system. In 2008, scientists made an alarming discovery. Based on the average movement of 7 mm since the last earthquake of 1751, it was found that the plates were jammed. On 12 January 2010 the stress of the surrounding rocks was finally overcome and the plates were released, resulting in a magnitude 7 earthquake.

The epicentre was 24 km south-west of the capital Port-au-Prince (home to 2 million people) and had a shallow focus of 13 km. The equivalent energy of an atomic bomb was transmitted outwards, violently shaking the entire country (Figure **2**). The primary and secondary impacts were ruinous (Figures **3** and **4**).

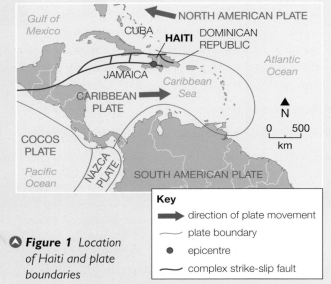

**Figure 1** *Location of Haiti and plate boundaries*

**Key**
→ direction of plate movement
⌒ plate boundary
● epicentre
⌒ complex strike-slip fault

**Figure 2** *Haiti earthquake intensity map using the Modified Mercalli Intensity (MMI) scale*

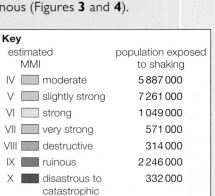

**Key**

| estimated MMI | | population exposed to shaking |
|---|---|---|
| IV | moderate | 5 887 000 |
| V | slightly strong | 7 261 000 |
| VI | strong | 1 049 000 |
| VII | very strong | 571 000 |
| VIII | destructive | 314 000 |
| IX | ruinous | 2 246 000 |
| X | disastrous to catastrophic | 332 000 |

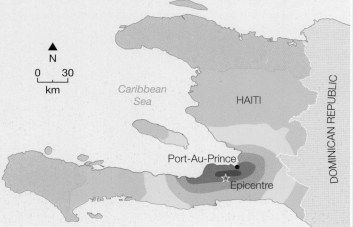

| Primary impacts | Secondary impacts |
|---|---|
| Much of Port-au-Prince was flattened in less than 60 seconds. Over 230 000 lives were lost. | Strong aftershocks were recorded, including a 6.1 magnitude earthquake on 20 January. |
| Fifty per cent of densely packed and poorly built concrete buildings collapsed, including key government buildings such as the police headquarters and the Palais Législatif (parliament building). The Palais National vwas damaged beyond repair (Figure **5**). | With the loss of hundreds of civil servants and destruction of ministries, the Haiti government was crippled. |
| Over 180 000 homes were damaged and 1.5 million people were made homeless. Over 600 000 people left Port-au-Prince – most to live with host families. Nearly 5000 schools were damaged or destroyed. | With the devastation of the police force and destruction of the main prison, much of the city became lawless. |
| Liquefaction on looser sediments caused building foundations to subside. Infrastructure was severely damaged – the main port subsided and became unusable, roads were cracked and blocked by building debris. | By the first anniversary of the earthquake, cholera had killed over 1500 and 1.5 million people were still homeless. |

**Figure 3** *Primary and secondary impacts of the 2010 Port-au-Prince earthquake*

**Figure 4** *High-density, poorly built concrete buildings destroyed in Port-au-Prince*

**Figure 5** *Palais National (National Palace) after the 2010 earthquake; the palace had to be demolished in 2012*

### Short-term (immediate) responses

◆ Rescue efforts – international search teams struggled within the dense and congested urban environment. Local people employed by the UNDP (United Nations Development Project) helped to find and rescue survivors out from the debris and to clear roads.

◆ Infrastructure – in Port-au-Prince, the US military took control of the airport to speed up the distribution of aid and reopened one of the two piers in the port.

◆ Security – 16 000 UN troops and police restored law and order, coordinated by a new UN/US Joint Operations Tasking Centre.

◆ Food – in absence of local food markets, the UN World Food Programme provided basic food necessities. Farmers were given immediate support before the spring planting season.

◆ Water – among other interventions, the UK Disaster Emergency Committee (DEC) provided bottled water and purification tablets for over 250 000 people.

◆ Health – emergency surgeries were established to perform life-saving operations. The DEC provided over 100 000 consultations and built 3000 latrines.

◆ Shelter – around 1.5 million homeless people were accommodated in over 1100 camps, in emergency shelter, mostly in the form of tarpaulins. Over 100 000 of these people were at critical risk from storms and flooding (Figure **6**).

## Think about

### Haiti – a multi-hazard environment

Haiti is currently the poorest country in the western hemisphere with a per capita income of less than one-tenth of the Latin American average – 54 per cent of the population live in abject poverty. It is a multi-hazard environment – it is seismically active and also lies in the middle of the hurricane belt (and so subject to severe storms from June to October). Flooding, landslides and periodic droughts further afflict a country that already experiences political instability, corruption, poor infrastructure, social inequality, exclusion and unrest.

Haiti suffered both economically and socially during the US-supported dictatorship from the 1950s until the 1980s. In fact, it took until 1996 for René Préval to be the first head of state to receive power peacefully from a predecessor in office and was also the first Haitian president since independence to serve a full term in office. His two terms in office were severely tested by food riots and Hurricane Gustav in 2008 and also by the earthquake and Hurricane Tomas in 2010.

▼ **Figure 6** Tent city – Haiti 2010

## Long-term responses

◆ Aid – the Haiti Relief Fund manages an US$11.5 billion reconstruction package with controls in place to prevent corruption. Reconstruction is due to be completed by 2020.

◆ Food – the farming sector was reformed to encourage greater self-sufficiency and less reliance on food imports.

◆ Health – a shift in emphasis to focus on follow-up care, including mental health, took place within local health care services.

◆ Buildings – hospitals, schools and government buildings were rebuilt to new life-safe building codes (Figures **7** and **8**). Local people were employed as construction workers. Slums were demolished and high-risk areas, such as unstable hillsides, were avoided when new settlements were built. The new homes are more affordable, safe and sustainable.

◆ Economy – some economic activities were moved away from Port-au-Prince to less earthquake-prone areas. A UN strategy was developed to create new jobs in clothing manufacture, tourism and agriculture, and also to reduce the effects of uncontrolled urbanisation.

▲ **Figure 7** *Iron Market, Port-au-Prince after the 2010 earthquake*

▲ **Figure 8** *Iron Market, Port-au-Prince after rebuilding*

## ACTIVITIES

1 Study Figures **1** and **2**. Describe and suggest reasons for the overall pattern shown by the Modified Mercalli Intensity scale map.

2 Describe the destruction shown in the photograph in Figure **4**.

3 Outline the impacts of the earthquake on the people living in Port-au-Prince.

## STRETCH YOURSELF

In November 2010, Hurricane Tomas hit Haiti. By May 2011 hundreds of thousands of people were still in camps and were highly vulnerable to the spread of cholera, and the impact of storms, flooding and landslides. Widespread unemployment and underemployment persist – more than two-thirds of the labour force do not have formal jobs.

Evaluating all evidence presented in this section, critically challenge the contention of some commentators that Haiti is `backward and so beyond hope'.

In this section you will learn about a recent seismic event – the 2011 Tōhoku earthquake and tsunami, Japan

> **Some natural disasters change history. Japan's tsunami could be one...**

(*The Economist*, 19 March 2011)

### The Tōhoku earthquake and tsunami, Japan, March 2011

On average Japan records 1500 earthquakes every year – around one-third of the world's total! So the biggest earthquake ever recorded in Japan had world-changing significance. And not just historical significance, it had immediate global impacts on the Earth. The Tōhoku earthquake and tsunami:

◆ moved the entire island of Honshu 2.4 metres closer to North America

◆ shifted the Earth's axis at least 10 cm

◆ made Earth days shorter by 1.8 microseconds

◆ calved 125 square kilometres of icebergs from the Antarctic coast

◆ caused visible waves in Norwegian fjords.

At 2.46 p.m. Tokyo time, on Friday 11 March, a magnitude 9.0 earthquake occurred under the Pacific Ocean, 100 km due east of Sendai on northern Honshu's eastern coast (Figure **1**).

A 400–500 km segment of the North American Plate, which was being dragged down by the subducting Pacific Plate, suddenly slipped upwards between 5 and 10 metres. The resulting sea water displacement caused a tsunami to spread in all directions – at hundreds of kilometres per hour.

Japan's tsunami warning system kicked in, but people along a 3000 km stretch of coastline had just minutes to escape. The first wave hit the north-east coast only 30 minutes after the earthquake.

There were ten waves, each about 1 km apart as they reached the shallower coastal water. Here they slowed and piled up, reaching a staggering 10 m in height.

They overwhelmed tsunami defence walls and surged up to 10 km inland (Figure **2**).

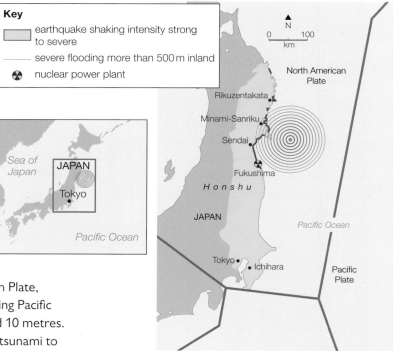

**Figure 1** The epicentre of the earthquake and the area hit by the tsunami

**Key**
- earthquake shaking intensity strong to severe
- severe flooding more than 500 m inland
- nuclear power plant

**Figure 2** The Japanese tsunami, 2011

## Primary effects

◆ Ground shaking caused buildings to collapse. Some were set ablaze by broken gas and petrol pipes.

◆ The tsunami swept inland, mainly along the north-east coast, causing devastation to everything in its path – boats, buildings, vehicles, trees.

◆ It flooded an area of almost 500 square kilometres.

◆ When the waters receded, whole cities were in ruins. Trains had vanished. Ships and boats lay tossed like toys.

◆ In Tokyo, skyscrapers had 'started shaking like trees'. But their earthquake-proof design meant damage was limited. In Ichihara, a commuter town of Tokyo, an oil refinery was engulfed in flames as fuel tanks exploded.

◆ In Sendai, areas near the sea were badly damaged, but the city centre, just inland, was largely unscathed. Rikuzentakata was almost completely submerged and was almost totally destroyed. In Minami-Sanriku, half the population of 17 000 died, and few buildings were left standing (Figure **3**).

◆ Over 18 000 were dead or missing, mainly due to the tsunami, although Japan's tsunami warning system saved many lives.

## Secondary effects

◆ Half a million people were homeless. For weeks, 150 000 people lived in temporary shelters.

◆ Over 1 million homes were left without running water and almost 6 million homes lost their electricity supply.

◆ There were shortages of food, water, petrol and medical supplies.

◆ In the two weeks after the earthquake, there were more than 700 aftershocks, causing concern and further damage.

◆ Explosions and radiation leaks at the Fukushima Daiichi nuclear power plant in the days after the earthquake spread fear around the world. The earthquake severed the power supply to the cooling system; the tsunami then destroyed the back-up generators. Workers struggled to prevent a meltdown (Figure **4**).

◆ Fears of a nuclear disaster caused panic-selling across global stock markets.

**Figure 3** *The destroyed town of Minami-Sanriku*

### Did you know?
Mobile phones and news helicopters mean it was the best recorded tsunami in history. Amateur footage of the disaster spread worldwide via social networking sites.

**Figure 4** *The Fukushima Daiichi nuclear power plant after the earthquake and tsunami. Reactors 1 to 4 are from right to left. Three of these reactors overheated, causing meltdowns that eventually led to explosions releasing large amounts of radioactive material into the air.*

## Immediate responses

◆ In freezing temperatures, survivors huddled in shelters and hoarded supplies as rescue workers searched the mangled coastline of submerged homes.

◆ Helicopter crews plucked survivors from rooftops and flooded farmland.

◆ 100 000 soldiers were mobilised to establish order, organise rescue work and distribute blankets, bottled water, food and petrol.

◆ Offers of aid poured in from other countries, including the USA and China.

◆ The UK sent 63 search and rescue specialists, two rescue dogs, and a medical support team to help. People were rescued after being trapped for several days (Figure **5**).

◆ An exclusion zone was set up around the Fukushima Daiichi nuclear plant. Homes were evacuated and iodine tablets, to prevent radiation sickness, were distributed (Figure **6**).

◆ The Fukushima Daiichi explosions prompted a government-ordered shutdown of the majority of Japanese nuclear power plants.

◆ There were no reports of looting or violence.

## Long-term responses

◆ Japan coped well with the earthquake. But the tsunami defences were inadequate against the extraordinary height and force of the water. Future contingency planning must consider whether defences should be built to defend the coast against a similar high magnitude, low frequency event.

◆ In 2013 Japan unveiled a new, upgraded tsunami warning system because many people had underestimated their personal risk and/or assumed that the tsunami would be as small as others previously experienced.

◆ The Japanese government set up an advisory body called the Reconstruction Design Council to plan a long-term growth in the Tōhoku region. Special Zones for Reconstruction were designated with relaxed planning regulations to encourage rapid rebuilding and tax incentives offered to promote new investment in industry and commerce.

◆ Prior to 2011, nuclear power provided 30 per cent of Japan's electricity – the shortfall was met through increasing its dependence on fossil fuels, particularly imports of oil and gas. Court orders brought about by anti-nuclear groups have meant that it was only in 2016 that reactors were issued licenses to restart.

▲ **Figure 5** *Japanese rescue workers in Ishiimaki carry a survivor who had been buried alive for five days*

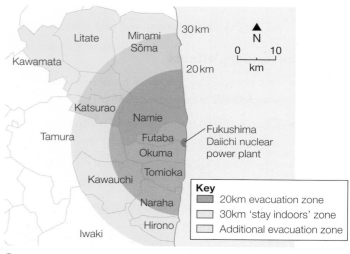

▲ **Figure 6** *Fukushima Daiichi evacuation (20 km) and stay indoors orders (30 km) were repeatedly revised and extended in the immediate aftermath of the explosions*

▶ **Figure 7** *200 000 children now suffer from precancerous thyroid abnormalities, primarily nodules and cysts*

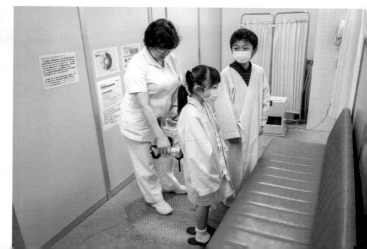

- Following the Chernobyl nuclear accident of 1986, the number of thyroid cancer cases among children started to increase rapidly after four to five years had passed. Some are of the opinion that, because of the Fukushima Daiichi disaster, a similar increase in the incidence of juvenile thyroid cancer may occur (Figure **7**).

- Five years after the Fukushima Daiichi explosions, 100 000 residents had still not returned home. In 2019 the evacuation order for one of the two towns where the nuclear plant is located was partially lifted.

- Radioactive rubble and refuse was still awaiting permanent disposal (Figure **8**). No Japanese prefecture was prepared to accept it.

- The total damages from the earthquake and tsunami are estimated at US$300 billion, making it the most costly natural disaster in history. Japan was already the most heavily indebted country in the industrialised world and the repair bill has had to be met by more government borrowing.

## ACTIVITIES

1. Study Figure **6**.
   a. Assess the effectiveness of the mapping technique adopted on this map.
   b. Describe the pattern shown.
   c. Suggest a reason that might explain extensions to the evacuation and stay indoors orders to the north and west.

2. The nature of tsunami damage varies considerably between areas. With reference to both the 2004 Indian Ocean (5.11) and 2011 Japanese tsunamis, discuss the human and physical factors that may be responsible for this variation.

3. For both HDEs and LDEs, which should come first – prediction or contingency planning? Explain your answer with reference to tsunamis or any other seismic hazard.

## STRETCH YOURSELF

'Those with least suffer most.' Critically evaluate this contention with reference to named seismic events studied.

### Think about

#### Challenges facing Japan

Japan is an island nation in the north-western Pacific Ocean. It is densely populated and arguably the most geographically hazardous nation on Earth – subject to the full range of tectonic hazards, in addition to late summer typhoons. Climatically varied, extensively forested, but chronically short of good agricultural land, its energy and mineral resources are scarce. Yet it transformed itself from a poor agricultural nation in the nineteenth century into a first-rank global power. Furthermore, out of the ashes of the Second World War, Japan has built one of the richest and most successful democracies in history.

But although outsiders may view the Japanese as diligent and disciplined and also culturally, economically and technologically sophisticated, Japan is not without its problems. The UN predicts that the country will reach an economic crisis point by 2050 when the dependent population will outnumber the economically active population. Japan's greatest resource may well be its remarkable people, but economic and social provision for this ageing, yet declining population will become ever more challenging.

⬣ *Figure 8* Workers haul a bag of radiation-contaminated leaves during a clean-up operation in the abandoned town of Naraha, just outside the exclusion zone surrounding the Fukushima Daiichi nuclear plant

In this section you will learn about the characteristics and causes of hazards associated with tropical storms

## What is a tropical storm?

The giant swirl of cloud in Figure **1** is instantly recognisable as a hurricane or tropical storm (cyclone). This is not just any hurricane – it is Hurricane Katrina, which killed over 1800 people and caused more economic damage in the USA than any previous natural disaster.

Despite the awesome power of a hurricane and the destruction that it can cause, it is a perfectly natural weather event and has to be expected in certain latitudes at certain times of the year. Indeed, hurricanes play an important role in the redistribution of heat across the world, transferring warm conditions from the tropics towards the poles. Without hurricanes the weather conditions and climate of the world would be very different.

## What are the characteristics of tropical storm?

A **tropical storm** is also known as a cyclone (India), hurricane (North Atlantic) and typhoon (south-east Asia), which can extend to 500 km in diameter and cause extensive damage and loss of life to coastal regions in many parts of the tropics. By definition, a tropical storm must have average wind speeds in excess of 120 km/h (75 mph).

⬀ **Figure 1** *Satellite photo of Hurricane Katrina*

Figure **2** shows a cross-section through a typical storm. Notice that there is a degree of symmetry around its central point or eye. The most powerful and damaging part of a tropical storm is the menacing bank of cloud that rings the central eye. This is called the *eye wall* and people who have experienced such a storm say that it 'hits like a train'! As you can see from Figure **2**, cloud and rain extend in a series of waves well beyond the eye wall. Within a tropical storm, tornadoes are commonly formed. Their highly localised nature makes them difficult to predict and their impact can be highly destructive.

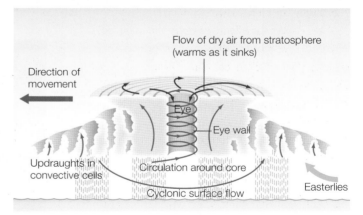

⬀ **Figure 2** *Cross-section through a tropical storm*

## Factors in the distribution of tropical storms

Figure **3** shows the distribution and frequency of tropical storms. Tropical storms do not occur randomly across the world. The vast majority are formed in the tropics, although they do extend beyond this region in places, for example, China, Japan and the eastern seaboard of the USA.

⬂ **Figure 3** *The distribution and frequency of tropical storms. The percentages show the proportion of the total global number of storms in each region.*

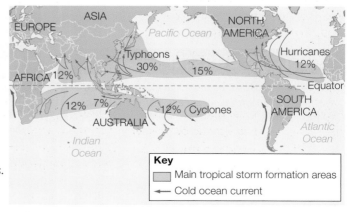

Several factors are important in affecting the distribution of tropical storms.

- Oceans – tropical storms derive their moisture (through the transfer process of evaporation) and energy (in the form of latent heat, see 2.3) from the oceans; there are clear links with the water cycle (see 1.2). This explains why tropical storms form and continue to develop over ocean areas and then peter out on reaching land.

- High temperatures – a sea-surface temperature in excess of 26 °C is required for tropical storm formation, which is why they are formed in low latitudes during the summer, when temperatures are at their highest.

- Atmospheric instability – tropical storms are most likely to form in regions of intense atmospheric instability, where warm air is being forced to rise. The ITCZ, where two limbs of the Hadley Cell converge to form low pressure on the ground, is a perfect spawning ground for tropical storms (see 1.3).

- Rotation of the Earth – a certain amount of 'spin' is needed to initiate the characteristic rotating motion of a tropical storm. The influence of the Earth's rotation on surface phenomena is called the Coriolis effect. This increases with distance away from the Equator and explains why tropical storms do not usually form in the region between 5 °N and 5 °S (close to the Equator).

- Uniform wind direction at all levels – winds from different directions at altitude prevent a tropical storm from attaining height and intensity. The vertical development is effectively 'sheared off' by the multidirectional winds.

Once a tropical storm has started to form, it will soon develop its distinct and clearly defined rotation. Warm, moist air rises rapidly in its centre, to be replaced by air drawn in at the surface (Figure **4**). A central vortex will develop as more and more air is drawn in and rises. The very centre of the storm (the eye) is often characterised by a column of dry, sinking air.

As the air rises, it rapidly cools. This leads to condensation and the formation of towering cumulonimbus clouds. Sometimes a number of isolated thunderstorms will coalesce to form a single giant storm. When condensation occurs, latent heat is released, which effectively powers the storm.

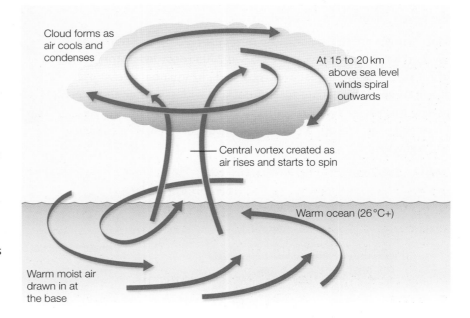

Cloud forms as air cools and condenses

At 15 to 20 km above sea level winds spiral outwards

Central vortex created as air rises and starts to spin

Warm ocean (26 °C+)

Warm moist air drawn in at the base

◮ **Figure 4** *The formation of a tropical storm*

A tropical storm continues to grow and develop as it is driven by the prevailing winds across the oceans. Only when it reaches land and the supply of energy and moisture is cut off will the storm start to decay. Should it move back out over the ocean, it will become reinvigorated.

## What are the hazards associated with tropical storms?

When a tropical storm makes landfall it brings with it a deadly cocktail of high seas, strong winds and torrential rain. With a large proportion of the world's population living close to the coast, there is a significant potential for loss of life and damage to property.

### Strong winds

By definition, a tropical storm packs a powerful punch, with average wind speeds in excess of 120 km/h (75 mph), although gusts of over 250 km/h have been recorded at the eye wall. The strong winds are capable of causing significant damage and disruption by tearing off roofs, breaking windows and damaging communication networks (Figure **5**). Debris forms flying missiles whisked up by the wind. Damaged power lines often lead to widespread electricity cuts (power outages) and occasionally even fires. Debris strewn over roads can cause major transport disruption.

### Storm surges

This is a surge of high water, typically up to about 3 m in height, which sweeps inland from the sea, flooding low-lying areas (Figure **6**). It is caused by a combination of the intense low atmospheric pressure of the tropical storm (enabling the sea to rise vertically) together with the powerful, driving surface winds.

Storm surges are the major cause of widespread devastation and loss of life. (Hurricane Katrina in the USA in 2005 recorded a storm surge of 7.6 m, one of the largest ever recorded.) In addition to loss of life, storm surges can inundate agricultural land with saltwater and debris, pollute freshwater supplies and destroy housing and infrastructure. Enhanced coastal erosion can lead to the undermining of buildings and highways.

### Did you know?

In 2004 tropical storm Catarina became the first recorded hurricane in the South Atlantic. It came ashore in the Brazilian state of Santa Catarina with 145 km/h (90 mph) winds and torrential rainfall. Tropical storms do not usually form in this region due to the lack of the Earth's 'spin' and the fact that high-level winds tend to shear off the storms, preventing their formation. Could this be a reflection of global warming?

**Figure 5** *Damaged caused by strong winds, Cyclone Winston, Fiji, 2016*

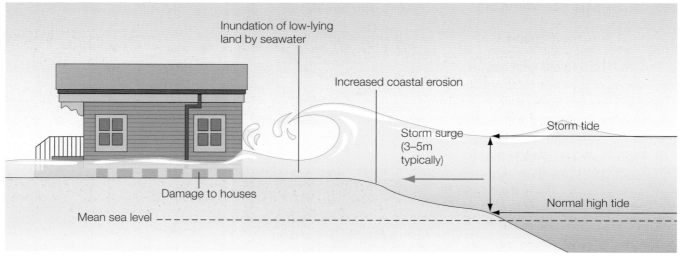

**Figure 6** *The impact of a storm surge on a coastline*

## Coastal and river flooding

The warm, humid air associated with a tropical storm can generate torrential rainfall, often in excess of 200 mm in just a few hours. This can trigger flash flooding at the coast, particularly in urban areas where surface water can overwhelm the drainage system. Urbanisation – with its system of drains, impermeable surfaces and high density of buildings – exacerbates the flood hazard by encouraging rapid overland flow and causing flash flooding.

As a tropical storm moves inland, it gradually weakens as its moisture (and energy) supply is cut off. However, it may still result in significant river flooding due to the intensity and sheer quantity of rain falling on the river basin. In August 2011, intense rainfall associated with Hurricane Irene caused widespread river flooding throughout New Jersey, USA, resulting in the evacuation of over one million people (Figure **7**). The damage was estimated at over US$1 billion.

▲ **Figure 7** *New Jersey after flooding by the Passaic River as a result of Hurricane Irene, 2011*

## Landslides

It has been estimated that up to 90 per cent of landslides each year are caused by heavy rainfall, and many are triggered by tropical storms. The intense rainfall increases pore water pressure (hydrostatic pressure within a slope), which weakens cohesion and triggering slope failure. The additional weight of water exacerbates the problem. In 1998 Hurricane Mitch – one of the North Atlantic's most powerful tropical storms – triggered multiple landslides that killed 18000 people in Central America (Figure **8**).

There is some evidence that load release caused by tropical storm-induced landslides may trigger earthquakes in tectonically stressed regions. For example, the 2010 Taiwan earthquake occurred just two years after Typhoon Morakot had dumped 300 mm of rain in just five days. Research carried out at the University of Miami found that 85 per cent of earthquakes of magnitude 6 and above occurred within the first four years after a very wet storm.

▲ **Figure 8** *Landslide caused by Hurricane Mitch, 1998*

## The Saffir–Simpson scale

Tropical storms be classified according to the Saffir–Simpson scale (Figure **9**), which was developed in 1971 as a means of enabling storms of different magnitudes to be compared. It is an absolute scale based on sustained wind speeds, and has five separate categories. Despite being widely used, particularly in the North Atlantic region, it is of limited value in assessing impact, as it does not take into account rainfall, or the area affected by a storm. So, a low-category cyclone that hits a densely populated urban area can be a far more damaging event than a high-category storm that makes landfall in a remote region.

| Saffir–Simpson Hurricane scale | | | | |
|---|---|---|---|---|
| **Category** | **Wind speed** | | **Storm surge** | |
| | mph | (km/h) | ft | (m) |
| **Five** | >156 | (>250) | >18 | (>5.5) |
| **Four** | 131–155 | (210–249) | 13–18 | (4.0–5.5) |
| **Three** | 111–130 | (178–209) | 9–12 | (2.7–3.7) |
| **Two** | 96–110 | (154–177) | 6–8 | (1.8–2.4) |
| **One** | 74–95 | (119–153) | 4–5 | (1.2–1.5) |
| **Additional classifications** | | | | |
| **Tropical storm** | 39–73 | (63–17) | 0–3 | (0–0.9) |
| **Tropical depression** | 0–38 | (0–62) | 0 | (0) |

▲ **Figure 9** *The Saffir–Simpson scale*

## Tropical storm frequency and magnitude

Currently, there is no clear evidence that the numbers or intensity of storms are increasing as global temperatures increase. Figure **10** shows no clear trend evident. In the last two decades there have been several years with a high number of tropical storms but the pattern is quite erratic – 2006, 2007 and 2009 recorded low numbers. There is also no evidence that hurricanes are becoming more intense (higher magnitudes).

Scientists would argue that a longer time period of study is necessary to see any long-term trends. There is some logic in expecting a warmer atmosphere to hold more moisture (and therefore result in more powerful storms), but this is countered by increased wind shear that acts as a negative feedback loop in nullifying tropical storm formation.

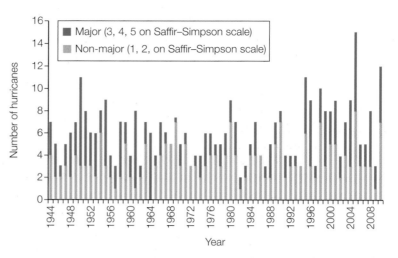

▲ **Figure 10** Hurricanes in the North Atlantic and Caribbean, 1944–2010

## Tropical storm regularity and predictability

To some extent tropical storms can be predicted – they are mostly restricted to the tropics and do not usually occur close to the Equator. They also mostly occur from late summer into autumn with a peak from August through to October. As Figure **11** illustrates, over a long period of time there is a degree of regularity (symmetry), although this does not mean that in any one year the pattern will be reflective of this.

Each year, NOAA (US National Oceanographic and Atmospheric Administration) publishes a prediction of hurricane activity for the forthcoming season. It uses a number of indicators, such as sea-surface temperatures, atmospheric conditions and short-term climatic cycles (such as El Niño and La Niña) to suggest the number of storms that might be expected. NOAA does not predict likely landfall or levels of activity for particular locations – these can only be predicted in the final few days before landfall happens.

For 2016, NOAA predicted a near-normal hurricane season, with four to eight hurricanes, with one to four predicted to be major.

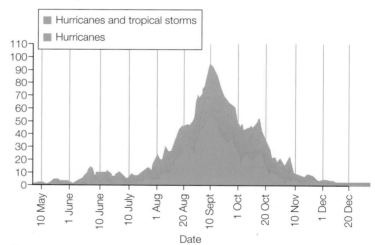

▲ **Figure 11** The seasonal occurrence of hurricanes and tropical storms in the Atlantic Basin. In this graph, the term 'tropical storm' is used to denote a storm of lesser intensity than a hurricane.

A hurricane return period (Figure **12**) is the frequency at which a certain intensity of hurricane can be expected within a given distance of a given location. For example, a return period of 20 years for a major hurricane means that *on average* during the previous 100 years, a Category 3 or greater hurricane passed within 100 km of that location about five times.

Once they have formed, tropical storms tend to follow similar tracks. However, this is an over simplification, as each storm develops its own characteristics and responds uniquely to the atmospheric and oceanographic conditions at the time. Whilst most tropical storms will erratically follow tracks similar to those in Figure **3**, some tracks are extremely eccentric, demonstrating the inherent dangers in attempting to make predictions (Figure **13**).

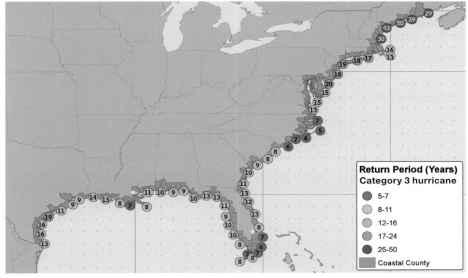

▲ *Figure 12* *Hurricane return periods (USA)*

**○** **Figure 13** *The unpredictable track of Hurricane Mitch, 1998*

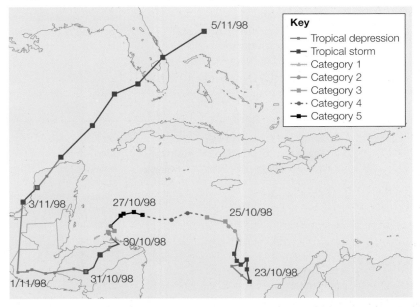

## ACTIVITIES

**1** Study Figure **3**.

   **a** With the aid of an atlas describe the main areas of tropical storm formation.

   **b** During the course of a year, which area spawns the highest percentage of tropical storms?

   **c** Use an atlas to identify a selection of major coastal cities or urban regions threatened by tropical storms.

   **d** Describe and explain the tracks of the tropical storms.

   **e** Why are no tropical storms formed at the Equator?

**2** Consider the hazards associated with tropical storms.

   **a** Why do you think a storm surge is the major hazard associated with tropical storms?

   **b** How can torrential rain inland combine with a storm surge to create major problems on the coast?

**3** Use the internet to find out more about flooding associated with Hurricane Irene. Refer to http://pubs.usgs.gov/sir/2013/5234/pdf/sir2013-5234.pdf

**4** Study Figure **10**. Describe the pattern of hurricane numbers since 1944. Comment on the assertion that hurricanes are becoming more frequent and more powerful as a result of climate change.

**5** Study Figure **12**. Use an atlas to identify the cities at greatest risk from hurricanes in the future. Discuss the value of the map in helping to predict and prepare for future hurricanes.

In this section you will learn about the impacts of and responses to hazards associated with tropical storms

## Impacts of tropical storms

In common with all natural hazards, it is possible to identify a range of impacts of tropical storms. These are summarised in Figure **1**. The detailed impacts of two recent tropical storms will be considered later in the chapter.

| Impact | Definition | Example |
|---|---|---|
| Primary impact | Initial and direct impacts of a tropical storm. These include strong winds, storm surge and heavy rain at the coast causing flooding. | Hurricane Sandy (2012) battered the eastern seaboard of the USA, causing extensive destruction. |
| Secondary impact | Impacts that happen as a consequence of the primary impacts, such as inland river flooding and landslides. | Landslides killed thousands of people as Hurricane Mitch tore through Central America in 1998. |
| Environmental | A range of impacts including inundation with saltwater (affecting freshwater habitats and aquifers), destruction of coastal environments (salt marshes, sand dunes, coral reefs) and pollution (e.g. sewage). | Cyclone Nargis (2008) caused widespread environmental damage in Myanmar. Vast swathes of productive cropland were ruined by saltwater and sewage pollution. |
| Social | Impacts on people, including death, injury and disruption to people's everyday lives. | Hurricane Katrina (2005) displaced over one million people from the city of New Orleans. Over 1800 people were killed. People were still being accommodated in trailer parks four years after the event. |
| Economic | The financial costs of a tropical storm to local people and governments. Increasingly, the financial burden is supported by the insurance industry. | Hurricane Katrina (2005) is the USA's costliest tropical storm disaster, with damages estimated at US$150bn. Property damage alone accounted for US$108bn. Some US$2.5bn was spent collecting and disposing of waste and debris. |
| Political | Tropical storms can lead to political issues of command and control. To what extent is it a local/regional issue or a national or even international issue? If a disaster receives national emergency status this often paves the way for greater personnel and financial support. Countries often ask for international aid from governments, trading blocs (e.g. EU) or NGOs (e.g. charities). | Cyclone Nargis (2008) created widespread destruction. However, the military Myanmar government did not encourage international support. This left many local communities having to cope themselves. A total of 80 000 people died. In the immediate aftermath of Hurricane Katrina (2005), President George Bush was criticised for remaining on vacation as people in New Orleans begged for food after terrible flooding. Some believe that this was the beginning of the end of George Bush's presidency. |

**Figure 1** Impacts of tropical storms

## Reducing the impacts of tropical storms

Although it is not possible to prevent a tropical storm from forming or making landfall, it is possible to take measures aimed at reducing the impacts associated with storm surges, wind damage and flooding.

These measures can be behavioural (e.g. increasing people's preparedness) or structural, which can involve small-scale building adaptations as well as larger-scale constructions, such as sea walls to prevent inundation by storm surges. In locations that are very prone to tropical storms, such as the Philippines and Bangladesh, people have had to adapt to the problems, making use of structures such as cyclone shelters.

## Preparedness

Most people living in areas at risk from tropical storms are aware of the potential dangers. This is through education and public awareness campaigns, using posters, radio and television to warn people of the dangers and provide instructions about what to do before, during and after a tropical storm event. This can involve making minor structural improvements to buildings (e.g. stronger doors and windows), preparing emergency supplies and planning evacuation routes. Property can also be insured against storm damage.

In Florida, USA, evacuation routes and cyclone shelters are clearly signposted. Individual families are encouraged to make a plan and have provisions ready in case they need to act quickly. There is no question that mass evacuation programmes have been immensely important throughout the world in saving lives.

It is now possible to use satellites and other technology such as radar to identify and track tropical storms. Computer models based on historical data enable scientists to predict the likely course or track of an individual storm. In the USA and the Caribbean a 'Hurricane Watch' is issued for those areas of land where hurricane-force winds are a serious possibility within 36 hours. This will be upgraded to a 'Hurricane Warning' when landfall is expected in the next 24 hours or less.

Look at Figure **2**. It shows the early track of Hurricane Gustav as it passed through the Caribbean in August 2008. Notice that a 'cone' of prediction has been plotted to show the area most likely to be affected by the hurricane. This results in warnings being issued to the general public and to civil authorities to enable them to act accordingly and prevent injury and loss of life.

The National Hurricane Centre in Miami uses a computerised model called SLOSH (Sea, Lake and Overland Surges from Hurricanes) to estimate storm surge heights and waves. It uses historical data and takes into account storm characteristics and shoreline features (water depth, bridges, bay configurations, etc.) to estimate the height and extent of a storm surge associated with a predicted hurricane. Data is uploaded every six hours and the model is refined accordingly. The prediction is accurate to +/− 20 per cent. Therefore, if a 2 m surge is predicted, the actual surge will be between 1.6 m and 2.4 m.

### Preparedness, mitigation, prevention, adaptation

There are several approaches that can be adopted to reduce the impact of a natural hazard.

- Preparedness – increasing people's awareness of the potential hazards associated with storms, wildfires, etc. and through their actions minimise the likely impact of the hazard.
- Mitigation – actions aimed at reducing the severity of an event and lessening its impact, which can involve direct intervention or post-event support in the form of aid and insurance.
- Prevention – actions aimed at preventing large-scale events from starting (if possible), for example wildfires.
- Adaptation – accepting that natural events are inevitable and adapting our behaviour accordingly.

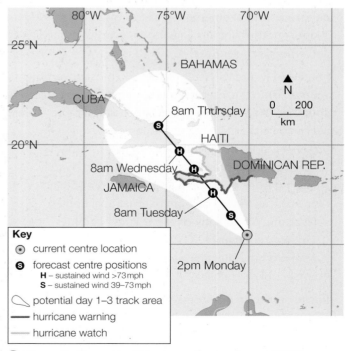

**Figure 2**  *The track of Hurricane Gustav, August 2008*

## Mitigation

Mitigation can involve a range of measures, including structural intervention, disaster aid and insurance cover.

### Structural responses

It is possible to offer some protection from storm surges by soft engineering schemes (planting trees and building up beaches) or hard engineering, such as constructing sea walls (Figure **3**). There is an increasing recognition of the importance of coral reefs in acting as a buffer in reducing storm surges. It is therefore important, for several reasons, to maintain the healthy reef ecosystems that fringe many tropical coastlines (see 6.10).

In South Carolina, USA, the South Carolina Department of Insurance administers the Safe Homes Program, which provides grant money to local homeowners to make their property more resilient to hurricane and wind damage. The principle behind the initiative is that less damage will result in fewer insurance claims and lower premiums. Increasing resilience includes reinforcing gable ends, strengthening roofs and installing stronger doors and windows.

 **Figure 3** *Sea wall flood defence at Corpus Christi, Texas*

### Disaster aid

Disaster aid can take two forms:

- ◆ immediate humanitarian relief in the form of search and rescue, food, water, medicine and shelter
- ◆ longer-term reconstructional aid that seeks to support recovery and reconstruction.

The first stage in any impending disaster is for the government to declare a state of emergency, as this often triggers federal/state support both financially and also in terms of mobilising armed forces and emergency services.

Immediately after disaster has struck, most governments seek support from the international community, particularly neighbouring countries and those with historical, political or economic links. Disaster aid can take the form of expert personnel (e.g. engineers, doctors, search and rescue), transport (e.g. helicopters) or relief supplies (e.g. food, water, shelter) (Figure **4**). Aid can also come from trading blocs such as the EU or from international bodies such as the World Bank or the United Nations. Charities and other NGOs also provide valuable support, often reflecting generous donations from members of the public.

Longer-term aid may come from a variety of sources and could take the form of either a loan to help a country rebuild or direct help, for example, involving charities working within stricken countries.

Disaster aid played an important part in both case studies (see 5.18).

**Figure 4** *Médecins sans Frontières workers distributing aid following the floods in Pakistan, 2010*

## Insurance

Insurance cover is widely used to mitigate the effects of tropical storms, particularly in HDEs. In the USA, people living in hurricane-prone areas are encouraged to take out insurance against wind damage and to follow certain building codes and regulations, for example on windows and doors. The fact that some car insurance policies only cover damage to windscreens and not side windows can result in apparent bizarre behaviour prior to a storm as residents seek to park their cars in the direction of the oncoming storm!

There are huge social issues regarding insurance – the rich can afford the high premiums, whereas the poorest in society cannot. Many of the poorest people in New Orleans who were most severely affected by Hurricane Katrina in 2005 did not have insurance; they remained behind in the city to safeguard their property, refusing to be evacuated to safety. This serves to illustrate how behavioural responses are determined to some extent by potential economic losses (Figure **5**).

## Prevention

In the past, scientists have attempted to use cloud seeding (dropping crystals into clouds to cause rain) in order to dissipate tropical storms and make them less powerful. However, this has not worked and scientists now focus on forecasting, together with mitigation and adaptation, in reducing the impacts of tropical storms.

While a tropical storm cannot be prevented as such, it is possible, to some extent, to mitigate some of its effects. For example, sea walls can be constructed to protect coastal developments from storm surges and river flood defences can help to protect property and land from flooding (Figure **6**). However, with low-return tropical storm probabilities for many coastal areas, such approaches can be seen as financially extravagant, with behavioural responses involving forecasting, evacuation and insurance being the more favoured options.

### Protecting Galveston, Texas, USA

In 1900 a devastating storm surge killed up to 12 000 people in the coastal city of Galveston, Texas, USA. In response to this event, the city authorities embarked on an ambitious plan to prevent such a tragedy occurring in the future. A 5 km sea wall was constructed to a height of 5 m to form an effective barrier against the sea. In addition, dredged sand was used to actually raise over 2000 buildings, some by as much as 5 m! In 2009 plans were put forward to extend the sea wall and install floodgates to cope with predicted rising sea levels and the increased threat posed by a storm surge.

**Figure 5** *New Orleans residents needed persuading to evacuate prior to Hurricane Katrina*

**Figure 6** *Sea wall in Havana, Cuba protecting the city from storm surge*

## Adaptation

Adapting to the threat posed by tropical storms is for many people around the world the most realistic option. Tropical storms cannot be prevented so people simply have to learn to live with the threat and do what they can to minimise the risks. Clearly, to some extent adaptation can also involve elements of preparedness and mitigation.

A good example of adaptation involves land-use zoning that aims to reduce the vulnerability of people and property at the coast. Most commonly this allows only low-value land uses (e.g. recreation) to occupy the coastal strip. In parts of north-eastern Florida, coastal properties are raised above the ground on stilts and have non-residential functions on the ground floor, such as a garage or storage area (Figure **7**). This shows how people have adapted the functionality of their houses to accommodate the threat posed by storm surges. Storm surge elevation markers help to give an indication of which buildings are at risk when a storm warning is issued (Figure **8**).

Following the devastation of the Australian city of Darwin in 1974, when Cyclone Tracy destroyed 94 per cent of the houses and left 40 000 people homeless, the city authorities recognised the need to adapt to the problem of strong winds when rebuilding the city. The use of improved wind-resistant structures in building design was mandatory, which has proved to be effective in reducing losses. Regular inspection and maintenance programmes help to ensure the effectiveness of this approach.

▲ **Figure 7** *Adaptation to coastal flooding, Cedar Key, Florida, USA*

⊙ **Figure 8** *A hurricane storm surge elevation marker on Tybee Island near Savannah, Georgia, USA*

### Adaptation in the Philippines

The Philippines is struck by some 20 tropical cyclones each year. In 2013 Typhoon Haiyan tore through the country leaving more than 6,300 dead. Winds reached 196mph and the coast was inundated by an 8m storm surge.

People are preparing for the another cyclone like Haiyan. Aid agencies are encouraging income diversification so fewer people are dependent on fishing and coconuts, sectors devastated by Haiyan. Mangrove restoration will help to create a buffer against any future storm surge. People are better educated about the dangers of a storm surge and warning systems have been established to encourage evacuation to higher ground.

## Reducing the impacts of tropical storms in Bangladesh

Bangladesh has been greatly affected by tropical storms in the past. In 1970 the Bhola Cyclone (named after the island that it devastated) killed up to 500 000 people. In 1991 some 138 000 people were killed by another powerful cyclone.

In response to these deadly events, a number of initiatives have been introduced:

◆ Tropical storms are monitored by the Bangladesh Meteorological Department and warnings issued over the television, radio and internet.

◆ In remote village communities, designated wardens help to spread the warnings and guide people to safety.

◆ Concrete cyclone shelters (Figure 9) have been constructed to provide a safe refuge for people threatened by flooding. These shelters, often raised up on stilts, serve as schools or community centres when not being used as a cyclone shelter.

In 2017 southern Bangladesh was struck by Cyclone Mora. Strong winds (over 110km/hour) and a storm surge affected farmland near Chittagong. Whilst 18 people were killed, some 500 000 were evacuated from the area following a multitude of tropical cyclone watches and warnings. This shows how recent initiatives have had a huge effect on saving lives.

**Figure 9** *Cyclone shelter in Bangladesh*

## ACTIVITIES

1 List the short-term and longer-term impacts of a storm surge.

2 Study Figure 5 in section 5.14. Describe the impacts of the strong winds associated with Cyclone Winston and suggest the short-term and longer-term impacts on the people of Fiji.

3 The secondary impacts of tropical storms have a greater impact on people's lives than the primary impacts. Do you agree with this statement?

4 Find out more about home improvement mitigation techniques as part of South Carolina's Safe Home Program by accessing the website at http://scsafehomes.com. Do you think this approach should be used more widely?

5 Comment on the effectiveness of the following approaches aimed at reducing the impacts of tropical storms:
   • Forecasting and issuing warnings
   • Construction of seawalls
   • Insurance
   • Adaptation

## STRETCH YOURSELF

In 1926 Miami was hit by a powerful and devastating hurricane that killed over 300 people. In 1992 Hurricane Andrew caused devastation just south of the city and, in 2005 Hurricane Katrina passed through the city before heading out into the Gulf of Mexico en route to New Orleans.

Miami is one of America's wealthiest and most rapidly growing cities, yet it is in the firing line for hurricanes. The low-lying barrier beach on which Miami Beach has been built is particularly vulnerable. Beach nourishment has been used to extend the width of the beach to provide a buffer to storm waves, and trees have been planted to stabilise the sand and form a partial barrier to the waves. Evacuation routes are clearly signposted and there is an elaborate system of early warning.

Use the internet to help you address the following aspects:

1 Why is Miami at risk from hurricanes? Consider the tracks that hurricanes take in the North Atlantic and the vulnerability of Miami itself (Figure **12** 5.14).

2 Briefly describe some of the historic events that illustrate Miami's future vulnerability.

3 What can and is being done to reduce the likely impact of a future hurricane? Consider both behavioural and structural approaches.

In this section you will learn about wildfires – causes, impacts and responses

## What is a wildfire?

**Wildfire** is the generic name used for an uncontrolled rural fire. They are known as bushfires in Australia and as brushfires in North America. Wildfires can affect different layers of vegetation – a *crown fire* spreads across tree canopies and affects forested areas; a *surface fire* burns across surface vegetation; a *ground fire* burns beneath the ground in layers of dry organic peat. The 'ladder effect' describes the process of fires from the forest floor spreading to the tree canopy (Figure **1**).

With the exception of Antarctica, every continent in the world experiences conditions favourable for the ignition of wildfires. As populations have grown and with more people moving into rural areas, the risk has increased. Not only are more people now living in vulnerable areas but there are also greater opportunities for ignition with the construction of power lines, use of machinery and the ever-present problem of arson.

## What conditions favour wildfires?

There are certain conditions that, when combined, are likely to result in a wildfire. These include a ready supply of fuel, usually in the form of dry vegetation, ignition sources (natural and human) and favourable climatic/ weather conditions. The behaviour of a wildfire (speed and direction of movement) is dependent on the type of vegetation (fuel), the climatic characteristics and the local topography (slopes, valleys, etc.).

## Vegetation type – the fuel characteristics

The type and amount of fuel (i.e. living and dead vegetation) influences the intensity of the wildfire (the output of heat energy) and the rate of spread (degree of threat). This is why grassland fires rarely produce the same intensity and degree of threat as forest fires. Moisture content is also a significant factor in wildfire formation. In Eastern Australia the fire season migrates from the north in the spring to the south in the autumn reflecting the seasonal variations in rainfall.

One of the reasons why Australia suffers from so many wildfires is the nature of the fuel. Common trees, such as the eucalyptus, are actually fire-promoting – during intense fires, highly volatile oils within the leaves can actually explode!

▲ *Figure 1* Types of wildfire

## The Black Saturday bushfires, Australia 2009

The Black Saturday bushfires were a series of wildfires that ignited in February 2009 in the Australian state of Victoria. It cost the lives of 173 people and injured over 400. Over 450 000 ha were burned and more than 3500 structures were destroyed. Power supplies were cut off for 60 000 people and around 11 800 head of livestock were killed along and vast areas of cereals, fruit and timber were destroyed.

In the week leading up to the fires, Victoria experienced the worst possible fire conditions – temperatures soared beyond 40 °C, vegetation was tinder dry and humidity was extremely low. Winds also gusted to over 125 km/h (80 mph). When the fires broke out, some ignited naturally by lightning but others due to human activity. During the weeks following fires raged throughout the state.

▶ *Figure 2* Black Saturday bushfires, Australia 2009

## Climate and weather conditions

Heatwaves, droughts and cyclical climatic events such as El Niño can create favourable conditions for wildfires. Most wildfires occur during or after prolonged dry periods when the vegetation (fuel) has become dry and combustible. This explains the occurrence of most wildfires in Australia and North America, as well as the fires that occur in savannah biomes during the extensive dry season (see 6.8).

Strong, dry winds blowing from continental interiors or deserts exacerbate the drying process – ideal conditions for lightning storms, which are a common form of wildfire ignition. Wind strength determines the rate of spread (Figure **3**). The Black Saturday fires in Australia in 2009 were driven by winds of up to 125 km/h (80 mph).

### The role of cyclical climatic events – El Niño

El Niño is a cyclical climatic condition that occurs on average every six to eight years. It involves the warming of the Pacific Ocean off the west coast of South America and affects global patterns of temperature and rainfall. Some places suffer from devastating floods, others are affected by prolonged periods of drought. The 2016 Alberta wildfire in Canada (see 5.17) was, in part, attributed to a strong El Niño event, as was Australia's Ash Wednesday fire in 1983.

Research carried out by NASA on the Indonesian forest fires of the 1990s has suggested clear links between El Niño warming events and carbon monoxide pollution plumes (associated with forest fires) (Figure **4**). During the El Niño phase there was a reduction in rainfall over Indonesia which coincided with an increase in forest fires (see 1.16).

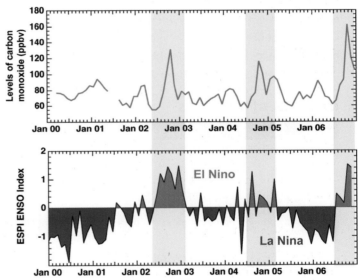

1. Desert winds originate from a clockwise flow of air around a high-pressure system east of the Sierra Nevada mountains.

2. Air from the mountains is compressed and warmed, becoming less humid. This lower humidity dries out vegetation and can fan any existing fires.

3. Winds squeeze through canyons with gusts between 65 and 95 km/h

4. Strong winds create turbulence and can make interstate travel difficult.

⊘ **Figure 3**  *The Santa Ana winds play a significant role in the development of and path taken by wildfires in southern California*

⊘ **Figure 4**  *El Niño warm periods and forest fire conflagrations in Indonesia (2000–2006); the top graph shows atmospheric levels of carbon monoxide associated with forest fires; the bottom graph shows the El Niño warm, dry phases (red) during this time period*

### The Indian Ocean Dipole

Australia's Black Saturday wildfire coincided with another cyclical event called the Indian Ocean Dipole (IOD) – an ocean and atmosphere phenomenon that affects the climate of countries that surround the Indian Ocean basin.

During a positive IOD period (cooler-than-normal water in the tropical east and warmer-than-normal water in the tropical west), there tends to be a decrease in rainfall over parts of central and southern Australia, thereby increasing the risk of wildfires.

Citing evidence that the wildfire season has lengthened since 1979, some scientists are predicting more frequent and devastating wildfires in the future as the climate warms.

## Causes of wildfires

Wildfires can be ignited by natural causes or they can involve human factors.

The vast majority of fires that threaten life and residential areas are the result of human actions, including discarded cigarettes and poorly controlled campfires. The most prone areas are in the so-called 'wildland-urban interfaces' – woodland that is close to large urban areas such as Sydney and Los Angeles.

Heat transfer processes (essentially radiation, conduction and convection) preheat vegetation ahead of the flames, preparing them for ignition and rapid spread of the fire (Figure **5**). These processes are most effective in preheating material that is above the fire (hot air rises!), thereby causing advance of the fire front vertically. This is why fires tend to advance more rapidly up a slope than on level ground. Experiments in Australia have shown that fires on a 20-degree slope advance at up to four times the rate of fires on level ground.

Burning fragments of vegetation (called *firebrands*) can be carried ahead of the fire front by convection currents and strong winds igniting isolated spot fires, their very randomness presenting a significant hazard. Gravity is also responsible for spot fires – firebrands can roll downslope and start fires some distance from the fire front.

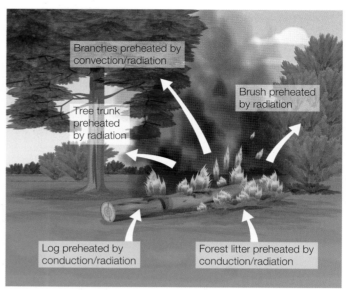

**Figure 5** *Heat transfer processes*

## Impacts of wildfires

**Figure 6** *Primary (immediate and short-term) and secondary (long-term) impacts of wildfires*

| | Primary impacts | Secondary impacts |
|---|---|---|
| Environmental | • Destruction of habitats and ecosystems<br>• Death and injury of animals, which impacts on food chains and food webs<br>• Short-term surge of carbon dioxide due to the burning of carbon stores (trees)<br>• Atmospheric pollution resulting from smoke and water pollution as toxic ash gets washed into water courses | • Lack of trees and vegetation causes depletion of nutrient stores, increased leaching and increased risk of flooding<br>• Increased carbon emissions impact on the greenhouse effect and climate change<br>• Effects on ecosystem development – secondary succession |
| Social | • Loss of life and injury<br>• Displacement – people being forced to temporarily live elsewhere<br>• Disruption to power supplies if power lines damaged by strong winds<br>• Damage to mobile phone stations and telephone exchanges affecting communications | • Possible need for new employment and income stream<br>• Behavioural adaptations based on wildfire experience – people may have to abide by new rules and regulations |
| Economic | • Damage/destruction of structures (homes, public buildings such as schools, fences and field boundaries)<br>• Financial loss (loss of earnings, damage costs)<br>• Destruction of businesses<br>• Loss of crops and livestock | • Costs of rebuilding or possible relocation<br>• Replacement of farm infrastructure, crops, fruit trees, livestock<br>• Cost of future preparedness and mitigation strategies |
| Political | • Actions of emergency services<br>• Responses of local and national government ('state of emergency' status etc.)<br>• Pressure on local authorities and emergency services to coordinate and prioritise responses in the immediate aftermath | • Develop strategies for preparedness and mitigation<br>• Decisions about replanting forests, compensation, future regulations, etc.<br>• Review laws/advice regarding use of countryside for leisure. |

## Wildfires and global systems

Wildfires can have a significant impact on global systems.

◆ Local ecosystems may be affected, with habitats destroyed, animals killed or displaced and soil nutrient stores depleted (see 6.5).

◆ Toxic ash can wash into water courses, adversely affecting aquatic ecosystems.

◆ The loss of vegetation will affect the water cycle, reducing humidity (less transpiration) and altering the relative significance of transfer processes such as surface runoff, evaporation and infiltration (see 1.7).

◆ At a local scale, nutrient cycles will be impacted as biomass and litter stores are burned.

◆ Burning will release carbon stored in trees, plants and peat. This will increase the amount of carbon dioxide in the atmosphere, enhancing the greenhouse effect and creating a positive feedback loop as increased temperatures increase the likelihood of wildfires (see 1.10).

◆ Wildfires will affect the development of vegetation successions, causing secondary successions to be initiated in forests and scrubland (see 6.9).

◆ Human systems associated with rural environments may also be affected by wildfires.

## Strategies for managing wildfires

In common with the management of natural hazards in general, there are four strategies for managing wildfires: preparedness, mitigation, prevention and adaptation (see 5.15).

### Preparedness

Increasing numbers of people who choose to live in rural areas at the edge of cities such as Los Angeles and Melbourne now find themselves at risk from the effects of wildfires and may not know how to reduce the hazards associated with wildfires – their everyday actions may even increase the fire hazard.

Community preparedness in the form of early detection and suppression of wildfires is an important element in reducing the spread and impact of wildfires. Many countries have rural firefighting teams staffed by volunteers.

During times of high fire risk, warnings are released, increasingly through social media, and fire bans may be introduced. In the USA, the National Weather Service issues warnings to alert fire departments and residents of critical weather conditions (low relative humidity, strong winds, dry fuels, the possibility of dry lightning strikes) that could lead to increases in wildfire activity.

◆ A Red Flag Warning is issued when weather conditions for extreme fire behaviour may be met within the next 24 hours. This is the highest level of warning and during this time residents are urged to exercise extreme caution.

◆ A Fire Weather Watch is issued when such weather conditions could exist in the next 24 to 48 hours.

People may be able to establish firebreaks or a 'defensible space' around their property (Figure **7**). However, in reality, a fire front is quite capable of jumping 30 ft (10 m). In the 2016 Alberta wildfire, the fire front jumped a one-kilometre wide river! Perhaps the best form of preparedness is to know when to evacuate and which routes to follow.

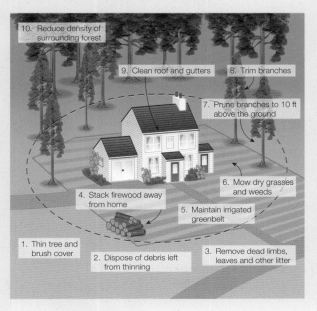

❯ *Figure 7 Reducing the wildfire hazard by establishing defensible space*

## Mitigation

Mitigation involves reducing the severity of an event by direct intervention as well as support following a disaster. Early fire detection is crucial. In the past this would have been by observation from high fire towers, but now satellites or infrared sensors can detect thermal variation. Cameras are also used to detect the early ignition of fires, particularly following lightning strikes.

NASA is developing drone technology for surveying vast areas, such as the Great Dismal Swamp on the Virginia–North Carolina border in the USA. A fully-equipped drone with a 2 m wingspan has a range of about 13 km and can fly for about an hour.

The sheer scale and unpredictability of wildfires can be overwhelming. Extensive wildfires, driven by strong winds and with multiple spotting events may have to be left to burn themselves out.

Figure **8** shows the main principles in controlling a wildfire, including removing the fuel (vegetation) – this is often achieved by deliberately back burning vegetation ahead of the fire front. It may also be possible to use natural barriers, such as rivers to control the spread of a fire. However, spotting, where embers are carried far from the active fire, can result in new fires being ignited elsewhere.

Disaster aid and fire insurance can mitigate the effects of wildfires. For example, disaster aid following the 1983 Ash Wednesday fires in Australia raised A$12 million for those affected. However:

- should everyone receive disaster aid whether or not they have private insurance?
- should insurance premiums be left to the free market or have some form of government control?
- should insurance premiums reflect attempts by householders and communities to reduce the wildfire hazard, say by following fire-safe building codes?

## Responses to the Black Saturday wildfires

In the aftermath of the Black Saturday wildfires, the Australian government contributed more than A$465 million towards reconstruction and recovery. It also announced a comprehensive recovery assistance package to address the psychosocial, economic, infrastructure, and environmental impacts. Assistance was provided to individuals, families, communities, businesses, primary producers and local governments.

In 2009, the Australian and Victorian governments jointly released the *Rebuilding Together: A Statewide Plan for Reconstruction and Recovery,* which granted A$193 million to replace major community facilities that were damaged or destroyed in the fires; support people to rebuild their lives and homes; and kick-start the economic and environmental recovery of these communities.

> **Figure 8** *Back burning to control the spread of a wildfire*

## Prevention

Wildfires need fuel. In forested areas, this is often dead vegetation that has collected as litter on the ground. Controlled burning reduces the amount of fuel but there is always a danger that they may get out of control. They also impact on the natural ecosystem and nutrient cycles by reducing the litter store, as well as releasing carbon dioxide into the atmosphere.

Public awareness is also important, especially in campsites and public areas where rules regarding the use of campfires and barbeques have to be strictly enforced. Many countries operate 'fire bans' during times of high risk.

Since 1944 Smokey Bear (Figure **9**) has been urging Americans to behave responsibly to fires, with good results – 96 per cent of Americans recognise him and 70 per cent can recall his message of fire safety. The average number of hectares lost to wildfires annually has fallen from 54 million in 1944 to 16.5 million today.

However, there is a school of thought that fires are a natural regenerative process within forest ecosystems and should be allowed – in certain circumstances – to take their course.

> **Figure 9** *Smokey Bear, increasing public awareness*

## Adaptation

At its most extreme, adaptation involves learning to live with the threat of wildfires and letting them take their course. Wildfires have a role to play in ecosystem development by burning away old and diseased wood enabling fresh growth, as well as directly stimulating germination of certain species. However, it is not entirely appropriate if we are to protect lives and safeguard human activities.

Planning regulations can be used to reduce the hazard associated with wildfires by restricting access to areas of risk during the fire season. Building design can be an effective form of adaptation – buildings need to be relatively simple, cheap and made of natural products that will not cause pollution if they burn down. Living with the possibility of fire damage is a risk taken by those who wish to live in fire-prone areas.

### STRETCH YOURSELF

Access the website wildfiretoday.com/tag/smoke to find out more about air pollution from wildfires. Find a recent map showing wildfires and add labels and annotations to describe the spread of the smoke. Which major cities were affected? Investigate wildfire events that have taken place since the Alberta fires in 2016 and assess their local and regional impacts.

## ACTIVITIES

1  Why has the wildfire hazard increased in recent years?

2  Assess the importance of fuel (vegetation), climate and topography in the behaviour of wildfires.

3  Look at the pages about savanna grassland (see 6.8). and use the internet to write a short summary about the nature, causes and impacts of wildfires in this global biome.

4  Consider the impact of wildfires on the water and carbon cycles.

**(S)**

5  Draw a revised version of Figure **5** to show how heat fluxes operate on a slope to increase the speed of an advancing fire front.

6  Discuss the assertion that short-term climatic cycles such as El Niño and the Indian Ocean Dipole have an important role to play in contributing to the wildfire hazard.

7  Use the internet to find out about the primary and secondary impacts of the Black Saturday fires in Australia, 2009. Use a colour code-system to sort the impacts into environmental, social, economic and political.

8  Use the internet to find out more about Red Flag Warnings in California.

9  What is meant by 'defensible space' and how effective do you think it is likely to be in reducing the fire hazard?

10  To what extent are publicity campaigns and advertising important in increasing public awareness about the wildfire hazard?

11  Use the internet to find out more about 'fire-adapted communities'. To what extent is adaptation to the fire hazard the most appropriate management strategy?

12  Discuss the assertion that wildfires should only be extinguished if people or human activities are at risk.

In this section you will learn about the impact of and responses to the 2016 Alberta wildfire at a local scale

Case study

### 'The Beast'

In May 2016 a huge wildfire (nicknamed 'the Beast') swept across parts of Canada's Alberta province, forcing the unprecedented evacuation of 90 000 residents of the city, of Fort McMurray as the fire devastated the city, destroying 2400 homes and businesses (Figure **1**). Remarkably no one was killed or injured as a direct result of the fire, which burned some 600 000 ha of land, roughly equivalent to the UK county of Norfolk. Analysts suggest that this fire could turn out to be Canada's most costly disaster.

### What were the causes and contributory factors?

The fire ignited in a remote forested area to the south-west of Fort McMurray on 1 May 2016, although the precise cause of the fire remains unknown. Initially the fire was under control, but a shift in the wind direction resulted in a blaze that tore into the outskirts of Fort Mc Murray, the largest settlement in this relatively remote part of Alberta.

Figure **2** is a thermally enhanced satellite photo showing the extent of the fire in Fort McMurray on 4 May. The fire front has advanced from the south-west but notice how erratic the fires are in and around Fort McMurray. This clearly illustrates the effect of 'spotting', where wind-carried burning embers ignite fires well ahead of the actual fire front. Firefighters faced enormous problems containing the blaze, which even jumped a one-kilometre river in places. Given the erratic spread of the fire, it is remarkable that nobody was killed and that 85 per cent of the city was saved.

Prior to the outbreak of the fire, the environmental conditions were extremely favourable. A lack of winter snowfall and an early snow melt in spring combined with warmer-than-average temperatures that dried out the ground. In late April temperatures soared and, combined with the very low humidity, vegetation in the area became tinder dry.

In the first few days of May, after the fire had started, temperatures exceeded 30 °C and winds increased. The intensity of the fire created its own weather patterns, including lightning, which led to the ignition of additional fires in the area. This positive feedback is extremely rare and testified to the intensity and ferocity of the fire.

⬈ **Figure 1** *Fire rages through Fort McMurray, Alberta, Canada 2016*

⬈ **Figure 2** *Thermal fire map of Fort McMurray, 4 May 2016. Active fires appear red; smoke appears white; burnt areas appear brown*

Climate scientists have linked the fire to a strong El Niño effect that may well have resulted in the unusually warm and dry early spring conditions. Links to climate change are tenuous at best, as the apparent increase in fires in recent decades could well be a reflection of better reporting. However, some climate change scenarios suggest that earlier springs in high latitudes could extend the fire season and thereby increase the wildfire hazard.

## What were the impacts of the wildfire?

The evacuation of an entire city caused considerable social and economic impacts. Figure **3** maps the worst affected areas in Fort McMurray and locates some of the places referred to in Figure **4**.

Figures **5** and **6** are two satellite infrared images showing Beacon Hill, a neighbourhood of Fort McMurray which was in the direct path of the fire (see Figure **3**). The images were taken a year apart, the second image just after the fire swept through the area. The devastation caused by the fire is plain to see.

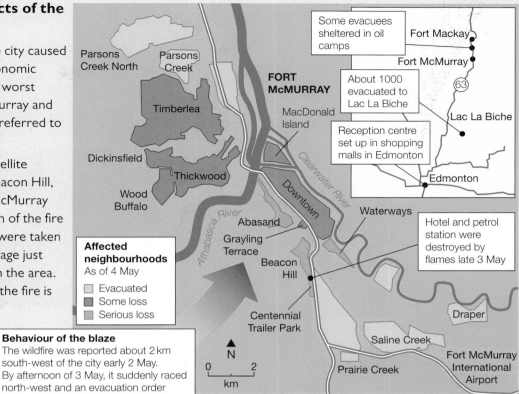

**Affected neighbourhoods**
As of 4 May

☐ Evacuated
■ Some loss
■ Serious loss

**Behaviour of the blaze**
The wildfire was reported about 2 km south-west of the city early 2 May. By afternoon of 3 May, it suddenly raced north-west and an evacuation order was extended to the entire city

Some evacuees sheltered in oil camps

About 1000 evacuated to Lac La Biche

Reception centre set up in shopping malls in Edmonton

Hotel and petrol station were destroyed by flames late 3 May

⊘ **Figure 3** *The neighbourhoods in Fort McMurray most affected by the wildfires*

⊘ **Figure 4** *Impacts of the Alberta wildfire 2016*

| Environmental | Social | Economic | Political |
|---|---|---|---|
| • The intensity of the fire severely affected the boreal forest ecosystem due to the scorched soil and burned tree roots<br><br>• The dry and scorched peaty soils could reignite at any time until the first heavy rainfall<br><br>• The fire will have released several million tonnes of carbon dioxide into the atmosphere<br><br>• Toxins including mercury, lead and organic compounds released from burning trees and buildings created air pollution as far away as the USA and the Gulf Coast<br><br>• Ash was washed into water courses after heavy rain, leading to water pollution and possible contamination of fish and other aquatic wildlife<br><br>• Huge quantities of waste (e.g. rotting food from freezers) and debris from the fire had to be disposed of. Much of this was toxic. | • 90 000 people forced to flee Fort McMurray<br><br>• 2400 homes and other building burned down in parts of Fort McMurray (Figure 3)<br><br>• Jobs and livelihoods were affected and movement in the area was restricted<br><br>• Increased levels of anxiety about the future, with some people suggesting fires of this kind may become more frequent as the climate changes<br><br>• Power supplies were disrupted<br><br>• Water supplies became contaminated as untreated water was deliberately introduced into the municipal water supply to assist firefighters | • Initial insurance company estimates suggested CAN$9bn of damage was inflicted upon Fort McMurray<br><br>• About a third of the 25 000 workers in the nearby oil sands industry (Figure 3) had to be evacuated from work camps. Production was halted in operations to the north of Fort McMurray – Shell Canada temporarily shut down its Albian Sands mining operation. Some 600 work camp units were destroyed by the fire. The fire is estimated to have cost the industry CAN$1bn.<br><br>• Transport in the region was seriously affected, including at the nearby international airport (Figure 3) | • The fire has fuelled political debate about the possible impacts of climate change and the increased vulnerability in the future, as the fire season may be lengthened by early spring melts<br><br>• Government officials had to oversee evacuation programmes and liaise with emergency services in implementing evacuation programmes<br><br>• The Alberta government had to oversee a phased and safe re-entry of they to ensure that they would be safe and secure<br><br>• Reconstruction programmes for buildings, services and infrastructure needed to be coordinated |

**Figure 5** *DigitalGlobe infrared satellite image of Beacon Hill North (see Figure 3), a neighbourhood of Fort McMurray before the wildfire, 6 May, 2015; the red areas are healthy vegetation*

**Figure 6** *DigitalGlobe infrared satellite image of Beacon Hill North (see Figure 3), a neighbourhood of Fort McMurray after the wildfire, 6 May, 2016; the red areas are healthy vegetation*

## What were the responses to the wildfire?

The initial response involved careful monitoring of the newly ignited fire using ground and satellite data. Meteorological information was used to forecast the likely direction of the fire's track, fire warnings were issued and emergency services mobilised. As the wildfire developed a sequence of responses followed.

◆ As the fire threatened Fort McMurray, a mass evacuation programme was implemented and some 90 000 residents of the city were escorted to safety. The lack of direct deaths and injuries is testament to the well-organised evacuation procedure. Aircraft were used to evacuate some of the oil sands workers as the fire threatened these areas to the north of Fort McMurray (Figure **3**).

◆ The Alberta government declared a state of emergency and this triggered support from the Canadian armed forces. Helicopters, water bombers and firefighters were brought into the area from neighbouring states and offers of help were received from many countries, including the USA, Australia and Russia.

◆ The Alberta government supported evacuees by providing CAN$1250 per adult and CAN$500 per dependant to help cover living expenses. By 9 May the Canadian Red Cross had received donations in excess of CAN$50 million.

◆ In nearby Edmonton, an online registry supported by local government and business organisations was created to help evacuees find accommodation and many landlords offered reduced rates to evacuees.

◆ In June, residents were gradually allowed to return to Fort McMurray to begin the long process of clearing up and rebuilding. The Canadian prime minister promised long-term aid to help support the rebuilding of Fort McMurray.

◆ At the end of June 2016 a benefit concert, 'Fire Aid', took place in Edmonton to raise money for those affected by the disaster.

## ACTIVITIES

1 Outline the favourable conditions that increased the fire hazard prior to the ignition of the Alberta wildfire.

2 To what extent did El Niño and climate change contribute towards the wildfire hazard in Alberta? Use the internet to find out more about this issue – you may find this website useful www.bbc.co.uk/news/science-environment-36212145

**(S)**

3 a With reference to Figure **3**, describe the location of the fires indicated by the thermal satellite map in Figure **2**.

 b In what way and for what reasons do the locations of the fires in Figure **2** present major problems for firefighters trying to contain and control the spread of the wildfire?

4 a Study Figures **5** and **6**. Describe the impacts of fire on the vegetation of the area.

 b Suggest why some areas of vegetation appear to have been unaffected by th fire. You may find it useful to refer to Google Earth.

**(S)**

5 With reference to other natural systems, describe how the Alberta wildfire may have significant short- and long-term environmental consequences.

6 Using carefully selected and annotated photos, outline the main social and environmental impacts of the wildfires.

7 What were the main challenges facing the residents of Fort McMurray in the weeks and months after the wildfires? What can the local and national governments do to support rebuilding and recovery?

## STRETCH YOURSELF

Use the internet to assess the success of the longer-term responses to the Alberta wildfire. Find out about the rebuilding and recovery that has taken place in Fort McMurray. Does the economic, social and political character of this community reflect what happened? To what extent is the city (and its people) better prepared if a similar event occurred in the future?

In this section you will learn about two recent storm events in contrasting areas of the world

## Hurricane Sandy, USA, 2012

On 29 October 2012 'Superstorm' Sandy (as it became known in the USA) swept ashore on the New Jersey coastline of the USA, bringing widespread destruction on a scale similar to that experienced when Hurricane Katrina struck New Orleans in 2005.

### What were the impacts of Hurricane Sandy?

All along the barrier coast, houses were destroyed by ferocious winds, storm surges and fires ignited from ruptured gas pipes. Roads were smeared with sand and debris, power lines were destroyed and many peoples' lives were disrupted (Figure **1**).

Fuelled by the warm waters of the Caribbean, Hurricane Sandy began life as a tropical storm in the Caribbean Sea south of Jamaica (Figure **2**). With sustained winds reaching over 160 km/h (100 mph), the storm passed through Jamaica, Haiti, Cuba and the Bahamas causing widespread damage and loss of life. In Jamaica 70 per cent of the population were left without electricity due to the strong winds, and in Haiti over 50 people were killed and 200 000 were left homeless. As the storm passed through Cuba, extensive flooding and strong winds destroyed 15 000 homes and killed 11 people.

Hurricane Sandy continued to travel northwards along the east coast of the USA (Figure **2**), with wind speeds still in excess of 140 km/h (90 mph). Unusually, the storm merged with a weather system moving in from the west, which transformed it into an 'extra-tropical cyclone', a rare but powerful storm system, hence the term 'superstorm'. By the time Sandy made landfall on 29 October it was no longer a true hurricane, as it was fuelled not by warm waters but by dramatic temperature contrasts associated with the merging weather system.

Figure **3** is a synoptic chart of Hurricane Sandy as it made landfall on 29–30 October 2012. The lines on the map are isobars, lines of equal atmospheric pressure measured in millibars. Notice that the hurricane is an intense area of very low atmospheric pressure. The very closely spaced isobars indicate strong winds – these are indicated in greater detail by the colour coding.

▲ **Figure 1** *Coastal flooding in New Jersey caused by Hurricane Sandy, 30 October 2012*

▼ **Figure 2** *The track of Hurricane Sandy 22–29 October 2012*

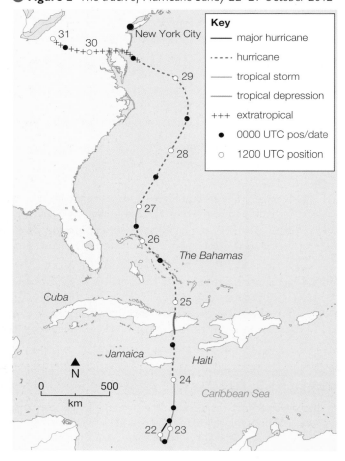

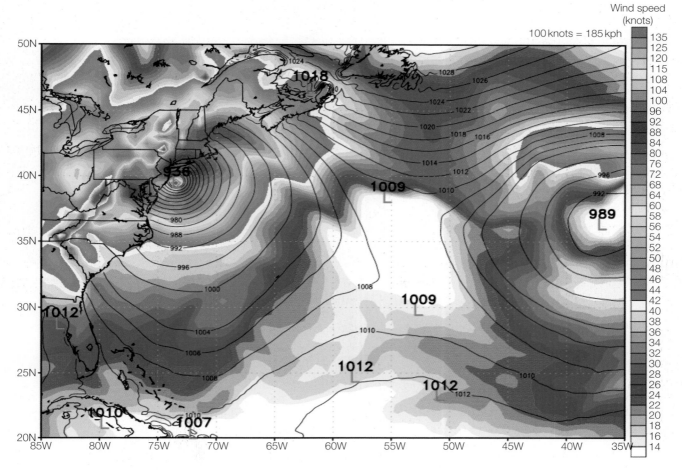

Wind speed (knots)

100 knots = 185 kph

<figure><figcaption>**Figure 3** *Synoptic chart showing pressure patterns and wind speeds (knots) for Hurricane Sandy (00hr 30 October 2012)*</figcaption></figure>

The strong winds (Figure **3**), torrential rainfall and powerful storm surges brought considerable devastation to much of north-east America. Thousands of homes were destroyed and millions of people were left without electricity. Some 160 people in the USA were killed by the storm, which affected a huge area that covered 24 states. Low-lying parts of New York City suffered particularly badly, as streets, tunnels and subway lines in the city were flooded and electricity was cut off to thousands of properties. Damage and loss of life was also reported in Canada.

Hurricane Sandy is the largest Atlantic storm on record – it extended over 1500 km in diameter. While it was less intense than Hurricane Katrina, its sheer extent resulted in widespread damage and loss of life. The storm resulted in the death of 233 people and total damages were put at US$75 billion, making it the USA's second most costly storm after Hurricane Katrina. It was the deadliest and most destructive hurricane of the 2012 Atlantic hurricane season.

## What were the responses to Hurricane Sandy?

As the storm approached the Caribbean and the USA, warnings were issued and preparations made to reduce the impacts on people and property.

◆ In Jamaica, schools and government buildings were closed and people spent time reinforcing the roofs of houses and stocking up on provisions. Kingston Airport was closed prior to the hurricane making landfall.

◆ In the USA, power companies hired additional contractors in preparation to repair power lines and restore electricity. The federal government signed emergency declarations to ensure that federal aid would be available to affected areas, and the National Guard and US Air Force were put on alert ready to assist when the storm arrived. Throughout the Eastern Seaboard, schools were closed, hurricane centres opened and people moved away from the coast and other flood-prone areas.

### Short-term responses

Short-term responses to the storm included the provision of food, water and shelter for those whose homes had been damaged or destroyed. The United Nations and World Food Programme sent relief supplies for 500 000 people in Cuba, one of the Caribbean countries severely affected by the hurricane.

In the USA, power companies gradually restored electricity and federal agencies and emergency services provided additional support. The American Red Cross supplied over 4000 volunteers to help people affected by the storm. In New York, the federal government provided emergency supplies of petrol.

**Figure 4** *In October 2013 Diana Nyad raised $103 001 for those who were still struggling a year after Hurricane Sandy struck. She completed a 48-hour swim in a temporary 36-metre pool, which was set up in Herald Square in New York.*

### Long-term responses

Several agencies in the USA raised money for the victims. A live telethon concert in November raised over US$20 million for the victims of the disaster and a 'Day of Giving' promoted by Disney-ABC Television Group raised a further US$17 million. In January 2013 the US government approved a relief aid package worth over US$50 billion to aid recovery in the affected areas.

Throughout 2013, rebuilding of communities continued with the help of financial support in the form of rebuilding payments (Figure **5**). Despite this support, by December 2013 less than half of those people who had requested support had received any help, and an estimated 30 000 residents of New York and New Jersey remained displaced. Some New Jersey hospitals saw a spike in births nine months after the disaster – such a post-disaster baby boom is quite common!

**Figure 5** *Rebuilding in Far Rockaway, New York, February 2013*

Politically, there was considerable debate in Congress about the impact of global warming on hurricane frequency and intensity, although as yet scientists have no clear proof either way. The storm hit the USA just one week before a general election and it led to discussions and debate around the issues of climate change and federal agency emergency response and relief provision. In the months after the disaster the US Army Corps of Engineers conducted a post-flood survey of flood defence provision. As a result, stricter building regulations were introduced in some areas.

# Cyclone Winston, Fiji, 2016

Fiji is an archipelago of some 300 islands in the South Pacific, about 2000 km to the north-east of New Zealand's North Island (Figure **6**). It has a population of about 860 000, most of whom live on the two islands of Viti Levu and Vanua Levu. In terms of development, Fiji ranks 90th in the world (just below China), compared with the USA, which ranks 8th, and the UK, which is 14th.

In February 2016 Fiji was struck by the Southern Hemisphere's strongest tropical cyclone in recorded history. The storm, which was classified as Category 5 on the Saffir–Simpson scale, had sustained wind speeds of over 230 km/h (145 mph – compare this with Hurricane Sandy, with gusts recorded at over 300 km/h (188 mph) in places.

Starting life as a small tropical storm to the east of Vanuatu, Cyclone Winston moved south and then north-eastwards towards Tonga. It strengthened over the warm seas, did a dramatic U-turn and headed in a westerly direction directly for Fiji, intensifying all the time. Its extraordinary circuitous track exemplifies the difficulty in predicting the course of some tropical storms (Figure **7**).

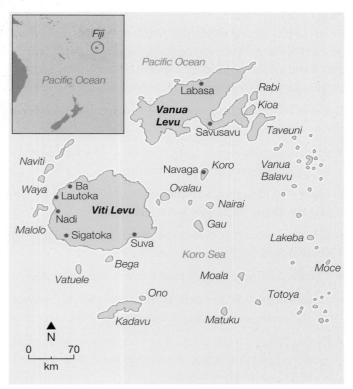

**Figure 6**  Location map of Fiji

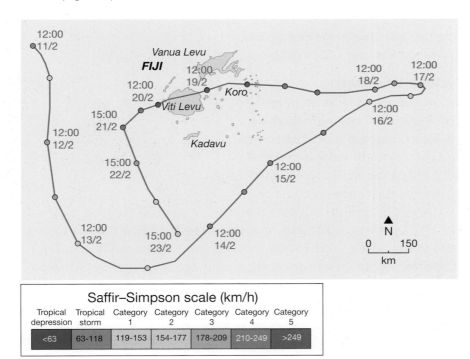

**Figure 7**  The circuitous track of Cyclone Winston, 2016. Figures in red show the location of Winston by time and date

## What were the impacts of Cyclone Winston?

Most of the damage on the islands was caused by the exceptionally strong winds that flattened houses, tore down power lines and destroyed agricultural crops. There was also some localised flooding and damage caused by storm surges, particularly along the southern coast of Vanua Levu.

Cyclone Winston caused extensive damage over a large part of Fiji.

◆ **Figure 8** *Destruction on Fiji*

◆ 350 000 people were significantly affected which was about 40 per cent of the population.

◆ Over 40 000 homes were damaged or destroyed (Figure **8**) and 44 people were killed.

◆ Communications between the islands were lost for several days.

◆ The total damage caused by the storm was estimated at US$1.4 billion.

◆ Approximately 80 per cent of the population lost power due to the extremely strong winds, which brought down trees and power lines.

◆ Destruction in some places was catastrophic, with whole communities being destroyed. For example, all buildings were destroyed, including schools, in Kade Village on Koro Island.

◆ Homelessness and the lack of shelter was a major problem throughout the islands, with 131 000 people in need of immediate shelter.

◆ Over 225 schools were damaged or destroyed.

◆ Food prices rose in the weeks following the storm due to the damage inflicted to crops by the storm – the agricultural sector lost over US$54 million.

## What were the responses to Cyclone Winston?

Prior to Cyclone Winston making landfall, monitoring and forecasting enabled warnings to be issued to the people of Fiji. Around 700 cyclone shelters were opened across the islands and people were encouraged to leave their homes and seek shelter. Fiji's military forces were put on alert to support the relief operation. Public transport was suspended and people were advised not to travel on the roads. These actions undoubtedly saved a great many lives.

### Short-term responses

With around 350 000 people affected by the disaster, many of whom were made homeless, the top priority was the provision of shelter along with food, water and medicine. The lack of electricity and damaged infrastructure left over 250 000 people in need of water and sanitation. Thousands of people took temporary shelter in the cyclone shelters and others turned to family or friends.

In the immediate aftermath of the passing cyclone, the Fiji government declared a state of emergency, which remained in place for two months. In the weeks that followed, relief aid was received from the international community, including governments and NGOs. This included financial assistance as well as relief supplies.

After two days, the international airport reopened, enabling emergency supplies to be airlifted in from countries such as Australia and New Zealand, and telephone services were also restored. After just a couple of weeks, almost all schools had reopened, although power supplies took three weeks to be restored due to the extensive damage caused by the strong winds. A huge amount of waste and debris had to be cleared to make way for reconstruction.

## Long-term responses

In February 2016 the Fijian government supplied US$9 million of financial support to over 40 000 families to help in reconstruction. In April, the government launched 'Help for Homes', a programme of house building aimed at low-income families who were unable to afford reconstruction costs. The organisation Empower Pacific provided psychological and counselling support.

International support was coordinated by Australia, New Zealand and France under the FRANZ Agreement, an agreement signed in 1992 pledging coordinated support in relief operations to South Pacific nations affected by cyclones or other natural disasters. Tens of millions of dollars of aid was received in total.

▲ **Figure 9** *Fiji Red Cross Society members talk to residents who lost their homes to Cyclone Winston – the most powerful storm in Fiji's history*

◆ New Zealand supplied personnel and relief totalling over US$3 million, including the deployment of aircraft and helicopters.

◆ Australia provided immediate relief worth some US$3.6 million and also sent air support and emergency personnel to support recovery and reconstruction.

◆ International aid was also received from countries that included China, India and South Korea as well as the EU and the Asian Development Bank and the United Nations.

◆ Many charities and NGOs raised money for people affected by the disaster, including the International Red Cross and Oxfam (Figure **9**).

## ACTIVITIES

1 Study Figure **1**. Outline the impacts of Hurricane Sandy on the USA and the Caribbean. Present your answer in the form of a table identifying environmental, social, economic and political impacts. Use the internet to find additional information including photos that can be annotated.

2 Study Figure **3**.
   a State the atmospheric pressure at the centre of Hurricane Sandy.
   b Where, in relation to the centre of the storm were the strongest winds experienced? Use the key to suggest the maximum wind speed and use an internet converter to give the equivalent speed in km/h.
   c Describe the pattern of the strongest winds.
   d Suggest the usefulness of this map in forecasting wind speeds and issuing warnings to people.

3 To what extent did the responses to Hurricane Sandy reflect the enormous wealth of the USA?

4 Describe the primary and secondary impacts of Cyclone Winston.

5 Outline the importance of international aid as a response to the Fiji disaster.

6 How does the scale of the disaster in Fiji compare with that of Hurricane Sandy?

7 To what extent did forecasting and public preparation help to reduce the impacts of the two tropical storms?

8 What are the advantages of coordinated international responses such as the FRANZ Agreement in supporting countries affected by natural disasters?

## STRETCH YOURSELF

Use the internet to find out more about the recovery and reconstruction that has taken place in Fiji in the aftermath of the cyclone. To what extent have people been able to rebuild their lives? Have there been improvements in housing and infrastructure? Is Fiji better placed to be able to cope with a future cyclone? You could present some of your findings in the form of an illustrated timeline, using annotated photos to support your written information.

The following are sample practice questions for the Hazards chapter. They have been written to reflect the assessment objectives of the Component 1: Physical geography Section C of your A Level.

These questions will test your ability to:

◆ demonstrate knowledge and understanding of places, environments, concepts, processes, interactions and change, at a variety of scales [AO1]

◆ apply knowledge and understanding in different contexts to interpret, analyse and evaluate geographical information and issues [AO2]

◆ use a variety of quantitative, qualitative and fieldwork skills [AO3].

---

**1** Outline contemporary theories that explain the movement of the Earth's plates. (4)

**2** Study Figure **3**, page 239, and Figure **7**, page 241. Analyse the relationship between distance from the epicentre and death toll. (6)

**3** Using Figure **1**, below, and your own knowledge, assess the influence of intra-plate vulcanicity at a global scale. (9)

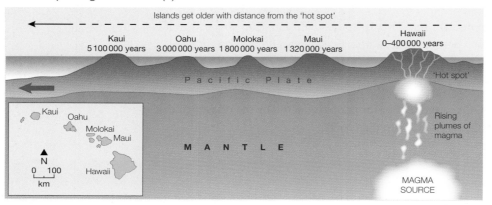

🔺 **Figure 1** The formation of the Hawaiian Islands

**4** Critically assess the relative impact of different types of seismic hazards on human life. (9)

**5** 'Economic and cultural geography determine the degree to which a potential natural event is perceived as a hazard worth planning for.' With reference to a local scale community you have studied, evaluate this statement. (20)

> **Tip** 💡
>
> Keep an eye on the time as this will help you prepare for the exams. Allow 25–30 minutes for a 20-mark question.

---

**6** For a multi-hazardous environment you have studied beyond the UK, explain the nature of the hazards faced. (4)

**7** Study Figures **2** and **3**. Analyse the data shown. (6)

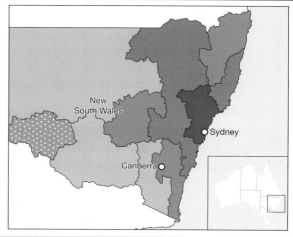

▶ **Figure 2** Fire danger ratings for New South Wales (November 2019)

**Key**
- ■ Catastrophic
- ■ Extreme
- ■ Severe
- □ Very high
- □ High
- ▨ Low-moderate

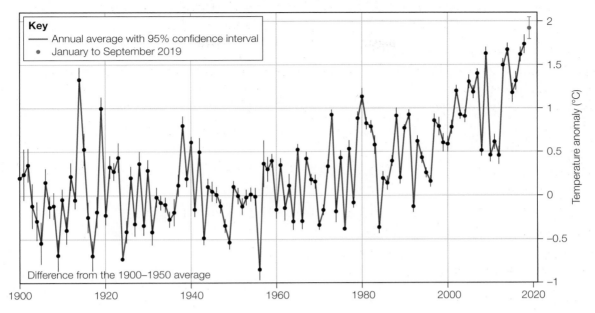

▲ **Figure 3** *The warming climate of New South Wales*

**8** Study Figure **4**. Using Figure **4** and your own knowledge assess the value of the use of new technology to inform human responses to seismic events. (9)

## ⚡ QUAKE UP! APP

★★★★☆ AlaskanGrizzly, 10/02/2019

**My one wish? That it were quicker**

I've been using this app for months now. I live in Alaska where we get a lot of quakes, and often. The most recent notable one was 7.0 in November 2018 (it was reported as 7.0 but where I live other sources suggest it was more like 8.0). Since then, we've had hundreds of tremors or aftershocks. I don't get the notification until between 6 and 10 minutes after it has actually hit us. For tsunami warnings and any other emergency updates, 6-10 minutes' leeway can be the difference between life and death — so why the delay?

**Response from Quake up!**

Quake up! sends alerts as soon as the USGS reports a tremor. The time it takes USGS to report a quake varies. In California, where they have many sensors, it only takes a couple of minutes. In other parts of the US, it takes USGS, on average, 8 minutes. Thanks for your feedback. I do hope this response is helpful.

▲ **Figure 4** *Quake up! app review*

**9** For one type of natural hazard, evaluate the importance of global governance for its successful mitigation and/or management worldwide. (9)

**10** To what extent is Park's model of human response applicable in understanding responses to storm hazards you have studied? (20)

> **Tip** 💡
>
> Illustrating an idea with a real world example is a great way to get an additional mark for 'developing a point'.

*Clean up process after an oil spill in July 2013 at Ao Prao Beach on Samet Island, Thailand. Think about the different impacts that this will have.*

## Your exam

Ecosystems under stress is an optional topic. You must answer one question in Section C of Paper 1: Physical geography, from either Hazards or Ecosystems under stress. Paper 1 makes up 40% of your A Level.

## Your key skills in this chapter

In order to become a good geographer you need to develop key geographical skills. In this chapter, whenever you see the skills icon you will practise a range of quantitative and relevant qualitative skills, within the 'Ecosystems under stress' theme. Examples of these skills are:

◆ Interpreting photographs and remotely-sensed images  6.9

◆ Using atlases and other map sources  6.6

◆ Drawing and annotating sketch maps  6.5

◆ Measurement and geospatial mapping  6.6

◆ Presenting field data and interpreting graphs  6.1, 6.3

◆ Analysing data, including applying statistical skills to field measurements  6.9

◆ Drawing and annotating diagrams of physical systems  6.3, 6.9

## Fieldwork opportunities

Small-scale, local fieldwork investigations relating to ecosystems are easy to fit into your busy course. They just require a bit of planning and thought about the equipment needed, as well as the best sampling technique to use.

*1. Collecting data in woodland areas*

Investigating the changes along a woodland transect, or comparing similar but different woodland areas (each area managed in different ways), could involve a measurement of the following at all sample points: light, soil pH, soil depth and temperature, and a species count. Also keep a record of percentage vegetation cover – a quadrat divided into 100 tiny squares lends a greater level of accuracy to this activity.

You would expect neighbouring areas of deciduous and coniferous woodland to differ in terms of light, soil pH and abundance of life – there are certainly fewer available food sources for nesting birds etc. in a coniferous forest. But what does your data tell you?

*2. Freshwater sampling*

Both kick sampling and the action of sweeping a net in a figure of eight can be used for freshwater invertebrate sampling. Choose one of these techniques – consistency of method between your sites is key. What is the impact of differing water quality on the abundance of these invertebrates and does water quality (turbidity) vary according to location within the alternating pattern of pools and riffles along a river?

In this section you will learn about the concept of biodiversity and recent trends at different scales

## The UK's ecosystem timebomb

In 2015 the results of the biggest survey ever of British wildlife was published. Researchers from the University of Reading and the Centre for Ecology and Hydrology studied records of 4424 species that were collected between 1970 and 2009.

Wildlife, trees and plants were analysed according to four functions they performed in the **ecosystem**: pollination, pest control, decomposition, carbon sequestration (see 1.9) and also species often recognised as being of cultural value, such as birds, butterflies and bees. They concluded that the marked decline in wildlife in the UK countryside is threatening core functions of the ecosystem that are vital for human wellbeing.

Groups such as hedgehogs, hoverflies, moths and birds are in the most serious decline, with individual species under particular threat, including the common red ant, red-shanked carder bee and the common banded hoverfly (Figure **1**). Climate change and habitat loss are leading to a reduction in biodiversity, with species that act as pollinators and natural pest controls most at risk.

## What is biodiversity?

**Biodiversity** has been described as 'the foundation of all life on Earth' because it is crucial to the functioning of ecosystems. Defined by the Convention on Biological Diversity (1992), and accepted internationally, the term 'biodiversity' means:

'the variability among living organisms from all sources including terrestrial, marine and other aquatic ecosystems and the ecological complexes of which they are a part; this includes diversity within species, between species and of ecosystems.'

The term 'biodiversity' has only been used widely since the 1980s. Since then, the term has been used increasingly within the scientific community and beyond, mainly because of concerns about the rapid loss of biodiversity at different scales around the world.

## Trends in biodiversity

Since the development of human civilisation, almost all of Earth's ecosystems have been dramatically transformed by human actions. Many animal and plant populations have declined in numbers, geographical spread, or both. African mammals, birds in agricultural lands, British butterflies, Caribbean corals, water birds and fishery species have all shown a significant decline in range or number. While species extinction is a natural occurrence, human activity has increased the extinction rate by at least 100 times compared to the natural rate.

⬆ **Figure 1** At risk: the common banded hoverfly

### Measuring biodiversity

Several indicators can be used to monitor trends in biodiversity:

- Species richness – essentially this is the number of different species.

- Population number – the number of genetically distinct populations of a particular species defined by analysis of a specific element of its genetic makeup.

- Genetic diversity – the variation in the amount of genetic information within and among individuals of a population, a species, an assemblage or a community.

- Species evenness – measurement of how evenly individuals are distributed among species.

## Local scale trends – the plight of the bees

At the local scale, the decline in bee populations has been linked to biodiversity decline (Figure **2**). Bees are extremely important pollinators, vital throughout the world in maintaining healthy and productive farming systems. In recent years their populations have declined. In a report conducted at the French National Institute for Agricultural Research (INRA), scientists discovered that bees that are fed pollen from a range of plants showed signs of having a healthier immune system than those eating pollen from a single type. Bees need a fully functional immune system in order to sterilise food for the colony. So, if biodiversity is decreasing, therefore reducing the variety of wild flowers, bees are restricted in their pollen sources and lose immunity as a result.

**▶ Figure 2** *Bees before and after colony collapse*

## Global scale trends – the Living Planet Index of biodiversity

The WWF's Living Planet Index (LPI) is a widely accepted measure of the state of the world's biological diversity. It is based on population trends of vertebrate species from terrestrial, freshwater and marine habitats. The index currently incorporates data on the abundance of 555 terrestrial species, 323 freshwater species and 267 marine species around the world. During the period 1970–2000, the index fell by some 40 per cent (Figure **3**). During this time, the terrestrial index fell by about 30 per cent, the freshwater index by about 50 per cent and the marine index by around 30 per cent.

Figure **4** shows the global distribution of threatened invertebrate species. In terrestrial ecosystems, those areas experiencing the most rapid decline in biodiversity include the Amazon basin, the Great Lakes region of Eastern Africa, Bangladesh, the Indus valley and parts of the Middle East and Central Asia. Much of this is due to deforestation, the expansion of intensive farming and desertification. In marine ecosystems, inevitably it is the waters closest to major populations, particularly in parts of Asia, where species face the greatest threats.

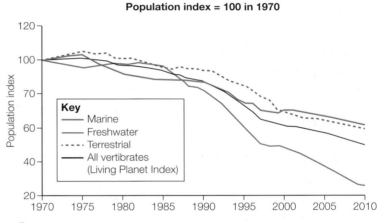

**Population index = 100 in 1970**

**▲ Figure 3** *The WWF's Living Planet Index*

**▼ Figure 4** *Global distribution of threatened invertebrate species , 2010*

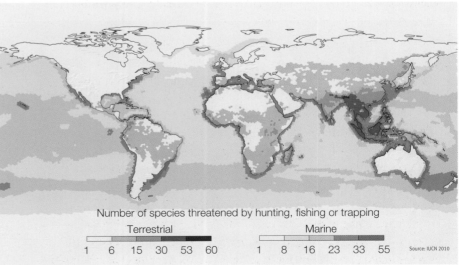

## Main causes of biodiversity decline

◆ More land was converted to cropland from 1950 to 1980 than between 1700 and 1850.

◆ Between 1960 and 2000, reservoir storage capacity quadrupled.

◆ Some 35 per cent of mangroves have been lost in the last two decades as coastal fringes have been developed for tourism or intensive fish farming.

◆ Roughly 20 per cent of the world's coral reefs have been destroyed by overfishing, pollution and, more recently, coral bleaching associated with climate change.

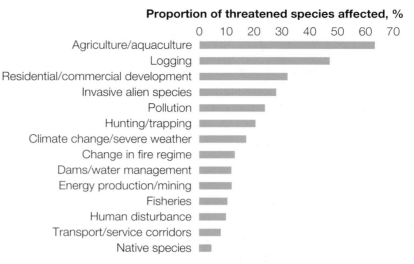

**Proportion of threatened species affected, %**

| | |
|---|---|
| Agriculture/aquaculture | |
| Logging | |
| Residential/commercial development | |
| Invasive alien species | |
| Pollution | |
| Hunting/trapping | |
| Climate change/severe weather | |
| Change in fire regime | |
| Dams/water management | |
| Energy production/mining | |
| Fisheries | |
| Human disturbance | |
| Transport/service corridors | |
| Native species | |

**Figure 5** *Causes of biodiversity decline*

## Recent trends

In recent years the decline has started to show signs of slowing down. In some regions this is due to the damage having already been done, with high rates of ecosystem conversion and species extinction. However, there is an increasing recognition in the value of maintaining (and increasing) biodiversity:

◆ The designation of protected areas, now covering nearly 13 per cent of the world's land area.

◆ Increasing recognition of indigenous and local community-managed areas.

◆ Adoption of policies and actions for managing invasive alien species – about 55 per cent of countries have legislation to prevent the introduction of new alien species and control existing invasive species.

◆ Regulations supporting sustainable harvesting, reduced pollution and habitat restoration.

◆ International financing for biodiversity conservation is estimated to have grown by about 38 per cent in real terms since 1992 and now stands at over US$3 billion per year.

'Coral reef decline will have alarming consequences for approximately 500 million people who depend on coral reefs for food, coastal protection, building materials, and income from tourism. This includes 30 million who are virtually totally dependent on coral reefs for their livelihoods or for the land they live on (atolls).'

(Global Coral Reef Monitoring Network: *Status of Coral Reefs of the World: 2008*)

**Figure 6** *Diving on St John's Reef, Red Sea, Egypt*

# The potential impacts of declining biodiversity

We have already seen how bee populations are declining and this could have a significant impact on pollination and subsequent agricultural productivity. Some scientists believe that in tropical rainforests some plant species become extinct even before they have been discovered! Many species in tropical **biomes** have valuable medicinal qualities and these could be wiped out. Imagine what medicines could be available if extinction does not happen.

Reduction in biodiversity can affect stores and flows such as decomposition rates, vegetation biomass production and, in the marine environment, fish stocks. For example, it is predicted that a reduction in marine productivity will mean that fisheries will not be able to meet the demands of a growing global population. Some scientists are concerned that at some point in the future a threshold may be crossed when positive feedbacks (see 1.1) lead to catastrophe on a par with global warming.

Biodiversity loss has negative effects on several aspects of human well-being, such as vulnerability to natural disasters (see 4.15), energy security, food security and access to clean water and raw materials (see *AQA Geography for A Level & AS Human Geography*, Resource security). It also affects human health, social relations and freedom of choice. Ecosystem modification frequently results in knock-on effects. So, for example, conversion to intensive farming can lead to a reduction in water availability. This may result in land degradation, water shortages for people, and even the need to construct reservoirs for water storage.

The practical and cultural value of ecosystems – collectively referred to as ecosystem services – cannot be traded like other commodities; as such it does not have a market value and is therefore ignored by financial markets. In order to emphasise the importance of ecosystems, attempts are being made to assign a monetary value that takes into account the benefits for recreation, water supply and cultural significance (Figure **7**). This may help to further slow down – or even reverse – biodiversity decline.

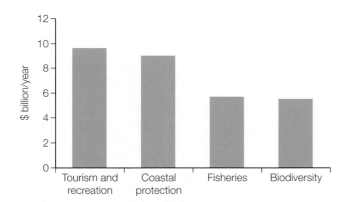

▲ **Figure 7** *Global annual value of coral reefs*

| Exploitation | 37% |
|---|---|
| Habitat degradation and change | 31.4% |
| Habitat loss | 13.4% |
| Climate change | 7.1% |
| Invasive species/genes | 5.1% |
| Pollution | 4% |
| Disease | 2% |

▲ **Figure 8** *Primary threats to LPI populations (2014)*

## ACTIVITIES

1 Consider the measures of biodiversity. Suggest the reasons why each one should be considered in assessing biodiversity decline. Species richness tends to be the most commonly used measure. Why is this and does it matter?

Ⓢ
2 Study Figure **3**.
 a Describe the trends shown in the graph.
 b Suggest reasons why freshwater species have declined at the fastest rate.
 c Why has the rate of decline slowed within certain biomes?
3 Use a method of your choice to present the data describing the main causes for biodiversity decline in Figure **8**.

4 Study Figure **7**. To what extent is it possible to give coral reefs an accurate monetary value for 'biodiversity'?
5 Assess the potential impacts of biodiversity decline. Focus on social, economic, political and environmental impacts. Consider variations in the rates that these impacts will be felt. Use the internet to assist your research and consider presenting your work in the form of an information poster.

## STRETCH YOURSELF

Use the internet to investigate further the impact of biodiversity decline on bee populations. To what extent is the collapse of bee colonies solely due to a reduction in the diversity of wild flowers or are there are other factors to consider? What is the impact of bee decline on natural ecosystems? What measures are being taken to address this issue?

In this section you will learn about the importance of ecosystems and their role in economic development and sustainability

## Why are ecosystems important for human development?

Since the beginning of human evolution, ecosystems have provided people with sustenance, shelter, medicine and energy. For thousands of years people have hunted in forests, gathered fruit, nuts and berries and caught fish in rivers, lakes and the sea. Trees have provided wood for fuel and buildings, and plants have been important sources of chemicals, both for medicinal purposes and as powerful poisons.

Look at Figure **1**. It shows the connections between ecosystem services (the functions offered by ecosystems) and human well-being (effectively human development). Notice the huge range of services offered by ecosystems and the highly complex way in which these interact with human systems. These interactions are liable to mediation by socio-economic factors, most obviously levels of development. Figure **1** should be viewed as a model for discussion.

**Ecosystem services**

**Supporting**
- Nutrient cycling
- Soil formation
- Primary production

**Provisioning**
- Food
- Freshwater
- Wood and fibre
- Fuel

**Regulating**
- Climate regulation
- Flood regulation
- Disease regulation
- Water purification

**Cultural**
- Aesthetic
- Spiritual
- Educational
- Recreational

Life On Earth–biodiversity

**Constituents of well-being**

**Security**
- Personal safety
- Secure resource access
- Security from disasters

**Basic material for good life**
- Adequate livelihoods
- Sufficient nutritious food
- Shelter
- Access to goods

**Health**
- Strength
- Feeling well
- Access to clean air and water

**Good social relations**
- Social cohesion
- Mutual respect
- Ability to help others

**Freedom of choice and action**
Opportunity to be able to achieve what an individual values doing and being

**Key**

**Arrow colour**
Potential for mediation by socio-economic factors
- ■ Low
- ■ Medium
- ■ High

**Arrow width**
Intensity of link ages between ecosystem services and human well-being
- — Weak
- — Medium
- ■ Strong

⬢ **Figure 1** *The relationship between ecosystems and people*

## Living in harmony with nature

### The Achuar, Amazon, Peru/Ecuador

The Achuar are a tribe of 6000 indigenous people living in small communities along the Ecuador–Peru border. They have been able to maintain their cultural identity in the face of progress and modernisation and live sustainably as part of their rainforest ecosystem, fitting into the natural ways and rhythms without creating any long-term environmental damage. Their lives depend on the ecosystem and they respect it.

The Achuar hunt monkeys, fish in the river (using a non-polluting, plant-derived poison to stun the fish) and gather roots such as yucca, sweet potato, plantains and squash. They mostly drink chicha, which is a slightly fermented drink made from manioc root. Their rich cultural folklore is embedded in the forest with which they have a powerful spiritual connection.

In common with many indigenous tribes, the Achuar's way of life and ecosystem are under threat from development. In 1964 oil was discovered in the Amazon. There have been violent conflicts as the oil companies have sought to gain access to the forest to exploit oil reserves. One result of this has been contamination of water supplies and land (Figure **2**).

⬢ **Figure 2** *Protestors including Achuar people protesting about toxic waste left by Occidental Petroleum on the lands of the Achuar tribe.*

Elsewhere in the world, large areas of tropical rainforests have been cleared to make way for cattle ranching to provide meat for fast-food restaurants, palm oil for industrial, catering and cosmetic uses, and new land for settlement. The burning of rainforests has wiped out many species of plants and animals and led to others, such as the orangutan becoming endangered. These rich and diverse ecosystems have been destroyed at an alarming rate to address the needs of an ever-expanding global population.

## Ecosystems, population growth and economic development

The last 500 years have seen an explosion in population growth reaching 7.7 billion in 2019 and projected to reach 10 billion by 2050. As population increases, agricultural productivity has to increase to avoid starvation and famines. This requires increased intensification (using chemicals and mechanisation) and the conversion of marginal 'natural' land into productive farmland. Almost inevitably, population growth, together with increased economic pressures such as resource exploitation, will lead to ecosystem degradation – pollution, deforestation, desertification and biodiversity reduction. Only through very careful management can this be avoided.

### The impact of population growth and economic development

A study of the impact of population growth on ecosystem services has been carried out for the community of Albemarle County and Charlottesville, Virginia, USA (population 279 642). The research focused on the effects of local population increase on a selection of ecosystem services – including water-related services (such as storm water retention) and air-related services (such as carbon sequestration and storage and air pollution removal).

The study found that, for a population increase of up to 125 per cent, degradation of ecosystem services is contained within the study area. The surrounding rural areas continue to function effectively. However, as population exceeds this threshold value, degradation becomes widespread, impacting all of the surrounding rural areas. The results of this study clearly indicate that planners have to balance the needs of the human population with local ecosystem health. A strong urban forestry programme, for example, in developing areas will go some way to assuaging the negative effects of development (Figure **3**).

▲ **Figure 3** *Aerial photo of Charlottesville, VA, USA*

### ACTIVITIES

1 Study Figure **1**.
   a Work in pairs to suggest additional Ecosystem services for each of the four categories.
   b What is meant by the 'potential for mediation by socio-economic factors'?
   c What is the significance of the role played by 'Freedom of choice and action'?
   d To what extent to you agree with the colours and proportionality of the arrows linking Ecosystem services to Constituents of well-being?

2 How and why do the Achuar tribe embody sustainable principles in their everyday lives?

3 Study the research conducted in Albemarle County and Charlottesville, Virgina, USA.
   a Assess the importance of small-scale community studies into the possible impacts of population and economic growth on ecosystem services.
   b What evidence is there in Figure **3** that the local authorities are concerned about maintaining ecosystem services as the city grows?

### STRETCH YOURSELF

Investigate the impact of population and economic growth on ecosystem services in the UK. Consider in particular how an increased rate of house building will impact on water management, drainage basins and flooding. Use Figure **1** to help you define the different ecosystem services and try to address as many of these as you can. Is the UK prepared to cope with population growth while maintaining ecosystem integrity?

# The nature of ecosystems – energy flows and trophic levels

**In this section you will learn about the nature of ecosystems and energy flows through the system**

Ecosystems are central to the study of biogeography. They can be defined as dynamic, ordered and highly integrated communities of plants and animals, together with the environment that influences them. Ecosystems occur at all scales, from the smallest patch of lichen on a rock surface to global biomes such as a deciduous forest (Figure **2**).

The whole of life on Earth itself is, in effect, an ecosystem! Their significance should never be underestimated. Humankind depends on ecosystems for survival because plants and animals are our basic food resources. But human activities are changing and threatening ecosystems. Understanding them is therefore essential if we are to manage them for a sustainable future.

**Figure 1** *The application of systems concepts to a woodland ecosystem*

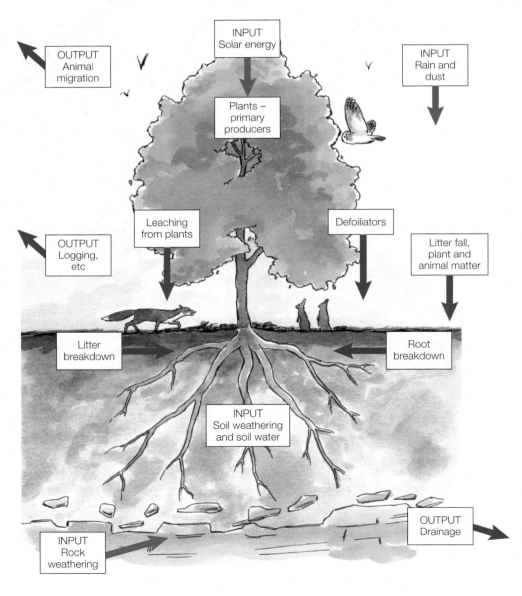

## The application of systems concepts in ecosystems

As the term implies, ecosystems are examples of natural systems, with inputs, outputs, stores and flows (transfers). They are open systems in that energy and living matter can both enter and leave. Look at Figure **1**. It shows the application of systems concepts to forests and the flows associated with energy, water and carbon together with the impacts of human activities (see 1.1).

The following processes are at work in a woodland ecosystem.

♦ Inputs – include precipitation, energy from the Sun and carbon dioxide. Additional inputs might include seeds blown into the forest and animals arriving from elsewhere.

♦ Outputs – include water lost through river runoff, overland flow, throughflow, groundwater flow and evapotranspiration. Nutrients leached through the soil also represent outputs, as does energy reflected back to space off the surface of leaves.

♦ Stores – nutrients held in the biomass, litter and soil are stores, together with water collected on leaves or as puddles on the ground.

♦ Flows – there are a great many flows in an ecosystem, primarily of energy, water and carbon.

♦ Positive feedback loop – one simple but very important feedback loop is the link between population and births – as the population of a species increases, the number of new individuals will increase (more births germinating seeds, etc). This accounts for the rapid spread of plant and animal species in certain ecosystems until, in the case of animals, they become regulated by the shortage of food (a negative feedback!).

♦ Negative feedback loop – animals have regulatory systems to maintain their temperature at about 37 °C. For example, blood vessels close to the surface of ears in desert animals enable them to lose heat. Shivering is a response to cold that warms the body.

## The structure of ecosystems

Ecosystems comprise living organisms in the *biotic* environment and inorganic non-living substances in the *abiotic* environment. Natural ecosystems will strive to achieve a balance between biotic and abiotic factors. This balance is often delicately poised and can be easily upset by sudden change, for example, natural wildfires or the actions of people, such as deforestation or conversion to farmland.

### The biotic environment

The biotic, living environment is made up of both plants and animals.

♦ Plants include all living vegetation, but also dead plant matter that is decomposing.

♦ Animals include all fish, birds, insects, microorganisms (such as bacteria) and mammals, including people.

**Figure 2** *A deciduous forest ecosystem in the UK*

### Biomass

The term *biomass* is used to describe the total weight of all biotic (living) organisms per unit area (e.g. one hectare). It is usually expressed as the mass of organically bound carbon that is present. Apart from bacteria, the total global biomass of plants and animals has been estimated at 560 billion tonnes of organically bound carbon. The vast majority of global biomass is terrestrial, with only about 5–10 billion tonnes of organically bound carbon stored in the oceans. Despite this imbalance of existing biomass, a similar amount of new biomass is produced each year on land and in the oceans.

♦ Plants have a much higher biomass than the animals that consume them – crops comprise about 2 billion tonnes of organically bound carbon, whereas domesticated animals comprise about 700 million tonnes.

♦ Antarctic krill form one of the largest biomasses of any individual animal species.

♦ Fungi make up about 25 per cent of global biomass.

♦ The total mass of bacteria may equal or even exceed that of all plants and animals.

## The abiotic environment

The abiotic, non-living environment includes all inorganic substances and other environmental influences. These are the chemical and physical components of an ecosystem. They include:

◆ minerals in the soil released by weathering of the parent rock

◆ water and gases (such as air) in the soil

◆ relief and drainage of the land

◆ climatic variables such as wind, light and seasonal patterns of precipitation and temperature.

## Energy flows

The source of all energy is the Sun. It is captured by the green pigment in plant leaves (chlorophyll) that converts carbon dioxide and water into their organic compounds through the process of photosynthesis. These so-called 'building blocks' of plants are tissue and food energy in the form of chemicals called carbohydrates.

The total amount of energy absorbed or 'fixed' by green plants is called the *gross primary productivity (GPP)*. Some of the GPP is lost through respiration as plants carry out their normal functions. The rest is used in the production of new leaves. This is known as the *net primary productivity (NPP)*. The productivity of plants is greatest under favourable conditions, i.e. high values of light, warmth, water and nutrients. As Figure **3** shows, tropical environments are highly productive compared with cold environments (tundra) and hot deserts.

| Producer | Biomass productivity (gC/m²/yr) |
|---|---|
| Swamps and marshes | 2500 |
| Tropical rainforests | 2000 |
| Coral reefs | 2000 |
| Algal beds | 2000 |
| River estuaries | 1800 |
| Temperate forests | 1250 |
| Cultivated lands | 650 |
| Tundra | 140 |
| Open ocean | 125 |
| Deserts | 3 |

⬆ **Figure 3** *Selected global producers of biomass productivity*

## Food chains and food webs

Plants form the basis of all nutrition and energy for the whole ecosystem. Carbohydrates contain amino acids, sugars, starches, proteins, fats and vitamins – all the organic materials needed by animals for growth, movement and reproduction. Food stored in the plants provides food for other organisms, which in turn feed others – this is known as a *food chain* (Figure **4**). Food chains trace single routes of pathways from one organism to another, for example in deciduous woodland:

plant ➜ insect ➜ toad ➜ snake ➜ fox

Simple food chains are useful in explaining the basic principles behind ecosystems. But if only one species occupied a particular trophic level (see below) then, as it was consumed, all the organisms in succeeding levels would be threatened by the reduction. In reality this will rarely be the case, not least because most animals have different sources of food. Also, any one species of plant or animal will be eaten by a variety of different consumers. In most ecosystems, therefore, such is the variety of plants and animals at each trophic level that a large number of food chains will be operating all at the same time and also interconnecting with each other. The resulting complex network of linked food chains is called a *food web* (Figure **5**).

⬇ **Figure 4** *Eastern garter snake eating gray tree frog*

## Marine ecosystems

Marine (aquatic) ecosystems illustrate food webs well (Figure **5**). When the basic requirements for life (water, nutrients, heat, light) are considered, the sea provides more favourable conditions for organic production than land. In the sea, there is no water shortage – oxygen is abundant and carbon dioxide readily available. Temperature variations are less marked than on land and the transparency of the sea allows a thicker photosynthetic zone for phytoplankton, algae and seaweed at the first trophic level. The second trophic level consists of herbivores such as zooplankton and the common periwinkle, which eat these primary producers. The third trophic level includes carnivorous and omnivorous fish, crabs and seals, with the fourth trophic level consisting of omnivores such as polar bears and humans.

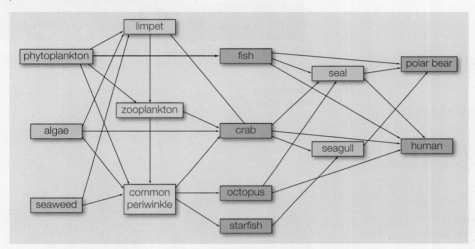

◀ *Figure 5  A marine ecosystem food web*

## Trophic levels

Plants represent the first **trophic** (or energy) **level** in the food chain (Figure **6**). They are called producer organisms (self-feeding autotrophs) – they produce their own food through photosynthesis. All other trophic levels are occupied by consumers, which include all animals – birds, fish, reptiles, insects and mammals including humans. We distinguish between primary consumers, which are vegetarian (herbivores), and secondary and tertiary consumers, which are meat eaters (carnivores).

So, the second trophic level is where herbivores eat producers directly. The third trophic level sees small carnivores feeding on the herbivores. Finally, the fourth trophic level is occupied by the larger carnivores. These top predators, known as omnivores, eat both plants and animals and so have two sources of food.

| Trophic level | Process | Examples | | |
| --- | --- | --- | --- | --- |
| **Level 1** – producers | self-feeding autotrophs produce energy by photosynthesis | grass | oak leaf | phytoplankton |
| **Level 2** – primary consumers | vegetarian herbivores eat plants | earthworm | caterpillar | zooplankton |
| **Level 3** – secondary consumers | meat-eating carnivores eat animals | house sparrow | blue tit | fish |
| **Level 4** – tertiary consumers | top predators (omnivores) eat smaller animals | tawny owl | hawk | great white shark |

▲ *Figure 6  Trophic levels and simple food chains*

## Reducer organisms

All food chains also include *reducer organisms*. These operate at all trophic levels. They complete the flow of energy through the chain by returning any remaining nutrients into the soil to support new plant growth. Reducer organisms fall into two groups.

◆ *Detritivores* are animals that eat dead and decaying organisms. Examples include lice, earthworms and vultures.

◆ *Decomposers* are organisms that cause the decay and breakdown of dead plants, animals and excrement (Figure **7**). Examples include bacteria and fungi.

▲ *Figure 7* Decomposers at work

## Energy pyramids

In any study of an ecosystem, nutrients should be thought of as being recycled while energy is lost. A staggering 90 per cent of energy is lost at each trophic level, mostly by animal respiration, movement and excretion. Consequently, energy that was input to the system by insolation is progressively lost along the food chain and at each higher trophic level fewer organisms can be supported. These energy losses are illustrated using energy pyramids and have been used to great effect in demonstrating the extravagant food and energy wastage, in terms of processing, involved in the majority of diets found in HDEs compared with LDEs (Figure **8**).

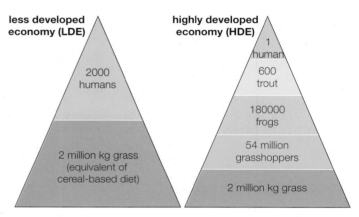

▲ *Figure 8* In a LDE 2 million kg of grass feeds 2000 people, whereas in a HDE the same amount of grass feeds just one person

## ACTIVITIES

**1** Ecosystems can be considered to represent 'the balance of nature'.
 **a** Explain what is meant by this statement.
 **b** Identify ways in which both natural events and human activities can destabilise this balance?

**2** Make a simple copy of Figure **1**. Use a colour key to show the different systems concepts, such as inputs and flows. Try to add some additional features to the diagram including feedback loops.

**3** Draw and label a simple food chain showing four trophic levels ending with humans.

**4** Figure **8** shows contrasting energy pyramids.
 **a** Explain why there must always be a greater biomass of plants than of animals in an ecosystem.
 **b** Compare and contrast the losses of energy at each trophic level.
 **c** Comment on the human significance of this comparison.

**5** Study Figure **9**, which shows a selection of organisms in a typical English deciduous woodland ecosystem.

Construct a four-column table listing all of the woodland organisms shown at their correct trophic levels. Complete the table to create a food web by drawing arrows linking each feeding interaction. (An example has been started beneath Figure **9**.)

| Level 1 producers | Level 2 primary consumers | Level 3 secondary consumers | Level 4 tertiary consumers |
|---|---|---|---|

oak tree ⟶ caterpillar ⟶ wood mouse
acorns ⟶ caterpillar ⟶ robin

decomposers – bacteria and fungi

**Figure 9** *Deciduous woodland ecosystem*

293

In this section you will learn about nutrient cycles

## What is a nutrient cycle?

Nutrients are plant foods. They consist of minerals and chemicals derived from precipitation, the weathering of rock and the decomposition of organic matter. Plants need nutrients in order to survive and grow and in the natural world, nutrients are readily available. However, house plants have to be fed regularly with commercially available plant foods. Even cut flowers are now sold with a sachet of plant food to be added to the water (Figure **1**).

⬆ **Figure 1** *Cut flowers require nutrient supplements*

The **nutrient cycle** constantly recycles the vital nutrients between the soil and the plants – it is vital in sustaining life in an ecosystem (see 4.2). It is commonly represented in a standard format developed by P.F. Gersmehl in 1976 (Figure **2**). Notice that the diagram takes the form of a simple system.

- ◆ *Stores* – the three stores (biomass, litter and soil) are represented by circles. The relative size of each circle approximates to its importance within a particular ecosystem or global biome.

- ◆ *Inputs* – the two key external inputs to the open system are nutrients dissolved in rainwater (or snow) and weathered parent material (bedrock or superficial deposits such as gravels or fluvial silts) dissolved in soil water.

- ◆ *Outputs* – nutrients may be lost from the system by runoff and leaching.

- ◆ *Transfers* – these include littering (e.g. leaf drop) and plant uptake from the soil. The thicker the arrow, the greater its significance within the system.

It is important to remember that the diagram is always drawn in this format and that the size of the circles (stores) and arrows (transfers) are drawn proportionately to the importance or significance within the system.

⬇ **Figure 2** *The Gersmehl Nutrient Cycle*

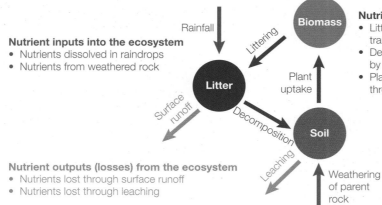

**Nutrient inputs into the ecosystem**
- Nutrients dissolved in raindrops
- Nutrients from weathered rock

**Nutrient outputs (losses) from the ecosystem**
- Nutrients lost through surface runoff
- Nutrients lost through leaching

**Nutrient transfers within the ecosystem**
- Littering – fallout from the plants, mostly leaf fall transferring nutrients to the litter
- Decomposition – decay of organic material in the litter by fungi and bacteria transferring nutrients to the soil
- Plant uptake – the uptake of nutrients from the soil through the plant roots

**Nutrient stores in the ecosystems (circles)**
- Biomass – contains all living plant and animal matter in the ecosystem
- Soil – contains minerals from the parent rock in addition to humus from decomposed plant and animal remains
- Litter – sits on top of the soil and contains both dead and decaying plant and animal material

Study the nutrient cycles for three major terrestrial biomes (large-scale ecosystems) in Figure 3. Take time to understand why the stores and transfers vary in size between the different biomes. Notice that the overriding controlling factor is climate.

▼ **Figure 3** *Nutrient cycles for three major terrestrial biomes (see Figure 2 for colour coding)*

**a** Tropical rainforest

High amount of precipitation

Very high biomass store as warm, wet conditions promote plant growth

Small stores as leaves are rapidly decomposed

High demand from lush vegetation growth

High runoff due to high rainfall

Warm, wet conditions are ideal for decomposers

Small store due to high uptake and leaching

High rainfall leads to significant leaching

Rapid chemical weathering due to high temperature and rainfall

**b** Coniferous forest

Low rainfall

Heavy fall of leaves in winter

Slow-growing trees

Leaves, especially needles, decompose very slowly

Relatively high uptake

Slow decomposition due to cold conditions (poor conditions for bacterial action) and waxy nature of leaves which slow down decomposition; also, the acidic conditions are poor for decomposers

Small store due to low inputs

Little chemical weathering due to low temperatures and low rainfall

**c** Grassland (prairie)

High amount of seasonal die-back of grass

Reasonable rainfall

High inputs of nutrients result in a large soil store

Rapid decomposition of grass

Moderate chemical weathering due to high summer temperatures and reasonable rainfall

## ACTIVITIES

1   Figure **2** shows the transfer of plant nutrients in an ecosystem. Briefly explain the nature of the nutrient transfer:
    **a**   from soil to biomass
    **b**   from biomass to litter
    **c**   from litter to soil.

2   Suggest how climatic factors affect the nature and significance of inputs and outputs in the nutrient cycle.

3   Study Figure **3**. Describe and account for the differences in the following for each biome:
    • biomass    • littering    • soil    • decomposition

## STRETCH YOURSELF Ⓢ

Consider one of the three biomes in Figure **3**. Suggest how human actions can lead to change and destabilisation of your chosen nutrient cycle. Draw a modified nutrient cycle diagram to show the immediate impacts of such a change. Then draw a second diagram to show how it adapts (and stabilises?) over time. Don't forget to include annotations on your diagrams.

In this section you will learn about the interrelationships between the factors affecting terrestrial ecosystems and how they respond to change

## What are terrestrial ecosystems?

There are a huge number and range of terrestrial (land-based) ecosystems (Figure **1**). You have already studied aspects of deciduous woodlands, lithoseres (developed on raised beaches), hydroseres (freshwater lakes) and will look at heather moorlands (see 6.9). Consider the many small-scale ecosystems such as field margins, hedges and ponds as well as all of the larger-scale ecosystems (biomes) such as deserts, tundra, tropical rainforests (6.7) and savanna (6.8) – you came across some of these earlier when studying nutrient cycles (see 6.4).

## Factors affecting terrestrial ecosystems

There are five main factors that determine the type and characteristics of an ecosystem. They are all interrelated.

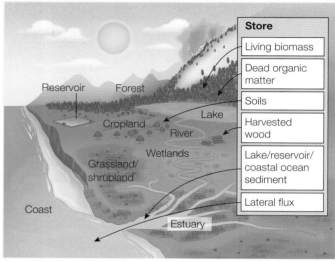

▲ **Figure 1** *Terrestrial ecosystems*

### Climate

Climate, essentially the temperature regime and precipitation type, amount and distribution, is the most important factor affecting the development of an ecosystem. To a large extent, the climate determines the plants that will be able to colonise a particular region. Most plants thrive in warm and wet conditions (e.g. the equatorial regions). However, desert and polar climates present huge challenges for plants, restricting species diversity and abundance. Even within the UK, plants that survive in the south, such as palms, would not survive the harsher conditions in Scotland.

### Topography

The relief of a landscape, its ups and downs, is its *topography*. Gently undulating hills, valley bottoms and steep, rocky outcrops all have slightly different controls on the development of soils, vegetation and the resultant ecosystem. Topography influences soil depth and drainage – valley floors accumulate deeper soils and are usually wetter than steep slopes. Topography can influence microclimates – cold air sinks into valley bottoms during calm, clear nights, leading to localised frost and fog. Altitude affects climate, which, in turn, affects vegetation – it tends to be cooler with increasing altitude.

### Soils

Soil is a mixture of inorganic minerals derived from weathered parent material, and decomposed organic material (humus) derived from plants, as well as water and air. Weathering is hugely influenced by the climate, being at its maximum in warm and humid conditions. Soil depth, texture and fertility determine the type and abundance of plants in an ecosystem. If the soils are thin, poorly drained and acidic, they will be unable to support a large biomass. If they are deep and fertile, they will be able to support lush vegetation growth.

### Vegetation

Vegetation is the primary food supply and also provides habitats. The type of vegetation (e.g. trees) has an influence on the plants and animals that thrive around it. For example, in a deciduous forest, bluebells thrive in dappled sunlight before the new leaves form a dense overhead canopy blocking out the sunshine from late spring.

### Time

Ecosystems can take hundreds of years to develop – the formation of soil and subsequent colonisation by vegetation is very slow. Yet, through human intervention (such as deforestation) or natural catastrophes (such as wildfires as in Alberta, Canada in 2016), an ecosystem can be harmed or destroyed in just a few hours, even minutes.

## How do these factors interact?

These five factors are all inextricably linked and connected with one another. Figure **2** shows a simplified profile through eastern Europe from north-west to south-east. Notice how climatic characteristics change across the latitudes, with harsh conditions being experienced at either end of the profile. The soil depth profile (a reflection of climate and also influenced by the amount of available organic matter) shows that the mid-latitudes experience the greatest depth of soil. The vegetation, largely a product of the climate, also varies considerably between the tundra (plants adapt to cope with frozen conditions, see 4.12) and deserts (deep water tables, searing temperatures and high aridity of hot deserts, see 2.2).

### Interrelationships on Exmoor

Figure **3** shows part of Exmoor in south-west England. This undulating landscape clearly demonstrates the role of topography on soil development and vegetation. Together, these factors influence small-scale variations in the ecosystems.

The rolling Exmoor uplands have relatively deep and acidic soils on which vegetation such as heather and bilberries thrive. These low bushes provide grazing for animals such as sheep and deer, which add nutrients to the soil through their faeces. The steeper valley slopes are well-drained, making them drier and suitable for bracken growth. With its dense fronds, bracken shades out other plants, restricting biodiversity.

The valley bottoms have thick soil deposits but they tend to be saturated and not very fertile. Reeds and grasses dominate these areas. The different vegetation types create a variety of different habitats and provide different foods, encouraging diversity in the fauna that inhabit these areas (Figure **4**).

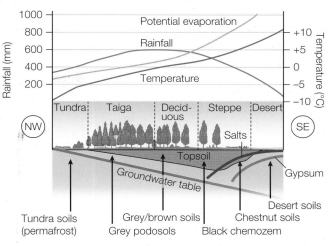

⌃ **Figure 2** *Profile through eastern Europe showing variations in climate, soils and vegetation*

⌃ **Figure 3** *The impact of topography on soils and vegetation, Exmoor, Somerset*

⌄ **Figure 4** *Typical slope profile, Exmoor, Somerset*

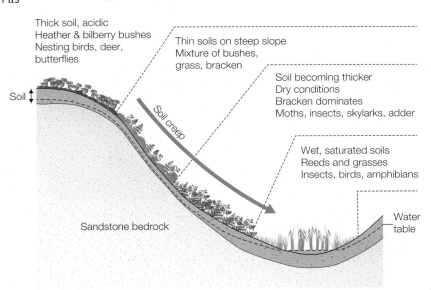

## Ecosystem interrelationships, Sierra Nevada , California, USA

The Sierra Nevada mountain range stretches for over 600 km from north to south through the state of California. The bedrock is mostly granite and much of the landscape reflects the glacial processes that were active during the last cold climatic period (see 4.2). The Sierra Nevada exhibits a distinctive altitudinal ecological zonation that shows clear interrelationships between climate, soils and vegetation (Figure **5**). This area is the subject of scientific study focusing on the rain/snow transition zone within the mixed conifer forest. It is here that signs of climate change might be most apparent and the impact on future water storage most detectable. The research is exploring the feedbacks operating at different scales.

The region as a whole experiences a Mediterranean climate – hot, dry summers and wet, often mild winters. Most of the region receives its rainfall (or snow at higher elevations) during the winter months. The snowpack at higher altitudes provides an important water store that enables some biotic and abiotic processes to continue through the summer. Future changes to the snowpack could have very significant impacts on the ecosystems of the region.

Each of the four altitudinal zones exhibit distinct soil and vegetation characteristics.

◆ Oak savanna – this ecosystem most closely reflects the regional mediterranean climate, characterised by oak trees, shrubs and grassland. Soils are generally deep and reasonably fertile as a result of active weathering and decomposition processes, a response to the high temperatures and presence of moisture, albeit seasonally.

◆ Pine/oak forest – mixed coniferous and deciduous forest reflects the slightly cooler conditions, with pine trees being well-suited to slightly harsher conditions (colder temperatures and snow). Soils remain reasonable thick but, with pine needles in the litter, acidity is slightly higher.

◆ Mixed coniferous forest – this zone comprises fairly continuous stands of trees and records the highest levels of productivity across the entire transect, due to the relatively high amounts of rainfall, moderate temperatures and long growing season. Soils are well-developed.

◆ Subalpine forest – here there are isolated copses of stunted conifers, which have developed on thin, patchy soils. There are a lot of bare rocky outcrops and some low-growing, closely packed tundra-like plants that retain heat and moisture. Seasonally cold temperatures limit net primary productivity (and therefore evapotranspiration) to short growing seasons. The soil is thin, probably due to the short annual interaction between vegetation and underlying rock. Incomplete recovery from glacial conditions (i.e. time) is another contributory factor.

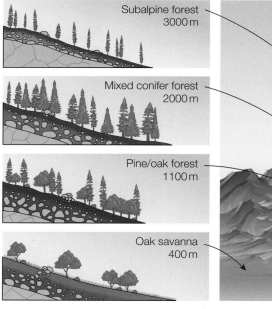

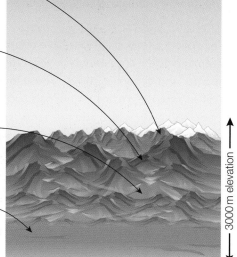

Subalpine forest
3000 m

Mixed conifer forest
2000 m

Pine/oak forest
1100 m

Oak savanna
400 m

3000 m elevation

Basin to regional

> **Figure 5** *Vegetation zonation in the Sierra Nevada, California, USA*

# How do ecosystems respond to present day change?

Ecosystem development is dynamic and will naturally adjust to change in one or more of its components. Some ecosystems are highly adaptable, robust and resilient, whereas others are more fragile and vulnerable to even the slightest change (see 4.12). Where growing conditions are amenable, there is often a high level of resilience. However, in hostile environments, such as tundra or desert, the implications of change are considerable.

The development of a balanced and stable ecosystem takes tens or even hundreds of years to be achieved. In this balanced state, slight variations such as a particularly cold winter or a period of drought are accommodated by subtle rebalancing – the concept of *dynamic equilibrium* (see 1.1). It is only when a significant and rapid change occurs that an ecosystem is likely to be tipped off-balance to such an extent that it suffers long-term damage. Figure **6** identifies some possible changes that may affect ecosystems.

**Figure 6** *Possible causes of ecosystem change*

| Component | Change | |
|---|---|---|
| | **Natural** | **Anthropogenic** |
| Climate | Natural climate change is generally extremely slow, enabling ecosystems to adjust. Short-term climate cycles such as El Niño can be more damaging to ecosystems in causing floods and drought. | The recent warming trend is undoubtedly affecting ecosystems through changes to the seasons and gradual latitudinal shifts in vegetation belts. Extreme weather events may be connected to climate change caused by human activity. |
| Vegetation | The vegetation succession results in changes to food supply and habitats. These changes affect wildlife but they happen very slowly, usually over hundreds of years. | The spread of invasive plant species (such as Japanese knotweed) and alien animal species can impact significantly on an ecosystem. Human management (drainage, rotational burning, deforestation) can be devastating to ecosystems. |
| Soil | Soils develop over very long periods of time, so are unlikely to trigger sudden change in an ecosystem. However, soil erosion or incidents of pollution could inflict more immediate change on an ecosystem. | Indirectly, management of vegetation impacts on soils. They might become compacted or eroded, for example. |
| Topography | The topography of an area develops extremely slowly in most regions, except where there are high rates of erosion or active processes of mass movement, for example in mountainous areas. | Terracing of slopes for agriculture or slope stabilisation affects drainage and soils and could therefore have an impact on ecosystems. |

## Impacts of invasive species on ecosystems

*Invasive* species are plants or animals introduced by people intentionally or accidentally that do not belong to a naturally developed (native) ecosystem.

'Through their impacts on species and ecosystem processes, alien invasive species can result in the fragmentation, destruction, alteration or complete replacement of habitats which in turn, has cascading effects on even more species and ecosystem processes.'
(UN Food and Agricultural Organisation)

In forestry, the introduction of alien invasive species, particularly insect pests and diseases, has had harmful ecological impacts, affecting the species diversity, gene pool, individual population numbers, habitats and food supplies.

◆ Miconia calvescens, a tropical American tree introduced to French Polynesia in 1937, has significantly altered the forests of French Polynesia and other Pacific islands by shading out all native plants and promoting erosion and landslides with its shallow roots.

◆ The invasive weeds Scotch broom and gorse are repressing the regrowth of the commercially important Douglas fir in British Columbia, Canada.

◆ Invasive grasses that are particularly fire-prone may lead to a permanent loss of forests.

◆ Australian acacia species, such as *Acacia cyclops*, have radically altered nutrient cycling regimes in nutrient poor ecosystems due to their ability to fix atmospheric nitrogen.

## Impact of mongooses on the Caribbean natural ecosystems

Mongooses are small carnivores about the size of a cat that are native to southern Europe and Africa (Figure **8**). They were introduced to islands such as St Lucia and Jamaica in the Caribbean in an attempt to control rats, which were damaging the sugar cane harvests. Rather than controlling the rats, the newly introduced mongooses hunted birds and bird's eggs. With no natural predators, they bred prolifically and their population has grown rapidly. They may have contributed to the extinction of some endemic vertebrates, and continue to cause livestock damage as well as pose a disease risk. The introduction of this alien species has disrupted the natural ecosystem, causing it to destabilise – at least for a while.

**Figure 7** *The Asian mongoose has had a detrimental effect on the local ecosystems*

## Impacts of wildfires on ecosystems

A **wildfire** is an uncontrolled fire in open country or wilderness. On average there are 60 000–80 000 reported wildfires each year, mostly in hot, dry environments. Some are started by natural causes such as lightning strikes (which may occur more frequently as a result of climate change), but the majority are started by people, either inadvertently or deliberately (see 5.15).

Wildfires can be an important and life-sustaining part of an ecosystem, clearing and converting dead wood to release important nutrients such as potash and phosphate and, in some ecosystems, triggering seed germination (e.g. the Ponderosa pine tree). However, if the frequency of wildfires increases, natural cycles will be disrupted and native species may start to suffer. Some invasive species are highly flammable. This acts as a positive feedback loop by encouraging further fires (see 1.1), which may wipe out more native species and lead to a proliferation of the invasive species. Wildfires can increase levels of carbon dioxide in the atmosphere, thereby strengthening the greenhouse effect, which may increase the likelihood of future fires (Figure **8**). With a reduction in ground cover, surface runoff may be encouraged, leading to soil erosion. Changes in the soil will then affect the biomass.

**Figure 8** *Californian wildfire (2015)*

## Impacts of conservation on ecosystems

Environmental concerns often lead people to intervene in order to preserve and conserve natural ecosystems. This can involve a variety of actions across a range of ecosystems at different scales.

**Figure 9** *Volunteers planting trees at Dundreggan Conservation Estate in the Scottish Highlands*

- Restoration of peatland involves filling in ditches to encourage water storage and reduce runoff. Soils become saturated and acidic. These environmental changes affect the vegetation and, in turn, the habitats for birds, insects and animals. Look back to 1.17 to read about peatland restoration on Exmoor.

- Tree planting restores local ecosystems affected by deforestation and is part of much larger-scale initiatives to respond to climate change. Trees affect water movement, soil development and the range of flora and fauna that can live and thrive in an environment. Fast-growing native tree species can be planted to encourage the regrowth of indigenous species – this is called assisted natural regeneration (Figure **9**).

- The trampling of vegetation and the exposure of bare soil leads to footpath erosion and habitat loss. Through careful management, vegetation can be reintroduced, which will, in time, reverse the damage to the ecosystem and return it to its native form. In the Lake District, the organisation Fix the Fells works to restore footpaths on behalf of public and private landowners, including the National Trust. They are currently working to restore 120 footpaths, mostly through replanting of native plant species (Figure **10**).

# How might ecosystems be affected by alternative futures?

So far you have studied the interrelationships between ecosystem components and the way that they respond to present-day changes. You have seen how the components of ecosystems are inextricably linked and how change can have significant effects on both the individual components and the ecosystem as a whole. Ecosystems are vulnerable and sensitive to change. How will they respond in the future to climate change and increasing resource exploitation?

## Climate change

Climate plays a dominant role in determining the development and characteristics of ecosystems. Therefore, any changes to the climate, particularly rapid ones, will have the potential to cause significant impacts, particularly if species are unable to adapt at the same rate at which change occurs. While climate change will bring many threats and challenges to ecosystems, it can also bring some economic benefits (Figure **11**).

**Figure 10** *Restoring footpaths at Scales Tarn in the Lake District National Park*

## Climate change and the United States Environmental Protection Agency (EPA)

'Climate change can alter where species live and how they interact, which could fundamentally transform current ecosystems. Impacts on one species can ripple through the food web and affect many organisms in an ecosystem. Mountain and arctic ecosystems and species are particularly sensitive to climate change. Projected warming could greatly increase the rate of species extinctions, especially in sensitive regions.'
(United States Environmental Protection Agency)

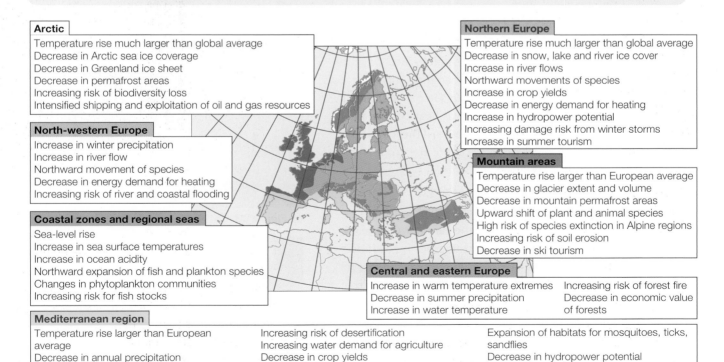

**Arctic**
Temperature rise much larger than global average
Decrease in Arctic sea ice coverage
Decrease in Greenland ice sheet
Decrease in permafrost areas
Increasing risk of biodiversity loss
Intensified shipping and exploitation of oil and gas resources

**North-western Europe**
Increase in winter precipitation
Increase in river flow
Northward movement of species
Decrease in energy demand for heating
Increasing risk of river and coastal flooding

**Coastal zones and regional seas**
Sea-level rise
Increase in sea surface temperatures
Increase in ocean acidity
Northward expansion of fish and plankton species
Changes in phytoplankton communities
Increasing risk for fish stocks

**Mediterranean region**
Temperature rise larger than European average
Decrease in annual precipitation
Decrease in annual river flow
Increasing risk of biodiversity loss
Increasing risk of desertification
Increasing water demand for agriculture
Decrease in crop yields
Increasing risk of forest fire
Increase in mortality from heat waves
Expansion of habitats for mosquitoes, ticks, sandflies
Decrease in hydropower potential
Decrease in summer tourism and potential increase in other seasons

**Northern Europe**
Temperature rise much larger than global average
Decrease in snow, lake and river ice cover
Increase in river flows
Northward movements of species
Increase in crop yields
Decrease in energy demand for heating
Increase in hydropower potential
Increasing damage risk from winter storms
Increase in summer tourism

**Mountain areas**
Temperature rise larger than European average
Decrease in glacier extent and volume
Decrease in mountain permafrost areas
Upward shift of plant and animal species
High risk of species extinction in Alpine regions
Increasing risk of soil erosion
Decrease in ski tourism

**Central and eastern Europe**
Increase in warm temperature extremes
Decrease in summer precipitation
Increase in water temperature
Increasing risk of forest fire
Decrease in economic value of forests

**Figure 11** *Threats, challenges and opportunities of climate change in Europe*

## Shifting ecosystem belts

One of the most striking global impacts of climate change will be the poleward shift in the latitudinal vegetation belts. In North America boreal (coniferous) forests are already invading the tundra zones, reducing habitats for the many unique species that depend on the tundra ecosystem, such as caribou, arctic foxes and snowy owls. As rivers and streams warm up, warm water fish are expanding into areas previously inhabited by cold water species. Cold water fish, including many highly valued trout species, are losing their habitats as the area of feasible, cooler habitats to which species can migrate is reduced.

In the UK, crops that are traditionally grown in southern Europe such as grapes and olives are already being introduced into parts of southern England. In the future, the cereal belt may shift into Scotland and coniferous forests in the UK may only survive in the far north. This northwards shift has ecological implications that affect food supplies and habitats, as well as economic and political implications.

Marine ecosystems will also respond to climate change. Many marine fish and shellfish species have certain temperature ranges at which they can survive. For example, cod in the North Atlantic require water temperatures below 12 °C. Even sea-bottom temperatures above 8 °C can reduce their ability to reproduce and for young cod to survive. In this century, temperatures in the North Atlantic are likely to exceed both thresholds. Figure **12** shows this northward trend of marine species.

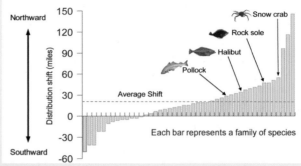

▲ **Figure 12** Trends in the distribution of marine species of fish and shellfish in the North Atlantic

## Changes to the fragile Arctic ecosystem

Declining Arctic sea ice has been well documented in recent years (see 4.13). As the ice declines in its extent and duration it leads to a reduction in the abundance of ice algae, which thrive in nutrient-rich pockets in the ice (Figure **13**). These algae are eaten by zooplankton, which are in turn eaten by Arctic cod, an important food source for many marine mammals, including seals. Seals are eaten by polar bears. So, you can see that a decline in ice algae can contribute to a decline in polar bear populations. Polar bears hunt close to the edges of sea ice; if the ice extent continues to decline, they will be restricted to the coast where food supplies are far less abundant.

## The Prairie Potholes, north-central USA

The Prairie Potholes is a vast area of small, shallow lakes, some of which temporarily dry out in the summer (Figure **14**). The lakes provide essential breeding grounds for many North American waterfowl species, as well as important habitats for fish (e.g. bass) and animals (including moose and deer). In the warmer, drier future that is predicted for this region, the lakes may dry out completely, which will be devastating for the many species that depend upon them.

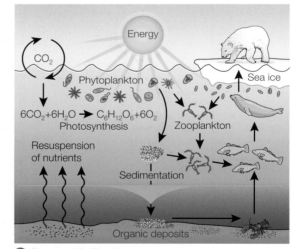

▲ **Figure 13** The Arctic ecosystem

▶ **Figure 14** Prairie Potholes, north-central USA

## Human exploitation of the global environment

For millennia human exploitation of the Earth's environmental resources was at a very low level, with little damage to the environment. As the population grew and civilisations expanded, resources were exploited at an increasingly rapid rate, often resulting in serious environmental degradation. Commercial agriculture and industrialisation have caused considerable change to global environments and ecosystems – the 'Dust Bowl' of the 1930s in central USA was caused by overcultivation and subsequent soil erosion.

Today there are very few 'untouched' wilderness areas – maybe parts of the Antarctic and Arctic and some tracts of rainforest and desert. Elsewhere, the world's natural ecosystems have been transformed; trees cut down, wetlands drained, grasslands replaced with cereal crops and expanses of land exploited for minerals.

Figure **15** shows how coastal ecosystems, both terrestrial and marine, have been affected by human exploitation. The destruction of mangrove forests and salt marshes to make way for development has massive ecological impacts on flora and fauna (see 3.11) Coral reefs are being harmed irretrievably by overfishing, pollution and global warming. Marine ecosystems have been harmed by pollution, oil spills and the discarding of waste, particularly plastics (see 6.10).

One of the most widespread and harmful forms of human exploitation involves the deforestation of tropical rainforests (see 1.11 and 1.15). Estimates suggest that half of the world's mature forests have been cleared by people. About 90 per cent of the world's species are found in tropical rainforests, yet these areas are amongst the most threatened on the planet (see 6.7).

**Goods & services**
Food
Recreation
Tourism
Biodiversity
Trapping sediment
Coastal erosion control
Mining
Shipping
Aesthetic landscapes
Culturally important places

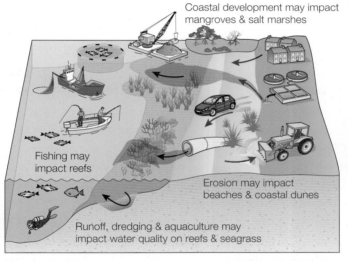

Coastal development may impact mangroves & salt marshes

Fishing may impact reefs

Erosion may impact beaches & coastal dunes

Runoff, dredging & aquaculture may impact water quality on reefs & seagrass

*Figure 15 The impact of human exploitation on coastal and marine ecosystems*

### STRETCH YOURSELF

Investigate the impact of weather-related events on ecosystems. You could study recent wildfires in California, Australia or Canada; or the 1987 'Great Storm' and how woodland ecosystems have responded since then; links between climate change and floods/droughts.

## ACTIVITIES

**(S)**

**1** Draw a spider diagram, centred on a sketch or diagram similar to Figure **1**, to outline the factors affecting the development and characteristics of terrestrial ecosystems.

**2** Study Figure **2**. Comment on the interrelationships between climate, soils and vegetation.

**3** With reference to Figures **3** and **4**, describe how and why topography can affect small-scale ecosystems.

**(S)**

**4** Use Figure **5** to draw an annotated sketch to describe the distinctive characteristics of the four altitudinal zones in the Sierra Nevada mountains. To what extent are there causal interrelationships operating between the different ecosystem components?

**5** In what ways can alien species cause change to native ecosystems.

**6** How does environmental concern and associated conservation measures result in ecological change? Is there an argument for not intervening but simply letting nature take its course?

**7 a** Study Figure **11**. Identify the threats to ecosystems posed by the changes identified. Is it all bad news?

  **b** Select one European region. Use the internet to help you to complete a short report to identify specific impacts of climate change.

**8** Describe the trend shown in Figure **12** and suggest how it will affect marine ecosystems.

**9** With reference to Figure **13**, describe how declining sea ice is affecting the fragile Arctic ecosystem. Refer to polar bears as an example.

**10** 'Human environmental exploitation represents by far the greatest threat to global ecosystems'. Discuss.

In this section you will learn about the concept of biomes and understand their global distribution

## What is a biome?

A biome is a large-scale ecological area with plants and animals that are well adapted to their environmental conditions. Biomes are generally defined by abiotic (non-living) environmental factors such as climate, relief, geology, soils and vegetation. They are categorised by their dominant vegetation, hence the terms tropical rainforest, savanna and tundra.

A biome is not technically an ecosystem, but it can be considered to operate like a large-scale ecosystem. There are terrestrial (land) biomes and aquatic (water) biomes (marine and freshwater).

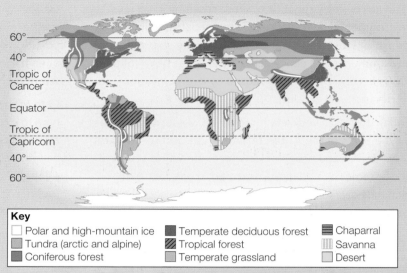

**Key**

☐ Polar and high-mountain ice
▨ Tundra (arctic and alpine)
▨ Coniferous forest
■ Temperate deciduous forest
▨ Tropical forest
▨ Temperate grassland
▤ Chaparral
▥ Savanna
▨ Desert

⬣ **Figure 1** *Distribution of terrestrial biomes*

## Distribution of global biomes

There is no definitive map of global biomes. An internet search will reveal considerable variation in the distribution and types of biomes. This is to be expected, as vegetation zone boundaries are inevitably indistinct as they naturally fade in and out. They are also irregular due to the local influences of, for example, geology, relief, proximity to the sea. You will also find variations in names, such as chaparral as opposed to Mediterranean or tropical grassland rather than savanna.

A global map of biomes is extremely generalised and does not take into account human activity. Consider the UK and much of Western Europe, which are classified as being 'temperate deciduous forest'. While this may be the natural **climatic climax vegetation** (see 5.9), it does not extend right across this region. In the UK, only 12 per cent of the land area is covered by forest, and much of this is coniferous plantation. So, the distribution of biomes (Figure **1**) is very generalised and somewhat theoretical.

## The role of climate in the distribution of global biomes

The pattern of global biomes more or less reflects the pattern of global climate zones. Figure **2** describes this link in more detail.

◆ The high rainfall and constant high temperatures in the equatorial regions creates ideal conditions for tropical rainforests, hence their broad swathe across South America, Africa and into south-east Asia (Figure **1**).

◆ The very high temperatures and low rainfall experienced at about 30 °N and 30 °S equates with the major world deserts.

◆ Further north, there are close links between climate zones and biomes, such as the Mediterranean region, the European temperate and coniferous forests and the sub-Arctic tundra.

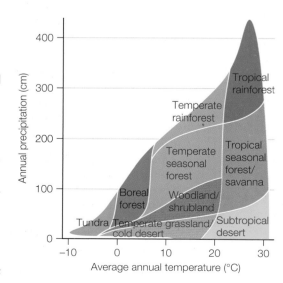

▶ **Figure 2** *The effect of climate on global biomes*

While it is possible to recognise roughly latitudinal belts of vegetation associated with climate zones, there are considerable variations within these zones due to continentality, relief and, at a more local scale, geology. For example, continentality and proximity to the sea influences rainfall and temperature ranges which will, in turn, affect the distribution of biomes. Look at the variation in biomes from the coast to the continental interior at the same latitude in either the USA or northern Europe (Figure **1**).

## The role of relief (altitude) in the distribution of global biomes

Relief (altitude) is another important factor affecting biomes (Figure **3**). As altitude increases, so temperature falls and rainfall tends to increase. This results in a zonation of vegetation (biomes) with altitude mirroring the latitudinal changes identified earlier. These variations will not always show up on a map of global biomes due to the scale involved, unless there is a significant horizontal extent of the different altitudinal zones.

Figure **4** shows the results of a scientific fieldwork study in the southern Himalayas. It shows an example of a place where local changes in altitude override the usual links between broad climatic zones and biome formation. In just a hundred or so kilometres, the vegetation changes from tropical rainforest to snowfield. (Find Butwal, west of Kathmandu, Mount Everest and Lhasa on an atlas map. The straight lines between them show the transect along which the data in Figure **4** was taken.)

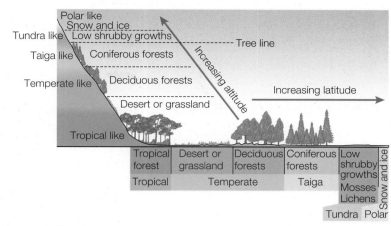

⊙ **Figure 3** *The effect of altitude and latitude on global biomes*

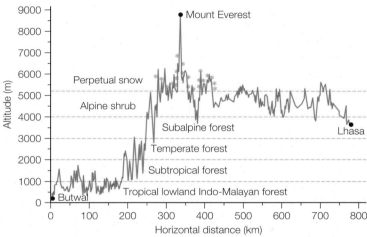

⊙ **Figure 4** *Vegetation distribution in the southern Himalayas from Butwal in Nepal to Lhasa on the Qinghai-Tibet Plateau*

## ACTIVITIES

1 Study Figure **1**. Describe and explain the distribution of the temperate grassland and coniferous forest biomes.

2 Study Figures **1** and **2**. To what extent does climate determine the distribution of global biomes?

3 Comment on the accuracy and value of the concept of global biomes.

4 Study Figure **4**.

Ⓢ

   a Draw a simple sketch to show the changes in vegetation with altitude along the line of section.

   b Describe and suggest reasons for the changes in vegetation with altitude between Butwal and Lhasa.

   c Use the internet to compare the biomes at Butwal and Lhasa. Use photographs to show the contrasts between the biomes at these two locations.

## STRETCH YOURSELF

1 Investigate the biomes (vegetation zones) in Australia. Find a detailed map of the biomes and explain why there is so much diversity. Consider factors such as climate, relief, proximity to the sea, ocean currents and wind directions. Try to discover why the biomes vary between the west and east coasts, despite the latitude being the same.

Ⓢ

2 Investigate the effect altitude has on the biomes on the slopes of Mount Kilimanjaro. There are some excellent satellite maps and examples of GIS on the internet. Describe in detail how vegetation changes with altitude and assess to what extent these vegetation zones are responding to recent climate change.

In this section you will learn about the characteristics of tropical rainforests. You will consider the controlling factors, the adaptations by flora and fauna and the issues associated with human activity.

## Characteristics and distribution of tropical rainforests

To many the tropical rainforest is a magical and unfamiliar place. In 1752, European explorers discovered large and beautiful flowers attached to bark, which they assumed were rooted inside the tree trunk.
They were wrong. So-called cauliflory trees are adapted to the needs of the tropical rainforest biome and bear their own flowers on the trunk (Figure **1**).

The rainforest biome is mostly located between the Tropics of Capricorn and Cancer. Tropical rainforests lie within the equatorial climate belt, an area 5 degrees either side of the Equator (Figure **2**). Tropical rainforests now cover less than 5 per cent of the Earth's surface – 200 years ago they covered twice as much. Yet they still support 50 per cent of all living organisms on Earth (see 1.15 and 1.16).

⬆ **Figure 1** The beautiful flowers of the cauliflory tree

### Tropical rainforest climate

The tropical rainforest (equatorial) climate results in the most productive biome on the planet. The two ingredients (inputs) necessary to support life in abundance – warmth (sunlight) and moisture – are ever present.

◆ Low diurnal (twice daily) temperature range – average daily temperature is about 28 °C; temperatures rarely fall below 22 °C at night. Cloud cover restricts daytime temperatures to 32 °C or lower.

◆ Annual temperature range as low as 2 °C.

◆ High annual rainfall of around 2000 mm – warm, moist and unstable air is forced to rise at the Inter-Tropical Convergence Zone (ITCZ) resulting in violent, afternoon convectional storms following intense daytime heating.

◆ A year-round growing season with no defined seasons – insolation is evenly distributed throughout the year with each day having around 12 hours of daylight.

◆ High humidity – rapid evapotranspiration from swamps, trees and rivers creates a sticky and oppressive heat.

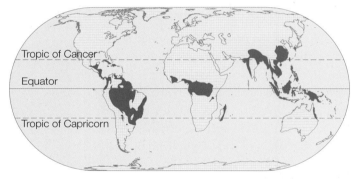

⬆ **Figure 2** Distribution of tropical rainforests

⬇ **Figure 3** The striking green layer of the forest canopy with isolated emergent trees

### Vegetation

The ideal growing conditions create a biome of high biodiversity. There may be up to 300 species of tree per hectare (there are only 5–10 in a forest in the UK) and 50 million species of animal in a tropical equatorial rainforest!

The permanent, rich, green landscape is deceptive as the trees are deciduous. However, the year-round growing season means that trees can shed their leaves at any time of the year. The constant fight for sunlight, which all plants need for photosynthesis, results in distinct stratification of the vegetation (Figure **3**).

## Soils

The hot and humid climate allows rapid chemical weathering of the bedrock and creates perfect conditions for the rotting and breakdown of the abundant leaf litter – important transfer processes in the system. Nevertheless, the soils are very fragile and depend on continuous leaf fall for nutrients. The rapid decomposition of the litter layer and the work of *biota* (flora and fauna), such as ants, result in a thin humus layer (Figure **4**).

Eighty per cent of the nutrients in a tropical equatorial rainforest originate from the vegetation. The nutrient cycle of the rainforest is extremely efficient at recycling these nutrients for sustainable growth (see 6.4). Where vegetation is removed, allowing heavy rainfall to dissolve and remove nutrients (leaching) the fertility of the soil is quickly lost.

## Adaptations of flora and fauna in tropical rainforests

With sunlight and nutrients at a premium in tropical rainforests, it is remarkable how many species thrive in this environment. They do so by forming close **symbiotic** relationships with one another and by evolving effective adaptations. This close relationship is well illustrated by the cauliflory tree (Figure **1**), the flowers of which produce a pungent night-time smell that attracts bats, which pollinate the plant.

Rainforests have a thin layer of fertile soil so most plants and trees have relatively shallow roots to maximise their ability to absorb nutrients and water, much of which is trapped by the umbrella of leaves in the canopy. The *soil water budget* (see 1.2) is high because rainfall exceeds potential evaporation, resulting in moist soils and lush vegetation, although leaching can be an issue.

A single hectare of rainforest may contain 42 000 different species of insect, up to 300 species of tree and 1500 species of higher plants. With two-thirds of all flowering plants living in tropical rainforest, it just proves how successful they have been in adapting to the challenging climatic and soil conditions.

Rainforests have layers, within which plants and animals adapt differently to the climate and soils (Figure **5**).

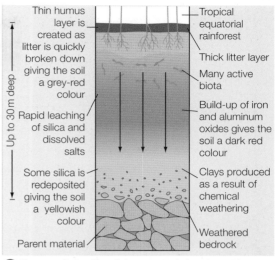

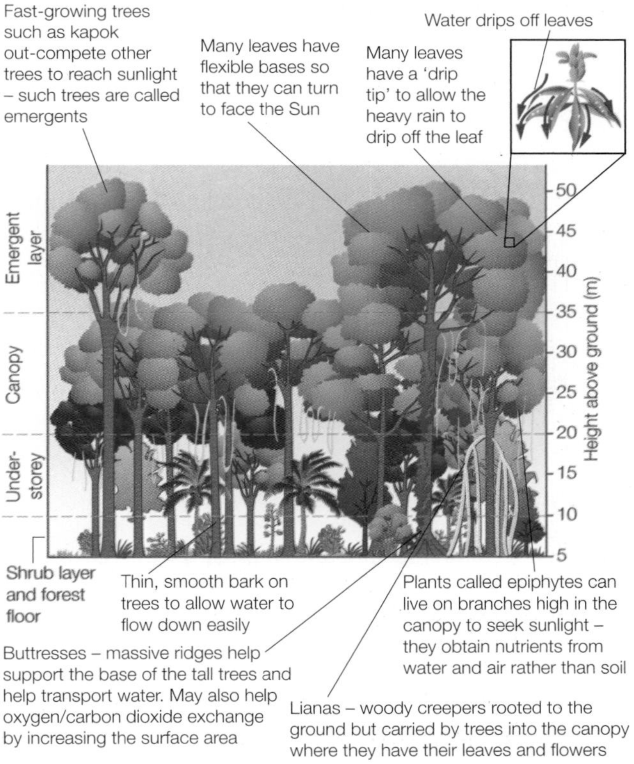

**Figure 4** *Profile of the iron-rich laterite soil commonly found in tropical rainforests*

**Figure 5** *Stratification and vegetation adaptations in a tropical rainforest*

### Emergent layer

Enormous trees tower above the canopy roof of the forest. They are particularly hardy, having to cope with the power of the Sun, high winds and heavy rain. One interesting adaptation of trees such as the kapok is that their seeds are fluffy, making use of the winds to carry them away. The only animals found here are bats and birds such as eagles (Figure **6**). It is an excellent place from which to survey the landscape for potential prey.

**Figure 6** *Harpy eagle in an animal sanctuary in the Brazilian rainforest*

## Canopy

This is the most productive part of the forest and where photosynthesis is at its greatest. Trees are carefully adapted to take gain maximum exposure to the Sun. Each mushroom-shaped crown has an enormous photosynthetic surface – the leaf area of a mature tree may amount to ten times the area of the ground beneath it! Lianas are climbing woody vines that have adapted by having their roots in the ground and climbing through the trees to seek sunlight in the canopy.

Daily heavy convectional rainfall has the potential to saturate leaves in the canopy that first dissolves then washes away nutrients from the leaf surface. Fungi, mosses and algae (all examples of *epiphytes* – see Figure **5**), may steal sunlight from the leaf surface. Around 80 per cent of canopy and emergent tree species grow *drip tip leaves* (Figure **5**) that help to shed water quickly and efficiently to stop the leaves rotting.

▲ **Figure 7** *The South American three-toed sloth*

The leaves of bromeliads form a vase or tank that holds water, which supports its own thriving ecosystem of bacteria, crustaceans, tadpoles, birds and frogs. These organisms provide nutrients through the decomposition of their faeces – a *symbiotic relationship*.

The canopy is the most heavily populated layer, with many species of birds and monkeys, often camouflaged to enable them to remain hidden from predators. The South American three-toed sloth has fur covered with green algae, which is perfect camouflage in the forest canopy (Figure **7**).

## Understorey

Slender and less substantial trees extend above the shrub layer. While the trees only receive the hazy light that breaks through the canopy, they are far more tolerant of shade than most trees and still manage to thrive. Most trees have thin, smooth bark – the high levels of humidity mean that they do not need thick bark to retain moisture.

The interlocking spindly branches of the trees form a 'spaghetti junction' of green corridors, along which lightweight animals can travel. The orangutan of Borneo and Sumatra could not be called lightweight but is able to swing from one bending branch to another (Figure **8**). Other animals that have adapted to live in this low light and relatively dry environment include bats, tree frogs and salamanders.

▲ **Figure 8** *Orangutan in the Borneo rainforest*

An almost unnatural gap exists between the top of the understorey trees and those above. Growth is restricted to several metres below the base of the canopy in order to receive a more even spread of sunlight and avoid some or all of the upper part of the tree (the crown) remaining in shadow.

## Shrub layer

Woody plants with many stems growing from their base, as well as younger trees characterise this layer. The majority of animals are unable to climb, so herbivores compete for fallen seeds, fruit, nuts and leaves. Omnivores have developed strategies to survive, including speed, stealth or camouflage. For example, the thick skin on the back of the neck of the Brazilian tapir, as well as its excellent swimming ability offer some protection from predatory crocodiles, cougars or jaguars (Figure **9**).

▲ **Figure 9** *Jaguar in the Peruvian Amazon rainforest*

## Forest floor

As little as one per cent of sunlight passes through the vegetation layers to the forest floor. The few plants that live here are mostly small herbs with large flat leaves that capture as much of the dappled light as possible. While there are few competing plants on the forest floor, such is the specialised niche that when light does enter, for example through the death of a large tree, the undergrowth is quickly killed by too much light.

Rising up to 2 m above the forest floor, thick buttress roots help to spread the weight of the towering trees over a wide area (Figure **10**). Other trees rely on stilt or prop roots for support that point downwards from the main trunk into the soil.

▲ **Figure 10** *Buttress roots providing stability for large rainforest trees*

The forest floor is teeming with insects and decomposing plants and animals. Large snakes and mammals, too heavy for the upper layers, roam this area. These include jaguars, leopards and tigers, which are all superbly camouflaged to remain hidden from potential prey. Tapirs and anteaters survive in low light conditions by developing a strong sense of smell and hearing.

## Human activity and deforestation in tropical equatorial rainforests

It is arguable that greed has driven the deforestation of tropical rainforest, in order to fund the economic development of many LDEs and meet our demands for cheap manufactured goods and food. Whether it is caused by greed or economic necessity, the effects of human activity in the tropical equatorial biome are widespread and affect us all.

Tropical rainforests are being cut down at a rate of about 32 000 ha per day and for a variety of reasons (Figure **11**) – most being related to money or the need to provide for people's families. At this rate, they could vanish in the next hundred years. Agriculture is the main driving force, where trees are chopped down to make way for ranching or commercial crop production.

| | Cause | Details | Examples |
|---|---|---|---|
|  | *To provide space for farming.* Most tropical equatorial rainforests are found in LDEs that have rapidly growing populations. Land has to be cleared to farm and to build additional settlements. | Traditional *slash-and-burn* shifting cultivation is very wasteful of space. *Cattle ranching* exposes pastures to soil erosion and produces low-quality meat, often used by TNCs for burgers sold in HDEs. *Plantations* of cash crops, such as oil palm and soya bean, require more and more land to be cleared. | Yanomami Amerindian tribes in western Brazil. Southern margins of Amazonia. Soya grown in Amazonia is used to feed chickens in the UK that then become chicken nuggets for fast-food chains. |
|  | *Logging* to provide timber. The increasing demand for valuable hardwood timber in HDEs provides a reliable and essential source of income. | The continued demand for furniture, building materials and other wooden products has encouraged felling of valuable equatorial hardwoods such as mahogany, teak and rosewood. | Japan alone accounts for a staggering 11 million cubic metres of equatorial hardwood used a year. |
|  | *Mining* of vast reserves of valuable minerals – bauxite, iron ore, tin, copper, lead, manganese and gold – but only accessible once the rainforest is cleared. | The cheapest method of mineral extraction is open-cast mining. This results in large-scale deforestation as trees and soil are stripped from the underlying rocks. | Carajás in northern Brazil is the location of the world's largest source of iron ore. 10 500 hectares of rainforest were cleared for the Juruti bauxite mine, also in Brazil. |
|  | *Road construction* supports the development of rainforest for other uses. | New private and government-funded roads allow people in and raw materials out, but cut broad swathes across the rainforest. | The Trans-Amazonian Highway extends 6000 km into Brazil's interior. |
|  | *Hydro-electric power (HEP)* provides cheap and plentiful renewable energy. | High rainfall gives potential for HEP generation. But the reservoirs that are created flood large areas of cleared forest. | Power for the Carajás iron ore mine is generated by the HEP station on the Tocantins River. |
|  | *Settlement growth* reflects the rapid population increase. | Population pressure comes from both natural growth and migration. The need for relocation cannot be ignored if other areas cannot support their populations. | Brazil has over 25 million landless people, so migration from the poorest parts, such as the drought-stricken north-east is encouraged. New roads have become 'growth corridors' for many of these 'colonists'. |

**Figure 11** *Key causes of tropical equatorial rainforest deforestation*

## Impacts of deforestation on the tropical rainforest biome

In the Amazon alone, some 17 per cent of the tropical rainforest has been lost in the last 50 years – almost half the size of continental Europe!

Refer back to the University of Leeds study (1.15) to see how deforestation affects climate and rainfall patterns.

◆ An estimated 20 000 tropical rainforest species become extinct each year, most without ever having been named (or even discovered!). Deforestation poses a huge threat to biodiversity (see opposite) and to the long-term sustainability of the tropical rainforest biome.

◆ Several animal species, including orangutans, have been decimated by loss of habitat in parts of south-east Asia. In Malaysia, more than half of river fish species have disappeared due to logging activities. Amphibian populations in tropical areas have declined because of habitat loss, pollution and disease.

◆ Without the protecting canopy above, fragile soils become exposed to direct rainfall. The resultant runoff leads to rapid soil erosion and leaching. Water courses may become silted, leading to localised flooding and soil erosion. Furthermore, with reduced rates of evapotranspiration, regional rainfall patterns may be altered (see 1.15).

◆ A tropical tree acts as a carbon sink. It absorbs around 22.5 kg of carbon dioxide each year (see 1.12). Conversely, deforestation accounts for around 20 per cent of global atmospheric carbon dioxide released each year, mostly through burning.

## The impacts of deforestation on biodiversity and sustainability

Tropical rainforests and coral reefs are the most endangered ecosystems in the world, and are vulnerable to a substantial loss of biodiversity. Deforestation depletes biodiversity by destroying habitat, by separating contiguous areas of rainforest from each other, by interfering with plant reproduction, and by exposing organisms of the deep forest to 'edge' effects. Logging does not simply remove a few trees from the forest. When canopy trees are cut, many smaller forest trees and plants, which are dependent upon them for shade, support or moisture, vanish. Animals dependent upon trees or other vegetation for food, shelter, water and breeding sites also disappear. Only those animals (generally the larger ones) that can migrate to contiguous forest areas are able to survive. Plants in the cleared areas often cannot be pollinated, or if they are, their seeds fall upon unsuitable open areas where they cannot survive. In addition, many rainforest species are restricted to relatively small areas and are found nowhere else. When the areas in which these species reside are logged or burned, they will disappear.

### Research on the changes in biodiversity in one area of Malaysian Borneo

One year after logging began in the area surveyed, there was no significant change in the total number of species. But some species, such as the Burmese brown tortoise, the rail babbler, and the four-striped ground squirrel, had disappeared from this area, and populations of ungulates (hoofed mammals), primates and hornbills (despite their protected status) were significantly lower. Hunting of many of these animals was far above sustainable levels. Squirrels thrived, since the removal of their larger food competitors left them greater food supplies (fruits, seeds). Some species – tree shrews, magpie robins, and bulbuls (species not previously found in the forest, but which are tolerant of disturbed areas) – appeared to replace the species dependent upon undisturbed forest.

Two to four years after selective logging, primate and mammalian species diversity was lower, except for ungulates. Hunting did not so much reduce the number of species present (although some species disappeared) as it did greatly diminish the abundance of animals. Generally, with the advent of selective logging and shifting cultivation, there were shifts in species, with edge-tolerant and coloniser species replacing primary forest species.

There are also unintended and unanticipated causes of forest removal. For example, flying foxes are the only known pollinators of some forest trees in Borneo and, as they are heavily hunted, some tree species will disappear. Similarly, civet cats are major seed dispersers necessary for forest maintenance and regeneration but, as they are also hunted to near extermination, the forests will be unable to sustain themselves. We know little of the eventual consequences of removing large animals from the forest ecosystem, and even less about forests' needs for various smaller animals, or insects or plants.

(www.rainforestconservation.org/rainforest-primer/3-rainforests-in-peril-deforestation/f-consequences-of-deforestation/3-loss-of-biodiversity-including-genetic-diversity)

## ACTIVITIES

1  Describe the main characteristics of the vegetation and soils in the tropical rainforest biome.

Ⓢ 2  Make a large simplified copy of Figure **5**. Add more details about the adaptations of plants and animals to the tropical rainforest biome.

3  To what extent is sunlight the driving force in the stratification of the tropical rainforest? Use examples to justify your answer.

4  Plants and animals of the tropical rainforest share a symbiotic relationship. To what extent do you agree with this statement? Include named examples of plants and animals in your answer.

5  Examine the links between the tropical rainforest biome and the water and carbon cycles. Consider the role of the biome and the implications of rainforest destruction.

## STRETCH YOURSELF

1  Use the internet to develop the Borneo case study above. Find out more about the species mentioned and consider in particular the impacts on food chains and food webs.

2  Find out more about the soil water budget for the tropical rainforest biome. How is it impacted by deforestation?

In this section you will learn about the characteristics of savanna grassland. You will consider the controlling factors, the adaptations by flora and fauna and the issues associated with human activity.

## Characteristics of savanna grassland

The savanna (tropical) grassland biome is characterised by vast rolling grasslands with scattered shrubs and isolated trees (Figure **1**). While there is enough rainfall to support grassland, there is insufficient to support tropical rainforest. These vast landscapes are often associated with safaris and animals such as giraffes, elephants and lions. Consider films such as *The Lion King* or some of David Attenborough's wildlife documentaries.

## Distribution of savanna grassland

The savanna grassland biome covers a little less than a third of the Earth's land surface. It is mostly located within the tropics, roughly between 15°N and 30°S, sandwiched between tropical rainforest and desert. As Figure **2** shows, the greatest concentration of savanna grassland is in Africa. There are also concentrations in South America, parts of south-east Asia and Australia.

This somewhat patchy distribution reflects the vagaries of the climate in these tropical latitudes, caused by variations in wind direction, ocean currents, relief and continentality (drier conditions tend to exist in continental interiors). Essentially, where the climate is partway between equatorial and desert, savanna grassland will tend to be the dominant vegetation.

⊙ **Figure 1** *The savanna grassland landscape*

⊙ **Figure 2** *The distribution of savanna grasslands (above each graph are figures for height above sea level, average annual temperature, annual rainfall)*

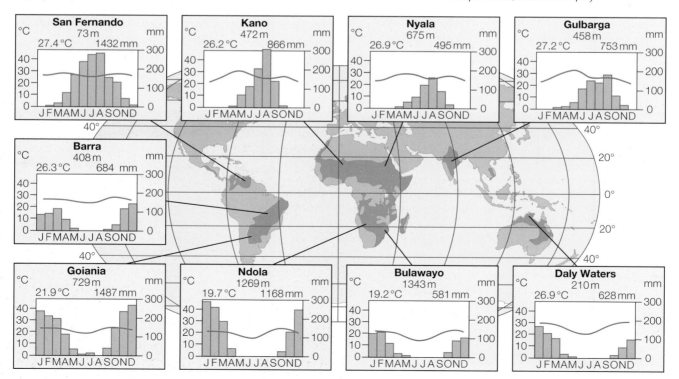

## Tropical savanna climate

The main feature of the tropical savanna climate is the presence of a clearly defined wet and dry season. This is shown by the climate graphs in Figure **2**.

In the Northern Hemisphere the wet season runs from roughly May to September, whereas in the Southern Hemisphere it runs from roughly November to March. This reflects the movement of the equatorial low-pressure belt that migrates north and south of the Equator with the 'overhead' sun. Associated with this low-pressure belt is the Intertropical Convergence Zone (ITCZ), a zone of intense rainfall and very unstable atmospheric conditions (see 1.3).

The reliability and amount of rainfall (a key input into the system) decreases with distance from the Equator, and this is reflected on the ground by a gradual change in the dominant vegetation – more trees in the wetter areas close to the Equator and more grassland in the drier regions away from the Equator.

So, within the savanna grassland biome, there is a natural variability in the type and abundance of vegetation as it reflects the climatic conditions. At the rainforest margins:

◆ precipitation is over 1000 mm per year, with one or two dry months

◆ the temperature ranges from 22 °C in the wet season to 28 °C in the dry season.

At the desert or semi-arid margins:

◆ precipitation is under 500 mm per year, with 9 or 10 dry months, and the reliability of the rainfall decreases with increasing latitude

◆ the temperature ranges from 18 °C in the wet season to 34 °C in the dry season.

Fires are common in savanna grasslands, often ignited by lightning strikes. They are an important input to the system and are part of the natural cycle, burning off dead vegetation and often stimulating new growth. Grass survives fire well with its root system safe beneath the ground, and most trees have thick barks to enable them to be fire-resistant.

After a fire, new shoots quickly appear on the surface nourished by nutrients released by the fire, such as phosphorus and potassium. Most trees in the savanna biome, such as the baobab (Figure **5**), have thick barks to enable them to be fire resistant. Most animals are able to move to safety or burrow beneath the ground.

Fires are used by pastoralists as a means of controlling the development of pasture. Widely practised in South Africa, controlled burning of grassland stimulates new growth providing a richer, more nutritious forage for animals compared with older, tougher grass.

### Savanna biome in transition

The savanna biome, as with other global biomes, is very much a transitional zone, varying enormously from north to south. In West Africa, for example, the southern edge of the biome fringes the tropical rainforest biome and has a high proportion of woodland as a result of the higher rainfall totals. With the more lush vegetation, animals are less inclined to take part in mass migration. In contrast, at the northern extent, the drier climate creates semi-desert conditions with bare soils and sparse, poor-quality grassland with few, if any, trees and shrubs.

The seasonal nature of rainfall as a result of the movement north and south of the ITCZ – has a profound impact on the growth of vegetation and therefore, the mass seasonal migration of animals – one of the main characteristics of the savanna biome. In East Africa, some two million wildebeest, gazelle and zebra migrate north in March/April from the Serengeti National Park in Tanzania to the Masai Mara National Park in Kenya. They make the return journey in October. Waiting for them are predators including lions, hyenas, leopards and cheetahs.

**Figure 3** *Migratory blue wildebeest crossing the Mara river, Masai Mara National Reserve, Kenya*

## Vegetation

The predominant type of vegetation is, unsurprisingly, grassland. Grass is ideally suited to the conditions – during the long dry season it can lie dormant with its network of roots below the ground surface, quickly bursting into life with the advent of the wet season.

The shrubs and trees will vary considerably from place to place according to local conditions. In Africa, the acacia tree is the one most commonly associated with this biome (Figure **1**), whereas in northern Australia it is the eucalyptus.

## Soils

As you have seen, the type of shrubs and trees vary from place to place depending on local controls. So also do tropical savanna soils – the local controls being, for example, geology, relief and climate. There can be pockets of wet, even swampy conditions where soils are saturated, whereas elsewhere they can be baked hard and have an almost impenetrable crust.

For the most part, the soils are quite porous due to the presence of the grass roots. This allows water to drain through the soil during the wet season. However, it does lead to the process of leaching (a transfer process), where nutrients are carried through the soil profile to be deposited (stored) further down. This results in a relatively infertile topsoil where any humus created from rotting vegetation is concentrated very close to the surface (Figure **4**). The high temperatures promote the action of *aerobic* (in the presence of oxygen) bacteria breaking down plant material and releasing the nutrients, thus limiting the amount of organic material stored within the soil.

The intense leaching causes the soils to have a characteristically red colour due to the presence of iron oxide. These soils are called *laterites*. Generally speaking, the more rain that falls in the wet season, the greater the intensity of leaching and the more impoverished the soils.

The soil water budget varies considerably during the year and also from place to place. It tends to fluctuate seasonally, experiencing a surplus in the wet season and a deficit in the dry season.

# Adaptations of flora and fauna in savanna grasslands

Plants and animals living in the savanna grassland biome have had to adapt to cope with potentially hostile and limiting environmental conditions. These include high temperatures throughout the year and a period of intense rainfall followed by a long period of drought and associated wildfires. Plants have to cope with infertile soils and a variable soil water budget, reflecting the seasonal nature of the rainfall.

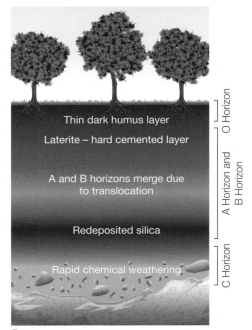

Thin dark humus layer

Laterite – hard cemented layer

A and B horizons merge due to translocation

Redeposited silica

Rapid chemical weathering

O Horizon

A Horizon and B Horizon

C Horizon

⏵ **Figure 4** *Typical savanna grassland soil profile*

## Duricrust

During the dry season evaporation is the dominant process, drawing water up to the ground surface from the soil below. Minerals dissolved in groundwater may be precipitated on the ground to form hard, silica-rich *duricrust*, which can vary in thickness from a few millimetres to several metres, depending upon the rainfall regime, the nature of the geology and the rates of evaporation. Below the ground, a hardpan soil may form in fine, acidic soils when there is a concentration of mineral deposition, typically iron and calcium, that effectively cements soil particles together. This hard pan soil restricts root penetration and hinders the growth and diversity of vegetation.

## Plant adaptations

Grass is well-suited to the conditions, dying back in the dry season and then regrowing quickly (up to 130 cm high) from its root nodules in the wet season. Some grasses are wiry and can turn their blades away from the Sun to reduce water loss. Others are more tussocky, enabling moisture to be retained close to the ground surface.

Shrubs growing in the savanna often have small, waxy leaves or thorns to reduce water loss. Deep tap roots enable them to seek water deep underground, while some have fire-resistant stems. Some plant seeds can remain dormant for years to survive prolonged drought (see 2.2).

Most savanna trees are deciduous, losing their leaves in the dry season. This helps them to reduce water loss. Evergreen trees tend to have small, leathery leaves that reduce the amount of transpiration (water loss). Two common trees with particular adaptations are the acacia and the baobab.

◆ Acacia trees (Figure **1**) have a low umbrella-shaped canopy that provides shade for the shallow roots, thus reducing soil water evaporation. They have small, waxy leaves to reduce water loss and they also lose their leaves (deciduous) in the dry season.

◆ The baobab tree is very characteristic of the African savanna, with its distinctive bulbous trunk (Figure **5**). It has a thick, fleshy trunk that can store water and thick bark to insulate and protect it from fire. Long tap roots seek water deep underground and it only bears leaves for a few weeks each year, thus reducing water loss.

## Animal adaptations

There are over 40 species of hooved animals (ungulates) that live on the African savanna, all coexisting successfully due to their highly specialised feeding habits. Some are grazers, living off the grass, whereas others are browsers, eating leaves from trees. Each graze/browse at different levels, eat different plants at different times of day and in different areas. This interspecies survival is the key to retaining biodiversity and sustainability in the savanna grasslands. For example, elephants eat shrubs and trees, which enables the grass to grow to support the many grazers.

Most animals have long legs that enable them to undertake long migrations as they follow the rains across the grassy plains to benefit from fresh grass growth.

⬣ **Figure 5** *A baobab tree in the Lower Zambezi National Park*

⬣ **Figure 6** *African wild dog, Madikwe Game Reserve, Botswana*

Some animals live in burrows to avoid the heat of the day and others lose heat through large areas of exposed skin, such as the ears of elephants.

The African wild dog is well adapted to life in the savanna grassland biome (Figure **6**). Its light body and long legs enable it to give chase to prey, its multicoloured coat enables it to be camouflaged and its large ears enable it to lose heat. They do not compete with hyenas and jackals because they only eat fresh meat; they are not scavengers. They do not compete with lions either, who chase larger prey.

Birds of prey such as hawks and buzzards use the isolated trees to nest, from where they have an excellent field of vision across the plains to spot prey.

# The impacts of human activity in savanna grasslands

## Overgrazing and ranching

Savanna grasslands are under threat from agriculture, and in particular overgrazing (Figure **7**). Nomadic herders have traditionally driven their domestic animals (cattle, sheep and goats) across the grasslands with little damage. However, with more permanent (sedentary) agriculture, overgrazing can be a real problem especially if exacerbated by extensive periods of drought. In the future, this situation might be made worse by climate change. Vegetation can be destroyed, exposing the soil to wind and water erosion, eventually leading to desertification, particularly at the desert margins (for example, the Sahel region in Africa, Figure **8**).

**Figure 7** *Domestic cattle grazing on savanna grassland in Kenya*

Overgrazing can be considered to be a positive feedback loop – it reduces the amount of vegetation, increasing the pressure on the remaining grassland, which, in turn, leads to more overgrazing and soil erosion.

In Botswana, the growth of ranching has not only put pressure on those areas demarcated for commercial grazing but has also led to a shrinkage of communal grazing land, resulting in overgrazing and the loss of biodiversity. This has been exacerbated by private landowners using the communal land as well as their own ranches. Landownership is clearly a major issue here and dual-grazing rights by private landowners will need to be eliminated if the system is to be sustainable.

In some places, grasslands are managed by people to create better land for grazing. This may involve the use of fire to clear areas of dead vegetation and stimulate fresh growth of grass. Such fires can kill young shrubs and trees, reducing biodiversity and the range of habitats for birds and animals. They can also burn uncontrollably if missmanaged.

Grazing by domestic cattle can result in some woody plants dying as their foliage is stripped bare. These plants then become replaced with animal-repellent trees and shrubs, such as thorny bushes, thereby reducing biodiversity.

**Figure 8** *Dry grazing land in the Sahel, Ethiopia*

## Tourism

Tourism – particularly safaris – has become popular in recent decades and this can lead to high concentrations of people in small areas and development pressures such as new roads, runways and hotels. Animals can become scavengers, eating human rubbish and being poisoned by toxic substances. The intensive use of vehicles in honeypot areas such as reserves and parks can damage the grass, leading to soil erosion (Figure **9**). Animals can be disturbed in their hunting and mating.

Poaching (e.g. for ivory) and hunting also pose threats to biodiversity and long-term sustainability.

◀ **Figure 9** *On safari in Ngorongoro Crater, Tanzania*

## ACTIVITIES

1  Study Figure **1**. Describe the characteristics of the savanna grassland shown in the photo. How sustainable do you think this landscape is?

2  Study Figure **2**. In what ways and for what reasons does variability exist within and between areas of savanna grassland? Consider the climatic variations and the implications for vegetation and animal habitat development. Attempt to draw a cross-section across the savanna in Africa to show latitudinal variability from rainforest-edge to desert-edge.

3. Make a copy of Figure **4** and add detailed labels to describe the main characteristics and controls on savanna soils. How does this typical soil type affect the ecology of the savanna grasslands?

4  Assess the role of fire in the savanna grasslands – is it predominantly positive or negative? How does fire fit into the savanna grassland system?

5  Make a detailed study of the adaptations of one named tree and one named animal to conditions in the savanna grasslands. Use the internet to seek additional information, and include an annotated photo for each.

6  Consider the impacts of commercial ranching on the ecology of the savanna grassland. What are the likely effects on biodiversity and how can commercial ranching be managed sustainably?

## STRETCH YOURSELF

Make a study of the savanna grassland biome in Australia using the excellent Savanna Explorer website www. savanna.org.au. What makes the Australian savanna distinctive in terms of its vegetation and animals? Focus on a selection of the plant and animal adaptations and consider the effects of human activities on biodiversity and sustainability. Attempt to represent the biome in the form of a systems diagram.

In this section you will learn about vegetation succession and climatic climax in the British Isles. You will also learn about the impact of human activity on vegetation succession.

## What is a vegetation succession?

Plant communities not only vary from area to area, but evolve and become more complex over time. This development and the associated changes in a plant community through time is called vegetation succession (seral progression) (see 3.7). Each distinctive stage in the progression can be identified as a seral stage (see 1.9). If allowed to continue undisturbed by human activities, the succession will end up in a state of perfect adaptation to, and in equilibrium with, the environment at the time. This is called the vegetation climax. A climax community of natural vegetation is sometimes called the *climatic climax vegetation* because climate is so often the main controlling factor. The UK's climatic climax is deciduous woodland, characterised by oak, birch and ash trees (Figure **1**).

There are two types of vegetation succession.

◆ Primary successions (*priseres*) occur on any surfaces that have had no previous vegetation. Examples include vegetation succession on bare rock exposed by glacial retreat, lava flows following volcanic eruptions and sand dunes constructed by onshore winds at the coast.

◆ Secondary successions (*subseres*) occur on surfaces that have already been covered by vegetation (colonised) but have since been modified or destroyed. Fire following lightning strikes, landslides or human activities such as deforestation can lead to a secondary succession.

**⬆ Figure 1** *Deciduous woodland, Hayley Wood, Cambridgeshire*

## Development of a vegetation succession

All vegetation successions pass through a sequence of seral stages, starting with a **pioneer community** (colony). This takes the form of a *ground layer* of hardy plants, such as lichens and mosses, because they can grow without soil. As weathering (an input) breaks down the rock and dead plant remains are decomposed (a transfer) into humus by bacteria, a *field layer* of herbs and grasses begins to grow in the immature soil.

The plants in the field layer are taller than the ground layer, so become dominant until fast-growing shrubs take over. Taller plants will always dominate smaller ones by capturing light for photosynthesis (an *energy transfer*) that would otherwise reach plants lower down. They will also provide shelter to allow other plants to become established. Consequently, each new seral stage will show an increase in both the height of plants and in the number of species.

**⬆ Figure 2** *Birdhouse Meadow, Ambleside, Cumbria – a plagioclimax community*

Over time, the *shrub layer* will ultimately be taken over by a *tree layer* of taller, slower-growing trees. Providing that the environmental factors do not change, the vegetation should eventually reach a state of equilibrium at this point – the climatic climax vegetation.

This whole sequence can take anything from a few decades to thousands of years, depending on the environmental circumstances. But in reality, human activities frequently alter the natural vegetation succession, whether by clearing the climax vegetation or preventing the succession getting to this stage. As a result, **plagioclimax communities**, such as meadows (Figure **2**), grasslands or heather moorlands, are more likely to exist in densely populated and highly developed regions.

Throughout the British Isles there are four generally accepted environments where primary vegetation successions at a local scale could progress to the climatic climax vegetation of temperate, deciduous oak–ash woodland. Lithoseres and psammoseres are found on land (vegetation successions on land are known as xeroseres) and haloseres and hydroseres are formed in salt and fresh water (Figure **3**). (Also see 1.9).

▼ **Figure 3** *Primary vegetation succession leading to climatic climax in the UK*

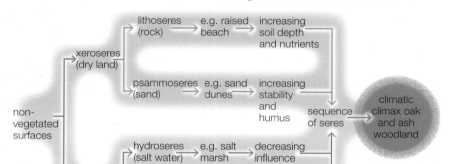

## Examples of vegetation successions

### Lithosere

A lithosere occurs when bare rock is progressively colonised as soils develop. Newly created bare rock can be formed by an event such as a volcanic eruption, or it can involve the exposure of a rocky surface previously covered by ice, water or vegetation.

Raised beaches are common features along the western coast of the UK. They were formed when sea levels fell after the end of the last glacial period some 8000 years ago. Free from the enormous weight of the ice, the land has risen slowly and erratically (isostatic rebound), resulting in the relative fall in sea level. Figure **4** shows a raised beach in south-west England.

▼ **Figure 4** *Raised beach at Prawle Point, Devon*

Look at Figure **5**. This shows the theoretical lithosere succession set in the context of a raised beach. Lichens, liverworts and mosses form a ground layer pioneer community encouraging soil formation. The field layer of herbs and grasses then follows, to be shaded out by the shrub layer of ferns, bracken and brambles as the soils mature. Small but fast-growing trees such as rowan can then establish themselves, to be followed by taller birch and pine trees. Finally, slower-growing oak and ash trees emerge as the dominant tree layer species of the climatic climax vegetation.

Look back at Figure **4** and notice that, despite the raised beach being exposed for some 8000 years, it is not covered in woodland. This is because people have converted and maintained the land as farmland, primarily grazing livestock. This is a good example of a plagioclimax, as the succession is prevented from progressing and reaching its climatic climax.

Consider the vegetation succession on the raised beach in terms of a system (see 1.1). The inputs would include energy from the Sun, rainfall, seeds brought in by the wind and human actions. Outputs would include nutrients leached through the soil or dead vegetation blown away by the wind. There are many stores of energy, carbon and water (see 1.2, 1.3, 1.8). The plants themselves are obvious stores together with the soil. Processes and transfers would include photosynthesis (see 1.9) and leaching (see 2.2). This system is being held in check, so human actions, in the form of farming, act as negative feedbacks. For example, grazing animals prevent the growth of shrubs and trees, thereby maintaining the dominance of grasses for them to graze. At the fringes, positive feedbacks may operate (soil formation leading to plant growth, decomposition, soil enrichment and so on), moving the succession through its seral stage.

⊙ **Figure 5** *Natural vegetation succession along a raised beach – a lithosere*

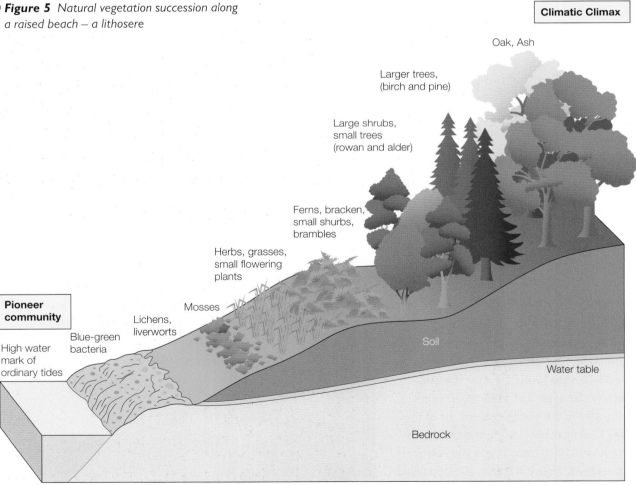

## Hydrosere

Hydroseres develop where freshwater environments such as ponds and lakes silt up over time (Figure **6**). The spores of algae and mosses, blown on to the water surface (an input), form a pioneer community of rafts of floating vegetation. Floating and submerged water weeds, including lilies, can then develop, trapping sediment at the water's edge to allow marsh plants such as reeds and rushes to establish. Continuing sedimentation of both silt and plant debris will slowly build up to eventually rise above the water level to produce a fen of small shrubs and trees, including willow and alder. Finally, the climatic climax of oak and ash trees can emerge, taking over the whole site once the pond silts up entirely (Figure **7**). Refer to the case study about Wicken Fen in 6.13.

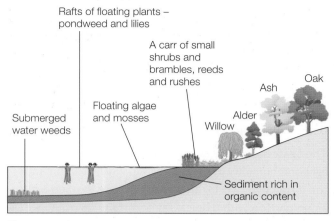

Rafts of floating plants – pondweed and lilies

A carr of small shrubs and brambles, reeds and rushes

Oak

Ash

Alder

Floating algae and mosses

Willow

Submerged water weeds

Sediment rich in organic content

⬇ **Figure 6** *Primary vegetation succession in a hydrosere*

⬆ **Figure 7** *Silted-up oxbow lake, Weaselhead Natural Area, Calgary, Canada*

## Temperate deciduous woodland biome

In the British Isles, the natural climatic climax is temperate deciduous woodland (see 6.6). This is the type of vegetation biome that would naturally occur if there was no land management or interference by people. Of course, so much of the British Isles has been developed for agriculture, infrastructure and urbanisation that vast swathes of deciduous woodland have been removed and only isolated pockets remain, almost all of which represent secondary successions (Figure **1**).

Temperate deciduous woodland is a high-energy biome with productivity second only to the tropical rainforest (Figure **8**). This is due to highly favourable growing conditions – plenty of rainfall, warm summers and cool winters. They do not have to cope with extremes of rainfall and temperature or with seasonal fires.

⬇ **Figure 8** *Organic productivity of the world's major biomes*

| Energy | Biome | NPP (g/m²/year) |
|--------|-------|------------------|
| High | Tropical equatorial rainforest | 2200 |
| | Temperate deciduous woodland | 1200 |
| Average | Tropical savanna grassland | 900 |
| | Temperate coniferous forest | 800 |
| | Mediterranean woodland | 700 |
| | Temperate grassland | 600 |
| Low | Tundra | 140 |
| | Hot deserts | 90 |

*Net primary production (NPP)* is a measure of the rate at which plants store energy as organic matter in excess of that used in plant respiration. It is expressed as an equivalent of grams of dry organic matter per square metre per year.

## The structure of temperate deciduous woodlands

The key characteristic of deciduous woodlands is that the trees shed their leaves in the autumn as temperatures fall – winter is not suitable for photosynthesis or leaf growth. There may be a shortage of water if it is frozen in the ground and leaves may freeze if exposed to very low temperatures. Losing their leaves helps trees to conserve water loss through transpiration. The weight of snow could also cause branches to snap.

The seasonal nature of leaf growth affects the levels of light and humidity in the woodland, which accounts for the distinctive layering that exists (Figure **9**). (See 6.7 for a comparison with tropical rainforests.)

◆ Ground layer – mostly limited to mosses and lichens growing among the thick layer of leaf litter, which is important in ensuring the deep, fertile soils that characterise this biome. For much of the year, this layer is heavily shaded and quite damp, with little movement of air – ideal for bacteria and fungi to thrive.

◆ Field layer – comprises brambles, bracken, ferns, grass and some flowering plants such as bluebells, which flourish in spring and early summer before the leaf canopy above has fully developed (Figure **1**). Plants living in deciduous woodland have to be able to cope with significant light (and humidity) variations during the year. From November until May they will be exposed to sunlight (albeit from a low-angled Sun for much of the time) and also direct rainfall. Woodland floors can become very wet and marshy during the winter. In the summer, dappled or dark shade will dominate as the full growth of leaves block out the Sun. Interception will reduce the amount of rainfall reaching the ground, although the lack of air movement will retain humid conditions.

⊙ *Figure 9* *Oak, birch and rowan trees in Killarney, Ireland*

◆ Shrub layer – this may include low bushes and smaller trees such as rowan, holly, hazel and hawthorn, all of which compete for light. Most of these trees are also deciduous, losing their leaves to conserve water in the winter.

◆ Canopy layer – dominated by oak trees and other tall deciduous species such as lime, elm, beech, chestnut, maple, sycamore and ash – all averaging 20 m in height. The trees develop extensive crowns of broad, thin leaves to absorb maximum sunlight during the summer. They shed their leaves in the winter primarily to conserve water.

Temperate woodlands are extremely biodiverse and they support a large range of animals, birds and insects that have adapted to cope with food shortages in the winter by migrating (e.g. deer, birds), hibernating (e.g. hedgehogs), storing food for winter (e.g. squirrels) or changing colour (e.g. stoats and weasels).

### Soils in temperate deciduous woodlands

The zonal soil in temperate deciduous woodland is brown earth (Figure **10**). These soils are characteristically deep, well drained and fertile and they are able to support abundant fauna. Decomposers break down the thick layer of leaf litter to create rich, dark brown humus, which earthworms and rodents aerate and mix into the layers (known as *horizons*) below. Leaching is variable, depending upon the amount of precipitation. However, any nutrient losses are usually balanced by gains from precipitation, weathering and decomposition, thereby maintaining high levels of fertility and creating slightly acidic soils that are ideal for plant growth. This helps to account for the high productivity of this biome.

⊙ *Figure 10* *Rich, crumbly brown earth soil*

## Impacts of human activity in temperate deciduous woodlands

According to the Wildlife Trust, ancient woodlands cover just three per cent of England's land area – 80 per cent are less than 20 hectares in size and half of these are less than five hectares. Much of the natural woodland has been lost or damaged through agricultural and urban developments. During the Second World War, much of the ancient broadleaved woodland was replaced by quick-growing conifers to supply timber for the war effort. In the 1950s and 60s, there was extensive planting of conifers on semi-natural habitats such as heathland, grassland, bog and wetland in order to secure timber supplies for the future. As a result, during the twentieth century, 40 per cent of England's ancient woodland was converted to plantations.

# The effects of human activity on succession

## What is a plagioclimax?

The final stage of a natural vegetation succession is called the climatic climax. In the UK as we have already seen, this is predominantly temperate woodland. But in reality, human activities frequently impact on plant succession, whether by clearing the climatic climax vegetation or preventing the succession ever reaching this stage.

For example, deciduous forest may be cut down and replaced with pasture, which, grazed by animals, is maintained at this artificial seral stage. A salt marsh succession may not reach its climatic climax if the land is drained and used for agriculture. Other human activities can include deforestation, afforestation, ploughing and clearance by burning. This artificial human-induced vegetation climax is referred to as a *plagioclimax*.

## Heather moorland plagioclimax

Heather moorland, such as the North Yorkshire Moors, is an excellent example of a plagioclimax community (Figure **12**). Despite being widespread and appearing to be a completely natural form of vegetation it is, in fact, heavily managed by people. Heather is the staple diet of red grouse and therefore the basis for the highly lucrative shooting industry.

Some 3000 years ago these areas were covered in climatic climax deciduous woodland on fertile brown earth soils. But the woodland was cleared for farming (growing crops and grazing), breaking the naturally efficient nutrient cycling of deciduous woodland and exposing the ground to heavy upland rainfall. Soils were both eroded and leached of their nutrients, leaving thin, acidic, less fertile peaty *podsols*, which are typical of temperate climates created by coniferous or heath vegetation.

As a result, the upland areas were colonised by more hardy plants including grass, bracken and heather, tolerant of a wide range of climate and soil. This mixed moorland vegetation could be maintained by sheep grazing. But in upland areas, such as the North Yorkshire Moors and the eastern Scottish Highlands, there is now deliberate management, using burning, to maintain a plagioclimax community of heather moorland.

## Nutrient cycle

Nutrient cycling (see 6.4) is very efficient in deciduous woodlands (Figure **11**). Deciduous trees are demanding of nutrients (hence the thick flow arrow) but the annual leaf fall together with high rainfall ensures a significant return of nutrients to the soil. The favourable conditions for decomposition promote a rapid transfer of nutrients back to the soil.

These ideal soil-forming conditions ensure that the deep roots of deciduous trees have easy access to moisture and nutrients with little danger from waterlogging. The agricultural potential of these soils is, therefore, enormous.

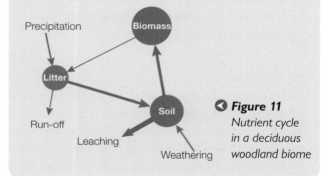

Precipitation
Biomass
Litter
Run-off
Soil
Leaching
Weathering

◀ **Figure 11**
*Nutrient cycle in a deciduous woodland biome*

**Did you know?**
The Glorious Twelfth in Britain refers to one of the busiest days in the shooting season. This is because 12 August sees the start of the shooting season for red grouse.

▼ **Figure 12** *Upland heather moorland, Goathland, North Yorkshire Moors*

## Managing heather moorland

Heather is a valuable evergreen forage plant – a major food source for hardy hill sheep and red grouse. Human management of heather moorland maintains this artificial seral stage, preventing the natural vegetation attaining its climatic climax.

Its natural cycle of growth follows four phases (Figure **13**).

**1** Pioneer phase – heather seedlings establishing themselves and the root system, growing more rapidly than the shoots.

**2** Building phase – the most productive and valuable. Vigorous growth and prolific flowering sees an increase in biomass to form continuous ground cover.

**3** Mature phase – the ground cover becomes discontinuous and the plants woodier.

**4** Degenerate phase – growth slows down. The heather's maximum height is reached and the oldest central branches die off. At this stage there are large gaps in the ground cover and the value of the heather is limited.

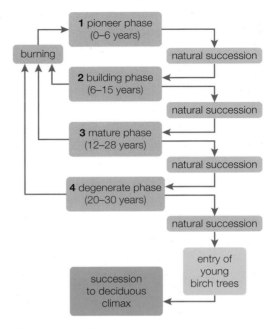

▲ *Figure 13* *The heather cycle*

Controlled burning keeps as much of the moorland as possible in the most productive building phase (Figure **14**), keeping the maximum amount of edible green shoots. The surface heather is burnt off in a 10–15 year rotation in small areas of about one hectare, which allows the burning to be carefully controlled. It also provides feeding areas for breeding grouse, with unburnt nesting cover nearby. Normally within each square kilometre there are about six burning patches, resulting in a patchwork quilt pattern of heather at various stages of regrowth.

If the burning was stopped the moorland would be invaded again. A natural secondary vegetation succession would follow of scrub, birch and, in the long term, reverting to a climax of deciduous woodland.

▲ *Figure 14* *Controlled burning of heather moorland, North Yorkshire*

## ACTIVITIES

**1** Define the following terms: vegetation succession, seral stage, climatic climax and plagioclimax.

**2** Study Figure **4**. What is the evidence that the vegetation communities on the raised beach represent a plagioclimax?

**3** Consider the hydrosere vegetation succession (Figure **6**) as a system. Make a simple copy of Figure **6** and use annotated labels to show inputs, outputs, stores and processes (transfers). Consider the presence of any feedback loops.

**4** Study Figure **7**. Comment on the development of the hydrosere in the photo. To what extent does this represent a plagioclimax?

**5** Using an atlas of thematic maps, compare and contrast global-scale maps showing climate zones, natural vegetation and soil regions. To what extent do these patterns overlap? Are there any regional anomalies? Suggest reasons why these might occur.

**6** Draw and annotate a sketch diagram to show vegetation stratification in a temperate deciduous woodland biome. Make sure that you clearly identify the ground, field, shrub and tree layers. Use the internet to help you develop and extend the detail of your annotations. Use labelled photos to illustrate your diagram if you wish.

**7** With the aid of simple sketches and/or photos from the internet, describe the management practises that retain heather as a plagioclimax vegetation community in moorland areas.

# How and why does vegetation change down a valley side slope?

To study the ecological characteristics of a steep slope of a valley on Exmoor, a group of A Level students collected data along a transect (Figure **15**). They expected to find variations in soil depth, soil water and vegetation. The slope is quite free draining (the bedrock is sandstone) and there is a small stream in the bottom of the valley. Figure **16** shows their predictions.

While their results (Figure **17**) were somewhat inconclusive in terms of changes in soil types, soil depth did decrease in the steeper middle section and increase towards the valley bottom. There were also significant changes in vegetation, with a dominance of grasses, heather and bilberry on the plateau, bracken and grass on the steeper middle section and soft rush, grasses and sphagnum moss in the valley bottom. This suggested that slope moisture was a key control and that this, in turn, was controlled by slope angle.

To extend their study, the students decided to consider more closely the possible relationship between slope angle and the abundance of bracken (calculated as a percentage cover using a quadrat). Bracken is a plant that prefers drier soils, so they predicted that it would be more abundant on the steeper slopes in the middle section of the slope. To test this relationship the students decided to use Spearman's rank to see if there was a relationship between bracken abundance and slope angle. They predicted a positive correlation.

▲ **Figure 15** *Students collecting data along a transect on a valley side slope, Exmoor*

▼ **Figure 16** *Cross-section of the south-facing valleyside side slope, Exmoor*

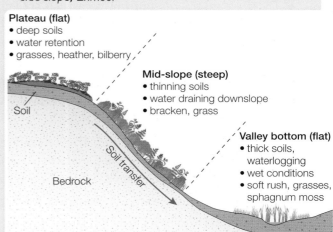

**Plateau (flat)**
- deep soils
- water retention
- grasses, heather, bilberry

Soil

**Mid-slope (steep)**
- thinning soils
- water draining downslope
- bracken, grass

Soil transfer

Bedrock

**Valley bottom (flat)**
- thick soils, waterlogging
- wet conditions
- soft rush, grasses, sphagnum moss

| Slope facet | Slope angle (degrees) | Rank (r1) | Bracken abundance (%) | Rank (r2) | d | d² |
|---|---|---|---|---|---|---|
| 1 | 3 | | 0 | | | |
| 2 | 6 | | 2 | | | |
| 3 | 9 | | 0 | | | |
| 4 | 22 | 4.5 | 11 | | | |
| 5 | 26 | 2.5 | 18 | | | |
| 6 | 18 | | 25 | 2 | | |
| 7 | 28 | 1 | 25 | 2 | −1 | 1 |
| 8 | 26 | 2.5 | 13 | | | |
| 9 | 22 | 4.5 | 23 | 4 | 0.5 | 0.25 |
| 10 | 21 | 6 | 21 | 5 | −1 | 1 |
| 11 | 18 | | 0 | | | |
| 12 | 17 | | 0 | | | |
| 13 | 15 | | 25 | 2 | | |
| 14 | 11 | | 19 | 6 | | |
| 15 | 6 | | 8 | | | |

▲ **Figure 17** *Table of measured data (n=15). Note that d = r1−r2*

## ACTIVITIES

**1 a** Use the data in Figure **17** to plot a scattergraph, with 'slope angle' on the horizontal axis and bracken abundance on the vertical axis. Draw a best fit line to show a relationship. What does the graph indicate?

**b** Now use the data to calculate the Spearman's rank correlation coefficient (ρ). Rank each column of data (r1/r2). The first six have been done for you. Calculate the difference (d) between r1 and r2 (note that the mean is taken for data of equal values). Use the formula to calculate the coefficient (n=15). Use the significance tables at www.york.ac.uk/depts/maths/tables/spearman.pdf to interpret your result.

$$\rho = 1 - \frac{6\, \Sigma\, d_i^2}{n\,(n^2 - 1)}$$

**c** Summarise your findings. Do they support the student's predictions? Comment on the reliability of your conclusion.

**2** The bracken is spreading further down the slope and is now more abundant in the valley bottom. Suggest possible reasons for this.

In this section you will learn about the characteristics and development of coral reef ecosystems and the threats posed by human activity

## Characteristics of coral reef ecosystems

A coral reef is a rocky ridge formed by the hard exoskeletons of millions of tiny coral animals. Some reefs are several million years old. Coral reefs are one of the richest and most biodiverse ecosystems on Earth (Figure 1) – they account for only about 18 per cent of the marine environment yet they are home to nearly 25 per cent of all marine species.

▲ **Figure 1** *Coral reefs – 'rainforests of the sea'*

## Why are coral reef ecosystems important?

◆ There are several reasons why coral reefs are important. They form one of the richest ecosystems on Earth, supporting many thousands of species of fish, plants and other organisms.

◆ More than 450 million people live within 60 km of coral reefs, with the majority directly or indirectly deriving food and income from them.

◆ With their stunning beauty (Figure 1), coral reefs are a popular tourist attraction. In the Caribbean alone, they are responsible for generating some US$10 billion annually.

◆ The global value of the world's coral reefs has been estimated at almost US$30 billion each year. In Hawaii alone, the benefits associated with tourism, fishing and biodiversity amount to US$360 million a year.

◆ Coral reefs form a natural barrier to protect the mainland from powerful waves and storm surges. This same principle applies to the less frequent but potentially much more devastating tsunami waves mainly generated by submarine earthquakes.

◆ Algae and sponges on coral reefs have valuable medicinal qualities that scientists believe may treat viruses and some cancers in the future.

◆ Coral reefs store carbon in the form of calcium carbonate – an important role in the carbon cycle (see 1.9). Carbon that has been weathered on land, dissolved and carried to the oceans by rivers is then used by corals to build their exoskeletons.

## Distribution of coral reef ecosystems

Warm coral reefs (there are some species that survive in colder climates) need a delicate balance of marine environmental conditions to survive and thrive. This explains why they are predominantly found within in the Tropics (Figure 2).

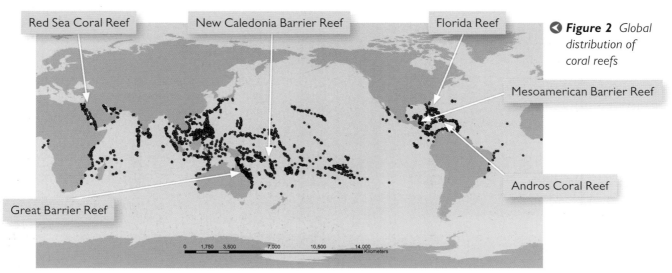

Red Sea Coral Reef

New Caledonia Barrier Reef

Florida Reef

◀ **Figure 2** *Global distribution of coral reefs*

Mesoamerican Barrier Reef

Andros Coral Reef

Great Barrier Reef

| 0 | 1,750 | 3,500 | 7,000 | 10,500 | 14,000 |
Kilometers

## Environmental conditions and coral reef development

Temperature – corals only live in oceans that have an average temperature of 18 °C and over. The ideal temperature is between 23 °C and 25 °C.

Acidity – corals thrive in relatively high levels of alkalinity (e.g. the waters of the Great Barrier Reef off the east coast of Australia have a pH of about 8.2). If the seawater becomes more acidic, for example by an increased absorption of carbon dioxide in seawater, it can stunt the growth of corals and even kill them.

Clear water – corals survive best in clear, unpolluted water. Sediment clogs their feeding structures and also reduces the amount of light. This is why coral reefs are not found at river mouths.

Salinity – corals can only tolerate salinity that is close to that of seawater. This explains why there are gaps in reef development at river mouths where freshwater is discharged.

Air – the vertical growth of coral is limited by exposure to air, which will kill them. So upward growth is limited to the level of the lowest tides.

Light – corals feed on algae, which need light to photosynthesise and grow. Coral reefs are found in relatively shallow water, usually less than 25 m in depth, where there is enough light for algae to thrive.

▲ **Figure 3** *Environmental conditions required for the development of coral reefs*

### Nutrient cycling in coral reefs – the role of algae in coral reef development

To appreciate the fragile nature of the coral reef ecosystem and to understand why they are vulnerable to changing environmental conditions, it is important to understand the coral reef nutrient cycle.

Corals live in nutrient-poor waters. It is only through very efficient nutrient recycling that corals maintain such a diverse ecosystem. At the heart of the recycling is a symbiotic relationship (both organisms benefit from an association) that exists between coral and algae.

◆ *Zooxanthellae* (plant-like algae) live within the tissues of the coral *polyp* (the tiny anemone-like animals that make up the coral). They are able to harness the light from the Sun, converting it into energy, just like plants, to provide nutrients to the corals.

◆ In exchange, the zooxanthellae benefit by having somewhere to live and having exclusive access to the waste-nutrients (nitrogen and phosphorus) produced by the coral, which fertilize the algae.

The most important nutrient is nitrogen. The symbiotic relationship between the algae and the coral captures and retains nitrogen very effectively – it is passed back and forth between the two organisms. In open waters, free-floating algae and marine animals lose nitrogen to the water, so nitrogen recycling is less effective.

The close relationship between coral and algae is supported by nutrients obtained from the water and from the consumption of microscopic prey called *zooplankton* (Figure **4**). Zooplankton obtain nutrients by consuming *phytoplankton*, a primary producer living in the ocean that converts light from the Sun directly into energy. Corals are also able to digest bacteria and edible detritus that enter the system by upwelling from the ocean floor. Phytoplankton is at the bottom of the food chain – without it the ecosystem would not survive.

Coral reefs providing food and shelter for fish. They also constantly excrete ammonia (a dissolved form of nitrogen) which is absorbed by corals and algae.

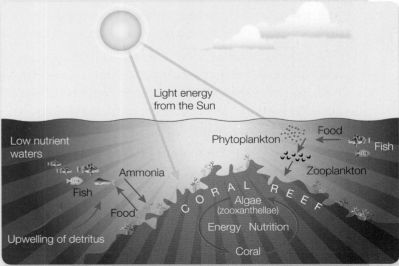

▲ **Figure 4** *Coral reef nutrient cycle*

## What are the threats to coral reef ecosystems?

Despite their global importance, coral reefs are one of the world's most threatened and endangered natural ecosystems on the planet. Some of the main threats include:

◆ Major drainage basin schemes, such as urban developments or deforestation can wash significant quantities of silt into the sea. This can clog the coral's feeding mechanism and smother it.

◆ Land clearance and onshore coastal developments (infrastructure, settlement, tourism) can also lead to more freshwater runoff and pollutants (e.g. sewage), which can alter the salinity and chemistry of the seawater, preventing the growth and development of coral.

◆ In some parts of the world, particularly the Middle East, coastal desalination plant discharges will have detrimental effects on coastal ecosystems by increasing salinity and temperatures, resulting in the accumulation of metals and potentially toxic chemicals. When discharge occurs in poor flushing areas (without strong currents to dissipate the discharge) concentrations can build up, with significant impacts on coral reef and other ecosystems.

◆ Nutrient-rich agricultural and sewage discharges can lead to the growth of algal blooms that can smother corals and block sunlight.

◆ Physical damage associated with fishing or tourism (e.g. boat anchors and people walking on the coral).

◆ Climate change is probably the biggest threat across the world. It can lead to acidification and coral bleaching, both of which could destroy vast swathes of coral reef in the future.

### Catchment pollution: Great Barrier Reef, Australia

The quality of the water is the main reason why the Great Barrier Reef is one of the most beautiful natural environments in the world. Climate change is the biggest threat to its future and is already having an effect. It also affects the quality of water from catchment runoff – the second biggest threat to the reef.

Thirty-five river basins drain an area of 424 000 km², supplying the inshore area of the reef with its greatest source of nutrients. Sediment inflow into the Great Barrier Reef has increased up to 10 times for some catchments since the middle of the nineteenth century and by 4–5 times as a whole.

The increased use of nutrients, pesticides and other pollutants on the land means more of these pollutants enter the waterways and subsequently the reef, reducing the quality of the water. This affects not only the coral, but also important habitats such as wetlands, seagrass beds and mangroves and the marine life they support. Associated industries such as tourism and fishing are also adversely affected.

### One Tree Island, Great Barrier Reef

A study of the Great Barrier Reef at One Tree Island in Australia found that by making the water less acidic (increasing its pH value) and closer to the value it would have been in pre-industrial days, the rate of deposition of calcium carbonate, which is used to grow the hard exoskeletons, increased.

'Our work provides the first strong evidence from experiments on a natural ecosystem that ocean acidification is already slowing coral reef growth. Ocean acidification is already taking its toll on coral reef communities. This is no longer a fear for the future; it is the reality of today.'

(Dr Rebecca Albright, the Carnegie Institution for Science, Washington DC)

🔻 **Figure 5** *Satellite image of the Great Barrier Reef, Australia*

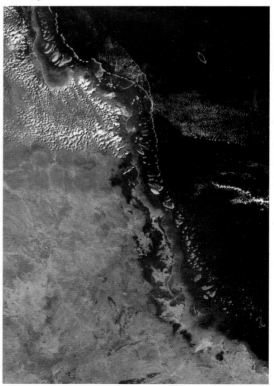

## Coral reef case study: Andros Barrier Reef, Bahamas

The Andros Barrier Reef is part of an extensive coral reef system in the Bahamas, off the south-east coast of Florida, USA (Figures **6** and **7**). The entire reef is the third most extensive coral reef system in the world. A 'barrier reef' is so-called because it forms a linear feature parallel to the shoreline separated from it by a wide lagoon.

The Andros Barrier Reef stretches for approximately 200 km and is separated from the land by a shallow lagoon with mangrove forests. The outer edge of the reef is marked by a steep drop to a depth of over 2000 m known as the 'Tongue of the Ocean'.

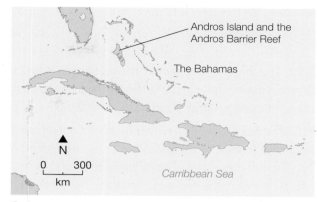

▲ **Figure 6**  *Location of the Andros Barrier Reef, Bahamas*

### Ecological characteristics of the Andros Barrier Reef

The warm tropical Bahamian climate is ideal for coral reef formation. The waters are relatively free from pollution and silt and there are no significant rivers discharging into the sea to upset the balance of salinity. The clear water enables maximum penetration of sunlight so that zooxanthellae and phytoplankton can flourish and photosynthesise effectively. Remember that the ecosystem would not survive without phytoplankton.

As a result of its favourable environmental conditions, the Andros Barrier Reef is extremely biodiverse and recognised as one of the healthiest reef systems in the world. Scientists estimate that over 160 species of fish and coral make up the reef community. They include red snapper, reef shark, rock lobster, sharp nose puffer and green turtle, together with many colourful species of coral (Figure **8**).

In common with other reef systems, the many different species are closely interrelated – the survival of one species is often dependent on the survival of several other species. Fish benefit from the safety and shelter of the reef for breeding and obtain their food from the plankton, crustaceans and other fish. Coral benefits from nitrogen excreted through the gills of fish and from detritus swept up by the swirling fish shoals.

▲ **Figure 7**  *The Andros Barrier Reef, Bahamas*

▶ **Figure 8**  *Diving with Caribbean reef sharks*

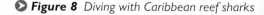

## What are the values of the Andros Barrier Reef?

The Andros Barrier Reef is important for several reasons.

◆ Coastal protection – acts as a buffer zone, providing vital shoreline protection from storms and tsunami, for example Hurricane Sandy in 2012. The shallow water above the reef forces waves to break early before reaching the islands. While this reduces coastal erosion and the risk of flooding on land, it can be quite destructive to the coral reef itself.

◆ Fish breeding grounds – creates sheltered conditions for the growth of mangrove forests, which themselves are important breeding grounds for fish. Fish are important commercially and for tourism, as well as being vital elements in the reef ecosystem. In the Bahamas, local and export markets for snapper, grouper, lobster and conch generate millions of dollars.

◆ Tourism – commercial and recreational activities such as fishing, sports-fishing, cruising, snorkelling and scuba diving bring in over US$150 million per year. The vertical wall of the 'Tongue of the Ocean' and sunlight penetration due to its east-facing aspect makes Andros Barrier Reef one of the most spectacular diving experiences in the region (Figure **9**).

◆ Healthy coral reef – widely recognised as being one of the healthiest reefs in the world, the Andros Barrier Reef is a superb outdoor laboratory for scientific research. In particular, it acts as a useful control in assessing environmental changes taking place in coral reef ecosystems elsewhere in the world.

**Figure 9** *Scuba diving the Tongue of the Ocean off the Andros Barrier Reef*

## Impacts of human activity on the Andros Barrier Reef

### Climate change

Perhaps the most significant, although indirect, human impact on coral reefs involves climate change. The increased quantities of carbon dioxide in the atmosphere have been counterbalanced to some extent by increased absorption (sequestration) in the oceans. While this may appear to be a good thing in helping to reduce climate change (less carbon dioxide in the atmosphere means less enhanced greenhouse effect), it has led to an increase in ocean acidification, which scientists believe is a key factor in the decline of coral reef ecosystems.

When carbon dioxide is absorbed in the ocean, it reacts with seawater to increase acidity. If the water becomes too acidic it dissolves the calcium carbonate corals that molluscs and creatures such as crabs and lobsters need to build their shells and stony skeletons. It has been estimated that in the Andros Barrier Reef, coral calcification may decline by as much as 10 per cent by 2040 due to ocean acidification.

Higher water temperatures resulting from climate change can trigger a stress reaction in corals, causing them to expel the zooxanthellae. This has a huge effect on nutrient flows and causes the coral to become 'bleached', literally turning white (Figure **10**). Eventually the coral dies. Projections from climate models suggest that the coral reefs in the Bahamas may experience sufficient thermal stress to lead to severe bleaching after about 2040.

 **Figure 10** *Coral bleaching*

## Overexploitation

Commercial fishing and intensive tourism can cause immense harm to the coral reef ecosystem. The depletion of herbivorous fish can lead to problems in that they help to keep competitors, such as seaweed, under control, preventing it from becoming invasive.

Grouper is a very popular fish to eat. However, their removal in some cases has led to an increase in damselfish, which groupers feed upon. Damselfish, in turn, create pockets in coral, ideal habitat for the algae upon which damselfish feed. In time, these algae can take over a reef, essentially smothering it. Overfishing of herbivorous (plant-eating) fish can also lead to high levels of algal growth.

Corals can be killed by physical contact with anchors, fishing nets, boat hulls and even people's feet. This is certainly an issue in the coastal waters around the Bahamas, where thousands of divers, fishermen and sightseers are attracted every year. Although the harvesting of sponges (Figure **11**) is an important local industry in the Bahamas (its origins stretch back to the 1840s when Greek divers first exploited the reef for its sponges), it can create a harmful imbalance in the ecosystem.

▲ **Figure 11**  *Sponge farmers trim the harvest for shipment, Andros Island, Bahamas*

## Pollution

Pollution can involve agricultural chemicals, sewage and silt eroded from hillslopes and discharges by rivers. Silt causes the water to become cloudy, restricting the penetration of sunlight used by zooxanthellae to photosynthesise. Coastal developments (particularly for tourism) have led to the clearance of vegetation, which has increased soil erosion and coastal silt deposition.

Between 1950 and 1980 there was considerable logging of the natural pine forests on North Andros Island, which led to significant quantities of silt washing into the sea, smothering some areas of the coral reef. Algal blooms in the Caribbean and the Florida Keys have also smothered parts of the coral and blocked the sunlight required by the zooxanthellae to complete photosynthesis.

Deep-sea fishing takes place off the Andros Barrier Reef. Marine-based oil and chemical pollution from trawlers and other ships can be harmful to both corals and fish.

## Hurricanes

The Bahamas is often affected by tropical storms and hurricanes, particularly from September to November, the so-called 'hurricane season'. While the Andros Barrier Reef provides important protection for the coastline, the coral can be severely scoured by strong currents and powerful waves and damaged by the snapping of branching corals and removal of sponges. Storms can also disturb the seabed sediments, clouding the water and potentially clogging up the corals' feeding systems.

Interestingly, hurricanes can cause an upwelling of cold water from the ocean depths, thereby cooling the surface waters and mitigating the impact of coral bleaching. In cooling the surface waters, this also makes subsequent hurricane formation less likely. This is a good example of a negative feedback loop.

## Management of the Andros Barrier Reef

Management of threats to coral reefs in the Bahamas is shared by the Department of Marine Resources (DMR) and the Bahamas National Trust (BNT). The government is committed to protect 20 per cent of its near-shore habitat by 2020. Several national parks and reserves have been established to help preserve parts of the valuable reef ecosystem (Figure **12**). The environmental quality of the reef ecosystem is frequently assessed for coral bleaching and to identify any harmful impacts from human activity.

- The Andros Westside National Park includes Andros Island and part of the coral reef and was designated in 2002 to balance long-standing traditions of the island, such as fishing and sponging, while also promoting resource conservation, recreational fishing and ecotourism. Developments in the National Park are strictly controlled.

- The North and South Marine Parks were established in 2002 on the eastern side of the island. Some activities are regulated or prohibited, such as fishing and collecting wildlife, mining, vessel anchoring, scuba diving and the discharging of materials.

- The Crab Replenishment Reserve was set aside to ensure a sustainable crab population for future generations.

- In the Exuma Cays Land and Sea Park, a coral nursery has been established. Here threatened species are conserved and monitored prior to being planted back in the coral reef.

- There are plans to establish a new national park to the north of Andros Island at Joulter Cays. This part of the reef is prized for its fishing and for its varied and extensive shallow water ecosystem. It is under pressure from excessive fishing, damage from boats and marine discards.

Several organisations monitor the environmental quality of the reef ecosystem to assess coral bleaching and to identify any harmful impacts from human activity. Fortunately, the Andros Barrier Reef remains one of the healthiest reefs in the world.

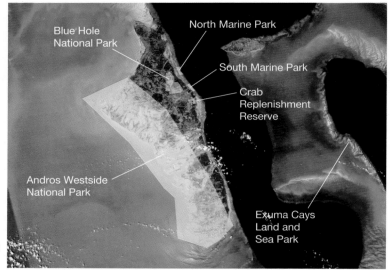

 **Figure 12** *Reef management and conservation areas on and around Andros Island*

## ACTIVITIES

**S**

1 Find a photo or diagram of a coral reef ecosystem and construct text boxes around your illustration to describe the six environmental requirements for coral reef formation.

2 **a** Study Figure **4**. What are the sources for nutrients in the coral reef ecosystem?

   **b** What is meant by the term 'symbiotic'?

   **c** Describe the symbiotic relationship that exists between the corals and zooxanthellae.

   **d** How does the coral reef ecosystem maintain such biodiversity in low nutrient waters?

   **e** Describe the important relationship that exists between coral and fish.

3 How does the Andros Barrier Reef benefit the economy of the Bahamas?

4 With reference to the Andros Barrier Reef and other coral reefs from around the world, assess the effects of river discharges (nutrients, silt and freshwater) on coral reefs.

5 Find out more about the effects of desalination plant discharges on coral reefs. This is one of the most localised human impacts on coral reefs and its effects are quite controversial. Decide whether you think desalination plants represent a threat.

6 To what extent is climate change the single most important threat to coral reefs?

7 How are management strategies addressing some of the issues associated with the Andros Barrier Reef?

# What are the future prospects for coral reefs?

According to the WWF, 27 per cent of the world's coral reefs have already been lost and, if present rates of destruction are allowed to continue, this will rise to 60 per cent over the next 30 years.

In 2011 an assessment carried out by the World Resources Institute reached the following conclusions.

◆ More than 60 per cent of the world's coral reefs are under local threat – such as overfishing and destructive fishing, coastal development, watershed-based or marine-based pollution.

◆ Approximately 75 per cent are rated as threatened when local threats are combined with thermal stress. This reflects recent rising ocean temperatures, which are linked to the widespread weakening and mortality of corals due to mass coral bleaching.

◆ If all threats are left unchecked, the percentage of threatened reefs will increase to more than 90 per cent by 2030 and to nearly all reefs by 2050 (Figure **13**).

**Figure 13** *The global pattern of threatened coral reefs. Corals use aragonite to build their skeletons. A saturation state of 4.0 or greater is best for coral growth, 3.0 or less is marginal. $CO_2$ levels in tropical waters are estimated to be around 450 ppm in 2030 and 500 ppm in 2050. In 2050 only a few areas may have adequate conditions for reef building.*

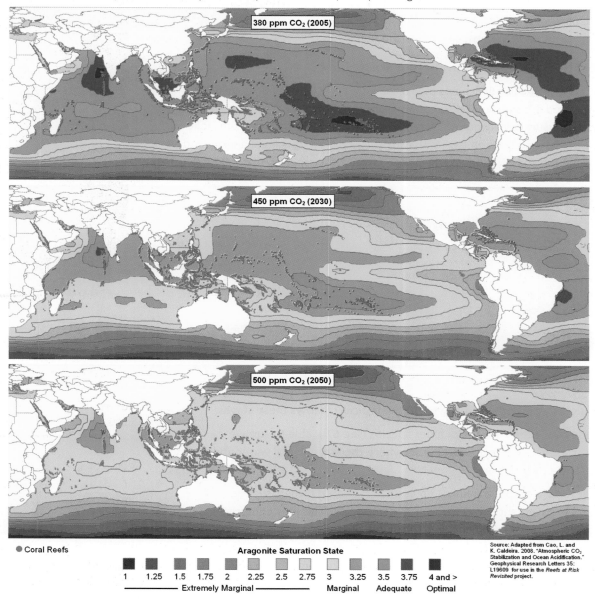

● Coral Reefs

**Aragonite Saturation State**

| 1 | 1.25 | 1.5 | 1.75 | 2 | 2.25 | 2.5 | 2.75 | 3 | 3.25 | 3.5 | 3.75 | 4 and > |

——————— Extremely Marginal ——————— | Marginal | Adequate | Optimal

Source: Adapted from Cao, L. and K. Caldeira. 2008. "Atmospheric $CO_2$ Stabilization and Ocean Acidification." Geophysical Research Letters 35: L19609 for use in the *Reefs at Risk Revisited* project.

In this section you will learn about the characteristics of urban wasteland ecosystems, the factors affecting their development and the impacts of change

## Characteristics of urban wasteland ecosystems

Urban wasteland is a general term used to describe abandoned land that has essentially been left for nature to 'take its course'. The term *brownfield* can be used to describe these sites that were used for urban or industrial purposes. They include former factories or routeways, such as railway lines or canals, dumping sites for industrial waste, building demolition sites and quarries (Figure **1**).

Urban wasteland potentially provides lots of different microhabitats due to the variety of surfaces and materials (e.g. bare soil, broken-up tarmac and concrete, piles of rubble, derelict buildings) and the varied topography, with flat open areas, hummocks, holes and depressions.

**Figure 1** *Nature taking its course on a derelict canalside factory in Hanover, Germany*

## Vegetation succession on urban wasteland

Despite the particular difficulties that these sites pose (which might include contaminated ground, limited soil depth, lack of moisture and nutrients) plant succession is surprisingly quick. Succession on abandoned industrial sites follows a lithosere-type succession (see 6.9), with adapted plant species changing (and improving) the soil conditions, which then favour other species that follow (Figure **2**).

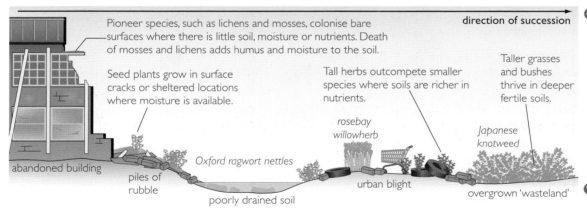

direction of succession

Pioneer species, such as lichens and mosses, colonise bare surfaces where there is little soil, moisture or nutrients. Death of mosses and lichens adds humus and moisture to the soil.

Seed plants grow in surface cracks or sheltered locations where moisture is available.

Tall herbs outcompete smaller species where soils are richer in nutrients.

Taller grasses and bushes thrive in deeper fertile soils.

*rosebay willowherb*

*Japanese knotweed*

abandoned building

piles of rubble

*Oxford ragwort nettles*

poorly drained soil

urban blight

overgrown 'wasteland'

**Figure 2** *Examples of plant succession on abandoned industrial sites or wasteland*

**Figure 3** *Purple flowering buddleia on waste ground*

Vegetation succession on a brownfield site typically follows these stages:

1 Mosses and lichen are the pioneer species (see 6.9), able to survive on surfaces such as bricks and walls where there is minimal water and nutrient supply.

2 Over time, a simple soil builds up in small sheltered niches, such as cracks in pavements or at the foot of damp walls. This may comprise dust blown by the wind and small amounts of humus and dead moss. Seeds may set and plants such as Oxford ragwort and buddleia may grow (Figure **3**). Buddleia – also known as the 'butterfly bush' on account of its attraction to butterflies – is particularly abundant alongside railway lines. Next time you are on a train, look out for it.

**3** In time, as soils develop further, plants such as rosebay willowherb, with its characteristic bright pink flowers, start to dominate. This plant, with its vigorous lateral *rhizomes* (subterranean stem from which roots and shoots grow), is well adapted to spread quickly across areas of relatively thin soil (Figure **4**).

**4** Grasses then follow and, as they die back, the soils become increasingly enriched and biodiversity increases.

**5** Invasive species such as Japanese knotweed may become established. These species can take over an area and smother other plants, leading to a reduction in species diversity.

**6** The final stage in the colonisation of an area of wasteland will involve the growth of small shrubs and trees such as brambles, willow and laburnum that exploit deeper cracks and crevices for their roots.

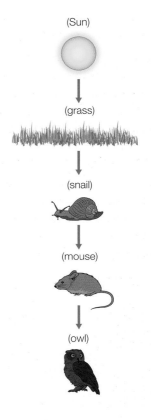

🔺 *Figure 4  Spreading rhizomes on a plant*

### Bugs and beasties associated with urban wasteland

Urban wasteland, with its derelict buildings and piles of rubble and industrial waste, forms a range of habitats for insects, birds and animals. There are many warm and sheltered niches for burrows or nests to be constructed. In addition to the many insects, butterflies, rodents and birds there will also be larger mammals such as urban foxes and badgers. Figure **5** shows a typical food chain and food web associated with an area of urban wasteland.

## Factors affecting the development of an urban wasteland ecosystem

There are several important factors that affect the development of an urban wasteland ecosystem including climate, soils and soil water budgets. The flora and fauna that live and thrive in such environments show ecological responses to these factors in the form of adaptations.

◆ *Climate*. Derelict buildings and piles of waste create several significant microclimates. There will be warm, sunny walls, sheltered sun traps, exposed and windy areas and places of variable humidity. The subtle variations in microclimate will be exploited by different plants and animals for growing or making their homes. For example, mosses thrive in damp areas, whereas spiders prefer drier habitats. Shrews prefer to live in relatively moist habitats with leaf litter and thick plant cover where they consume foliage-living insects, such as grasshoppers (Figure **5**). Flowering plants such as buddleia will tend to colonise sunny walls (Figure **3**) and in the summer they will often be covered by butterflies.

◆ *Soils*. Soils will be largely absent or, at best, thin and infertile. Mosses and lichens cope well under these conditions and plants with rhizomes (such as rosebay willowherb and couch grass) spread out laterally through the thin soils. Where soils develop in cracks, deeper-rooted plants will flourish and even trees such as laburnum.

◆ *Soil water budget*. Urban wasteland sites are often dry and dusty, with thin, rubbly soils. This results in low moisture retention and negative soil water budgets, with potential evaporation exceeding precipitation. Plants have adapted to these conditions in various ways. Some grasses and plants, such as rosebay willowherb, have shallow roots, enabling them to obtain moisture quickly following rainfall events. Others will exploit deep cracks in the ground where water may collect. Mosses act rather like sponges, retaining water to enable them to survive dry periods.

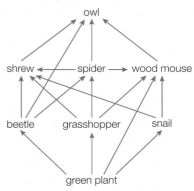

🔺 *Figure 5  Urban wasteland food chain and food web*

335

## Issues of change in an urban wasteland ecosystem

Urban environments are subject to constant economic, social and political change, which creates urban wastelands. However, as circumstances change, so wastelands themselves become threatened. They may be wiped out to make way for new developments or conserved and managed for the benefit of local communities or to protect endangered species. They may become transformed by the introduction of invasive species. It is interesting to consider how areas that were once 'good for nothing' have in many cases become highly valued beacons of green in a predominantly grey urban landscape.

## Urban redevelopment

By definition, an area of urban wasteland is just that, an abandoned area that has no immediate economic purpose, although it may be used informally for recreational activities such as biking. However, should this site be redeveloped in the future the ecosystem that has developed in the meantime will be subject to significant change and may even be completely destroyed. Ponds may be drained or infilled, land may be flattened and bulldozed and waste materials together with vegetation removed (Figure **6**).

## Invasive plant species

Invasive plant species such as Japanese knotweed and Himalayan balsam can have a devastating impact on the development of urban wasteland ecosystems. Japanese knotweed (Figure **7**) was brought into the UK in Victorian times as an ornamental plant and is now a recognised invasive weed. In London alone the plant has doubled its coverage in the last 20 years or so and is present along most of the capital's waterways.

Japanese knotweed is resistant to treatment – its stems regrow, float downstream and subsequently disperse at a new site. It can only be effectively treated by spraying with strong herbicides and then completely disposing of the weed and roots, by burning, for example. Cleared sites need to be regularly monitored so that it does not return.

**Did you know?**
On a global scale, exposed soil in brownfield sites is recognised as an important sequester of carbon, helping to offset the build-up of $CO_2$ in the atmosphere. For example, carbon sequestration in the UK's brownfield soils has the potential to meet 10 per cent of the UK's annual $CO_2$ reduction target.

**Figure 6** *Removing a derelict factory and clearing wasteland at High Wycombe, Buckinghamshire*

**Figure 7** *If left untreated, Japanese knotweed will consume local biodiversity and is a real threat to conservation areas*

## Conservation and amenity use

In the UK, government policy focuses on a 'brownfield first' approach, targeting new developments on available sites within urban areas. While this may seem like a good idea in principle, in that it conserves 'greenfield' sites outside the urban area, it does not take account of the often thriving ecosystem that has developed.

The National Planning Policy Framework allows for the protection of biodiversity, even on brownfield sites, but even so they are often seen as 'useless' areas attracting antisocial behaviour such as fly-tipping. Several pressure groups and charities such as Buglife, the Wildlife Trust and the RSPB have been active in the identification of brownfield sites that require conservation rather than development. As a result, several sites have become Nature Reserves or even Sites of Special Scientific Interest (SSSI), affording them protection from development.

Increasingly, areas of urban wasteland are being claimed by local communities as valuable ecological assets amid the urban landscape. Former wastelands have become important green areas for leisure, recreation, education and even as urban farms.

While such conservation developments retain a green footprint in the urban environment, they do represent a degree of change to the naturally developing wasteland ecosystem. Should brownfield sites be left alone to develop 'naturally'?

### Canvey Wick SSSI, Canvey Island, Essex

In 2005, Canvey Wick became the first brownfield site to be designated a Site of Special Scientific Interest (SSSI) specifically for its invertebrates, after a campaign by the conservation organisation Buglife and local residents to protect it from development. This former oil refinery now supports an outstanding invertebrate assemblage of national importance including the shrill carder bee (Figure **8**) and two species thought to be extinct, the Canvey Island ground beetle and the Morley weevil.

As an SSSI, Canvey Wick is closely monitored and strictly managed to retain the important habitats that will enable the rare species to survive. There are very strict planning controls and no harmful developments will be allowed.

Current management at Canvey Wick includes:

◆ Extensive vegetation and invertebrate surveys.

◆ The maintenance of critical ecological habitats such as bare ground and scrub through periodic disturbance and scrub clearance, retaining drought-stressed brambles and some scrub bordering tarmac bases.

◆ Improving grassland management to reduce the cover of rough grassland from 20 per cent down to 10 per cent. This will allow an increase in short sward grassland from 10 to 15 per cent, and herb-rich grassland supporting species, such as wild carrot and narrow-leaved bird's-foot trefoil, from 10 to 30 per cent.

◆ The annual clearance of 10 per cent of ditches to diversify successional and vegetation types.

The site was originally an oil refinery, which was decommissioned in 1973 and, bizarrely, was never used (Figure **9**).

▲ **Figure 8** The shrill carder bee

▼ **Figure 9** The remains of the Occidental Petroleum jetty at Canvey Wick SSSI, Canvey Island in Essex

### New Ferry Butterfly Park, Wirral, Cheshire

Situated next to Bebington railway station on the Wirral, New Ferry Butterfly Park is described by the Cheshire Wildlife Trust that manages the site as 'an oasis of green tranquillity in a densely populated area'. This urban nature reserve has developed on a former railway coal yard, goods yard and water-softening plant. Despite the thin, nutrient-poor soils, carpets of wild flowers have developed, including wild carrot and bee orchids.

The former uses of the site have been put to good use. Calcareous grassland has been created using the lime waste from the water-softening plant, creating a thriving community of flora and fauna that have adapted to this type of habitat.

Drifts of coal dust have been transformed into acidic grassland now home to common bent, sheep's sorrel and bird's-foot trefoil. Up to 26 species of butterfly have been recorded here, 18 of which breed on site (Figure **10**).

Through the work of its many volunteers, the Cheshire Wildlife Trust aims to identify and preserve threatened habitats and species in a wide variety of ecological settings. It provides advice to local government and landowners, and directly manages a number of sites including New Ferry Butterfly Park. It organises public events and open days and seeks to educate local people and in particular school children, so that they learn to value their natural environment and appreciate their role as custodians of the future (Figure **11**).

▲ *Figure 10* Gatekeeper butterfly at New Ferry Butterfly Park

▼ *Figure 11* Open Day at New Ferry Butterfly Park, 2015

## Prinzessinnengärten, Moritzplatz, Berlin, Germany

An area of urban wasteland in Berlin that had been left abandoned for over 50 years has been transformed into an area of productive gardens called Prinzessinnengärten (Princess Gardens). Beginning in 2009, the group Nomadisch Grün (Nomadic Green) worked with local communities to clear away rubbish from the site at Moritzplatz in Berlin Kreuzberg to construct organic vegetable plots and in effect create an urban farm (Figure **12**).

This shows that there is alternative to building on open space within large cities. The resident-led project not only created a biologically diverse and sustainable microclimate within a large city, it also gave a sense of community that added to the quality of life for those that created and use the gardens.

 **Figure 12** *Development of an organic urban farm at Prinzessinnengärten, Moritzplatz, Berlin*

## ACTIVITIES

1 Study Figure **1**. To what extent does this area of urban wasteland provide 'lots of different microhabitats'?

**Ⓢ 2** Outline with the aid of a simple diagram the typical vegetation succession on a patch of urban wasteland.

3 Describe the species diversity in an urban wasteland ecosystem.

**Ⓢ 4** With the help of the internet, describe the adaptations of flora and fauna to the environmental conditions (such as climate and soils) experienced in an urban wasteland ecosystem. Use annotated diagrams (such as Figure **4**) and photos to support your account.

5 To what extent is the plant buddleia an invasive species? Use the internet to find out more about this species. Find out why it is such a successful colonising plant on wastelands.

6 Use the internet to find out more about the important role played by the holly leaf miner in the urban wasteland food chain. Refer to the Field Studies Council website at www.field-studies-council.org/urbaneco/urbaneco/introduction/feeding.htm.

7 Outline the conservation strategies employed at Canvey Wick and New Ferry Butterfly Park. Assess the need for intervention in the management of urban wasteland.

8 Should urban wasteland be conserved or is there an argument for leaving it to develop naturally without any human interference?

## STRETCH YOURSELF

Use the internet to investigate one of the brownfield ecological case studies on the Buglife website (www.buglife.org.uk/brownfield-case-studies). For your chosen case study, consider the former use of the land, the main characteristics of the ecosystem and the management strategies that have been adopted. Find out if there are any similar projects close to where you live. An investigation into urban wasteland ecosystems would make an interesting fieldwork investigation.

In this section you will learn about the ecological changes that have taken place on Exmoor, their causes and responses

## Exmoor National Park and its ecological characteristics

Designated a national park in 1954, Exmoor is an area of upland moorland covering about 700 sq km of west Somerset and north Devon in south-west England (Figure **1**). Its name is derived from the River Exe, which has its source and headwaters in the hills.

There is a wide range of natural habitats (and associated flora and fauna) on Exmoor including heaths, coastal marshes, ancient woodlands and upland peat wetlands (Figure **2**). There are several protected areas, including the South Exmoor SSSI, which is home to rare species of trees such as sessile oaks, and nesting populations of birds such as redpoll, wheatear and stonechat.

About 25 per cent of Exmoor is uncultivated heath and moorland mainly used for sheep grazing or recreation. Herds of red deer roam the moorlands and hundreds of species of birds and insects including peregrine falcons, merlin and curlew.

The peat wetlands (locally called *mires*) of upland Exmoor are of great ecological importance and contain a range of habitats and a rich biodiversity. They have an important role in maintaining water quality and act as a carbon sink, helping to reduce carbon dioxide concentrations in the atmosphere. Additionally, the peat wetlands have preserved archaeological artefacts stretching back 10 000 years.

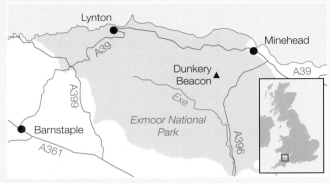

▲ **Figure 1** *Location map of Exmoor National Park*

▲ **Figure 2** *Moorland landscape in Exmoor National Park*

Evidence of environmental conditions in the past are preserved in the soils and sediments. Pollen grains, plant fossils and the remains of insects and other organisms provide a link to climatic conditions in the past. They are also invaluable to scientists studying the possible effects of climate change.

## Ecological change on Exmoor – its nature and causes

There are three main drivers of ecological change on Exmoor – agriculture, tourism and climate change. Collectively, these factors illustrate how the economic, social and political character of the community reflects Exmoor's ecological setting.

### Agriculture

Historically, agriculture has posed the greatest ecological threat on Exmoor, particularly in the upland peat moorlands where farmers have dug ditches to drain the land in an attempt to make it more productive. This has caused peatlands to dry out in places and become more vulnerable to soil erosion by wind and rain. Their importance in maintaining water quality and sequestering carbon from the atmosphere has been diminished and, in altering the wetland habitats, biodiversity has been reduced.

Currently, farming on Exmoor is dominated by sheep and beef cattle. While much of the low-lying land is cultivated, the higher moors are generally only used for rough grazing. Diversification has increased markedly in recent years to supplement farm incomes; about 50 per cent of farmers are involved in off-site diversification and 30 per cent in on-site diversification. Much of this is concerned with small business enterprises associated with tourism – bed and breakfast, camping and farm shops. While these economic and social developments to some extent reflect Exmoor's ecological setting, they also impact upon the ecology by disturbing wildlife and altering habitats.

## Tourism

Tourism brings both advantages and disadvantages to Exmoor. It is estimated that some two million people visit Exmoor each year, attracted by the rich cultural and wildlife heritage as well as the many outdoor recreational activities. Tourism on Exmoor brings in over £105 million to the economy and the equivalent of over 2000 full time jobs (approximately 20 per cent of Exmoor's entire population!). Local authorities and county councils take decisions to promote tourism to underpin the economy of the region and ensure its long-term sustainability. Therefore, Exmoor's ecology can be seen to directly affect the community's economic, social and political character.

However, tourism can lead to ecological damage, particularly at honeypot sites, roadsides and car parks. Footpath erosion destroys vegetation, which leads to soil erosion and increased rates of surface runoff, for example at popular sites such as Dunkery Beacon (Figure 3). Exmoor's roads are narrow and windy, resulting in congestion and occasional conflicts with local people. Parking on verges can damage wildlife. Politically, there are many hard decisions to be made in terms of prioritising conservation or economic development; striking a balance must be the priority.

▲ **Figure 3**  *Well-trodden footpaths on Dunkery Beacon*

## Climate change

It seems increasingly likely that warmer summers, a longer growing season and rising $CO_2$ levels will increase the productivity of commercial woodlands in Exmoor National Park. However, this would be balanced by drier summers, wetter winters and more extreme weather events which could cause stress in some tree species and increase soil erosion on thin soils and susceptible slopes.

◆ Species such as oak, beech and sweet chestnut are likely to find the climate tolerable, but some, such as common alder, small-leaved lime or black walnut, may suffer stress.

◆ Some of Exmoor's upland woodlands may become more like those in parts of lowland England with a greater mixture of broadleaf tree species.

◆ Many conifer species should grow well, although some (such as larch) may find it difficult to adapt.

◆ Other woodland flora would be affected, such as lichen communities and ground flora, some of which may reduce in extent.

◆ There could be a greater risk of drought and woodland fires, with which many species would not be able to cope.

In a warmer climate, non-native pests and diseases may be able to survive milder winters and extend their range, putting our trees and woodlands under further stress. Non-native species could expand their range (e.g. rhododendrons, laurel and Himalayan balsam), so existing measures that manage these species may need to be increased in intensity.

## Responding to ecological change

In 2011 ENPA published a collaborative study, *Exmoor Moorland Units*, that considered the issues, opportunities and management strategies for selected moorland blocks within the National Park. The study involved local landowners as well as organisations such as the Forestry Commission, Natural England and the RSPB. Community involvement and support was considered essential. The extent of this collaboration and cooperation between different bodies serves to illustrate the political implications of managing an ecological region such as Exmoor National Park.

### Dunkery Beacon

In addition to the impressive views from the highest point in Exmoor National Park, Dunkery Beacon (Figure **4**) provides important evidence of Bronze Age and medieval settlements that demonstrate the area's suitability for habitation during previous warmer climatic conditions. Ecologically, the diverse vegetation (moors, heaths and mires) provide a range of habitats for flora and fauna, including the rare heath fritillary and breeding birds such as the Dartford warbler.

Its popularity with tourism has led to footpath erosion and unauthorised vehicular damage to the moorland. Other issues include the spread of gorse, scrub and rhododendrons, damaging archaeological sites and the invasion of bracken on some slopes, reducing biodiversity.

The National Park is addressing the issues by:

- reviewing paths and provision of visitor access
- encouraging heather management, to include *swaling* (controlled burning) and grazing
- restoring mire sites by filling ditches to prevent runoff
- continuing to proactively manage the heath fritillary
- controlling and removing gorse, scrub and bracken where it has no ecological value, but protecting bird-breeding areas.

### The role of Exmoor National Park Authority

Exmoor National Park Authority (ENPA) has an important political role in striking a balance between conservation and supporting the community's economy, particularly its farmers. It also has a responsibility to promote access for the public for leisure and recreation. With tourism representing an increasing proportion of the region's income, ENPA has to work strategically with local and national political bodies, including local authorities and county councils, as well as national organisations such as the RSPB and English Nature. ENPA manages but does not own the land, so it needs to work closely with private landowners as well as planning authorities and pressure groups. This need for political cooperation and collaboration in the management of Exmoor's unique ecological setting is critical in promoting and supporting both conservation and continued economic development.

⬤ **Figure 4** *Dunkery Beacon, at 519 m it is the highest point in Exmoor National Park*

### Upland peat/lowland peat

Blanket bogs in the UK are most commonly associated with upland areas, primarily in Scotland. In cool, wet climates, bog mosses and other plants break down very slowly to form a layer of peat up to 3 m of more in depth. Blanket bogs are most likely to form on gentle waterlogged slopes where there is poor drainage.

In lowland areas, raised bogs are more likely to exist. These unique ecosystems started to form in shallow depressions some 8000 years ago at the end of the last glacial period. Lowland peat bogs comprise mostly of sphagnum moss, which can hold twenty times its own dry weight of water. Peat formed from decaying moss builds up slowly (about 1 mm/year), raising the bog above the surrounding land – hence 'raised bog'. In some raised bogs, peat has formed to a depth in excess of 10 m preserving important clues about climatic conditions in the past. The surface of a raised peat bog is very acidic, ideal for plants such as cranberry. The shallow pools are ideal breeding grounds for dragonflies and damselflies.

## Exmoor Mires Project

The Exmoor Mires Project (EMP) is an integrated management plan that addresses a diverse range of issues associated with deteriorating peatlands (mires) on Exmoor. It is financially supported by South West Water and works with individuals, local communities and partner organisations such as ENPA, Natural England and the Universities of Exeter and Bristol.

The overall aim of EMP is to restore the hydrological function of the peatlands in the upper catchment of the River Exe by keeping the peat wet in order to withstand the effects of climate change. It also aims to preserve the landscape and the historic information it holds for all to enjoy and benefit from in the future (see 1.17 and 1.18).

After its launch in 2010, EMP conducted a great deal of scientific research involving mapping, remote sensing and empirical data collection. This enabled restoration plans to be drawn up for each site containing details of ecology, historic environment, landscape, access, land management and ditch-blocking areas. It also estimated timings and costings. All restoration plans are consulted and agreed upon by the appropriate government bodies, landowners and farmers.

Subsequent action has largely involved blocking the drainage ditches constructed in the past by farmers (Figure **5**). This has reduced surface runoff and has resulted in the rewetting of the peatland. Increased water storage in the peatlands helps to control runoff into the Exe, thereby attenuating the annual flood hydrograph and reducing the flood risk downstream. It also improves water quality and helps to fix carbon storage.

Restoration work is carried out between August and April to avoid the ground nesting bird season. Untreated timber is used to block the drains; in time this will biodegrade naturally, by which time the ditches will have silted up and peat will be re-establishing. Any areas that have been disturbed by the works are covered by vegetation of bales of purple moor grass.

Between 2010 and 2015, 133 km of ditches were successfully blocked and 1139 ha of peatlands restored. Researchers at the University of Exeter found that the height of the water table had risen by an average of over 2 m and there was a marked increase in species richness and diversity of mire flora and fauna. Funding has been secured to enable the project to continue until 2020.

### ACTIVITIES

1. Outline the ecological, hydrological, historical and cultural importance of wet peatlands (mires) on Exmoor.

2. Study Figure **2**. Describe the landscape of Exmoor and suggest its value as an economic and social resource.

3. How does farming and tourism on Exmoor reflect its unique ecological setting?

4. Study Figure **4**. What are the arguments for and against encouraging wider access to Exmoor wild places?

5. What are the implications of climate change on Exmoor's ecology? To what extent can restoration projects such as the Exmoor Mires Project mitigate the future effects of climate change?

6. How does the management of a National Park such as Exmoor create a unique set of political challenges? In answering this question, consider the role of political collaboration and cooperation in the successful management of Exmoor National Park.

7. Dunkery Beacon is one of several 'moorland blocks' identified by Exmoor National Park Authority as requiring special management. Search the internet for the pdf file titled 'Exmoor moorland units'. Find out more about another moorland block of your choice. Present your work in the form of an information poster.

⬆ **Figure 5** *Blocking ditches to restore peatlands on Exmoor*

### STRETCH YOURSELF

Use the internet to find out what work has been undertaken by the Exmoor Mires Project since 2015. Include up-to-date statistics of the restoration and consider the impacts on the ecology and hydrology of the peatlands. Assess the importance of the collaboration between the political and environmental organisations, as well as the local community. Why is it important to ensure political consensus if a project like this is to be successful? The Exmoor Mires newsletters published by South West Water will be of particular use and interest.

Wicken Fen, Cambridgeshire, UK

In this section you will learn about the characteristics and management issues associated with Wicken Fen, a wetland ecosystem in Cambridgeshire

## Where is Wicken Fen?

Wicken Fen is a 758 ha wetland nature reserve 15 km to the north-east of Cambridge (Figure **1**). This unique landscape of natural lakes, meadows and fen wetlands has been managed by the National Trust since 1899.

Since the seventeenth and eighteenth centuries in particular, the East Anglian Fens have been drained to exploit the fertile peat soils for agriculture, although not parts of Wicken Fen, making it one of the last remaining natural fenland habitats in Europe. The term *fenland* applies to a low-lying wetland with grassy vegetation, usually forming a transition zone between land and water.

Despite its conservation status, Wicken Fen is facing development threats from housing, transport and commercial agriculture, much of it driven by the rapid growth of Cambridge, one of the fastest-growing cities in the UK.

## Wicken Fen – a biodiversity hotspot

Wicken Fen comprises multiple habitats (scrub, sedge fields, woods, reeds, ponds and ditches) each with its own distinct ecosystem, interacting with each other and forming larger ecosystems – a biodiversity hotspot.

The complexity of the food webs and the interlinking nature of the ecosystems at Wicken Fen can be illustrated by focusing on the water channels known locally as *lodes* (Figure **2**) – straight, raised waterways dug by the Romans, which formed important trading routes linking Fenland villages to the River Cam and the coast. Today, the lodes support different habitats that interlink with each other. This accounts for the phrase 'biodiversity hotspot'. There is a complexity of food webs.

- The slow-flowing, clear waters provide perfect conditions for submerged aquatic plants (such as some species of stonewort, water millefoil and water lilies) to photosynthesise, using energy from the sun to create new tissue.

- Herbivores, such as water snails, nibble the aquatic vegetation.

- Small carnivorous fish (roach and minnows) are prey for larger, freshwater carnivores (pike).

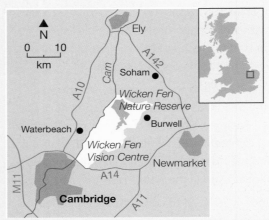

⬥ **Figure 1** *Location of Wicken Fen Nature Reserve and Wicken Fen Vision Area, 'one of the most important wetlands in Europe – an iconic habitat, supporting an estimated 9000 species of plants, insects, birds and mammals' (National Trust)*

⬥ **Figure 2** *Wicken Fen – a 'biodiversity hotspot'*

- Some species such as dragonflies and damselflies move between aquatic and terrestrial ecosystems, forming an important component in two food webs. Within the aquatic food web the larvae feed off other larvae and water snails but are prey to carnivores such as water boatman and fish.

- On emerging from the water, they become part of the terrestrial ecosystem, feeding on midges, mosquitoes, butterflies and bees (and each other!) but falling prey to birds, spiders, wasps, frogs and small mammals.

## Sustainable development – challenges and opportunities

While the National Trust has worked hard to protect species diversity in the wetland, development pressures (intensification of agriculture, climate change and developments due to the growth of Cambridge) have had detrimental impacts on this small and isolated reserve, affecting the hydrology and its biodiversity. The sharp environmental gradient between the preserved wetland and the commercial world beyond means that there are no 'soft' transition zones or ecological buffers. This means that developmental pressures (e.g. intensive agriculture) have a greater impact as there is dramatic land-use change directly adjacent to the reserve.

⬆ **Figure 3** *Konik horses, well-suited to wetland conservation grazing*

### The National Trust's Wicken Fen Vision Plan

In 1999 the National Trust launched an integrated management plan, the Wicken Fen Vision Plan, to extend the nature reserve to a maximum of 5300 ha and restore its fen and wetland habitats to create a 'sustainable landscape-scale space for wildlife and people' (National Trust).

The aim of the Vision Plan is to create a mosaic of natural wetland habitats, including wet grasslands, reed beds, marsh, shallow ponds and ditches, as well as establishing natural chalk grassland and woodlands where the soil and topography are appropriate. The Trust recognises the need to integrate wildlife conservation with the needs of local people, the economy of the area and tourism. This holistic approach is receiving widespread support from the local community.

One of the main challenges (and opportunities) is climate change, which may result in wetter winters but warmer summers in this part of eastern England. This will have an impact on the soil water budget, with more water available in the winter but less in the summer (less rainfall and higher rates of evaporation). Water will need to be stored during times of surplus in the winter – fenland environments are ideal environments for this storage to take place.

In addition, the National Trust plans to rewet and conserve the peat soils in the south to reduce their friability and promote greater biodiversity (erosion rates have exceeded 2 cm/year). Peat is an efficient carbon store and its destruction increases carbon emissions into the atmosphere enhancing the greenhouse effect. Rewetting will help to stabilise the peat and cut carbon emissions (see 1.7).

Some of the drier grassland will be grazed by domestic herds of large herbivores such as cattle, horses and deer. A herd of hardy Konik horses (Figure **3**) has been introduced from the Netherlands to help to keep the land open by grazing on weeds, reeds and grass and give plants, birds and insects the chance to settle in the area.

Through light-touch management, the National Trust hopes to have restored this whole area of natural fenland by the end of the century, ensuring biodiversity and public access in one of Europe's most important and endangered environments.

### ACTIVITIES

1  Study Figure **2**. Identify the different habitats shown in the photograph and use this to account for Wicken Fen being described as a 'biodiversity hotspot'.

2  To what extent does Wicken Fen represent a plagioclimax?

**S**

3  Attempt to draw a diagram to show the food web associated with water channels (lodes) in Wicken Fen. Show the interconnections between the aquatic and terrestrial ecosystems.

4  Outline the hydrological challenges that lie ahead for the management of Wicken Fen.

5  Examine the links between the carbon cycle and the draining and subsequent rewetting of the peatlands at Wicken Fen. Consider the presence of feedback loops.

6  Find out more about Konik horses and the role that they play in conservation and biodiversity enhancement at Wicken Fen.

7  In what ways and for what reasons does the National Trust's Wicken Fen Vision Plan offer a sustainable landscape-scale vision for the future of Wicken Fen?

### STRETCH YOURSELF

Find out more about the Wicken Fen Vision Plan – how much land has been purchased and how is the National Trust managing its sustainable development? Is it successful in addressing the needs of conservation, tourism and the local economy? Use the strategy planning document at www.nationaltrust.org.uk/wicken-fen-nature-reserve/documents/wicken-fen-vision-strategy-document.pdf

# Practice questions

The following are sample practice questions for the Ecosystems under stress chapter. They have been written to reflect the assessment objectives of the Component 1: Physical geography Section C of your A Level.

These questions will test your ability to:

◆ demonstrate knowledge and understanding of places, environments, concepts, processes, interactions and change, at a variety of scales [AO1]

◆ apply knowledge and understanding in different contexts to interpret, analyse and evaluate geographical information and issues [AO2]

◆ use a variety of quantitative, qualitative and fieldwork skills [AO3].

1 Explain the distinction between food chains and food webs. (4)

2 Complete Figure **1** below and then analyse the result. (6)

$$X^2 = \sum \frac{(O-E)^2}{E}$$

where:

$X^2$ = Chi-squared
$\sum$ = sum of

$O$ = observed frequencies
$E$ = expected frequencies

| Sampling point | Distance from high-water mark (m) | Number of marram grass plants observed within a plant quadrat (O) | $\frac{(O-E)^2}{E}$ where E = 70/10 = 7 |
|---|---|---|---|
| 1 | 10 | 3 | $(3-7)^2/7 = 2.29$ |
| 2 | 20 | 12 | $(12-7)^2/7 = 3.57$ |
| 3 | 30 | 15 | $(15-7)^2/7 = 9.14$ |
| 4 | 40 | 18 | |
| 5 | 50 | 9 | |
| 6 | 60 | 7 | |
| 7 | 70 | 3 | |
| 8 | 80 | 2 | |
| 9 | 90 | 1 | |
| 10 | 100 | 0 | |
| | | 70 | |
| | | | $X^2 = \sum \frac{(O-E)^2}{E}$ = |

| Degrees of freedom (the size of the sample minus 1) | Significance (confidence) level | |
|---|---|---|
| | 0.05 (95%) | 0.01 (99%) |
| 8 | 15.51 | 20.09 |
| 9 | 16.92 | 21.67 |
| 10 | 18.31 | 23.21 |

**Tip**

In Geography a 95% confidence level is adequate. 99% confidence, however, is desirable because we can state with certainty that there is only a 1% chance of error.

⬆ **Figure 1** The Chi-squared test

3  Using Figure **5**, page 320, and your own knowledge of a hydrosere you have studied, compare the processes leading to climatic climax ecosystems. (9)

4  Evaluate the concept of ecosystem services with reference to a specific ecosystem at a local scale. (9)

5  'Almost inevitably, population growth, together with increased economic pressures such as resource exploitation, will lead to ecosystem degradation – pollution, deforestation, desertification and biodiversity reduction. Only through very careful management can this be avoided.'
To what extent do you agree with this statement? (20)

6  Explain the term 'symbiotic relationship'. (4)

7  Study Figure **2**, page 297. Analyse the interrelationships between climate, soils and vegetation. (6)

8  Using Figures **1** and **2**, page 304, and your own knowledge, assess the extent to which climate determines the distribution of global biomes. (9)

9  To what extent do you agree that savanna grasslands have always been shaped by human activity? (9)

10  'Luxuriant, dynamic and vulnerable, tropical rainforests inspire stronger passions, more superlatives and expressed concerns than any other biome on Earth.'
To what extent do you agree with this statement? (20)

11  Outline indicators used to monitor trends in biodiversity. (4)

12  Study Figure **3**, page 283. Analyse the data shown. (6)

13  Study Figure **14**, page 324. Evaluate the controversy surrounding this management practice. (9)

14  With reference to a region experiencing ecological change, assess the response of the community and the long term sustainability of this coexistence of people and plants. (9)

15  Study Figure **2**. To what extent do you agree that the value of biodiversity must be viewed in terms of ecosystem services in the future? (20)

### Tip

**Evaluate** is a command word asking you to consider different options, ideas or arguments before coming to your own conclusion/ opinion. Your conclusions must be supported by evidence.

# What is the future of global biodiversity?

'I think we need to view biodiversity in two ways. Given the pressure on land for urbanisation, food and fuel, we'll never get beyond 15 per cent [of the world's plant and animal species within] protected areas. So we view those [reserves] on one side and that's where some biodiversity will remain. Elsewhere, the biodiversity that we have to maintain is the biodiversity we identify as being important for us and for human wellbeing.'

▲ *Figure 2*  *Statement by Kathy Willis, Director of Science, Royal Botanic Gardens, Kew (2015)*

347

### Stages in a geographical investigation

Like all other scientific investigations, a good fieldwork investigation follows a series of logical stages. This helps to ensure that the investigation is accurate, thorough and stands as a valid piece of research.

In *Component 2: Human geography and geography fieldwork investigation*, you could be asked questions on any aspect of a fieldwork enquiry, as described in Figure **2**, which you may find useful as a checklist when you are revising. Would you be able to answer a question on every aspect?

With regard to fieldwork, get involved at every stage! At AS you should experience a minimum of two days of fieldwork. Over the course of these two days you should undertake fieldwork relating to both physical and human processes in geography. Always be clear on the theory or idea your investigation sets out to test. You will need to go back to it towards the end of your analysis and consider how your results relate to it.

**Figure 1** *Taking readings at Spielboden Glacier meltwater channel, Saas-Fee, Switzerland*

❏ Preparation for fieldwork, including:
  o background reading
  o drawing up aims and objectives for the enquiry
  o planning research in the field and from secondary sources
  o using data-sampling techniques
  o carrying out health and safety procedures.

❏ Collection of primary data in the field and using secondary data sources.

❏ Processing and presenting data using relevant graphical and cartographical techniques.

❏ Analysing data, including using statistical techniques where relevant.

❏ Drawing conclusions related back to the original aims and objectives and linking these conclusions to both the place studied and the general ideas forming the basis of the enquiry.

❏ Reviewing the success, or otherwise, of all stages of the enquiry.

❏ Considering how the enquiry could be further developed.

**Figure 2** *AS Geography: stages of a fieldwork-based enquiry*

### Using different kinds of data

You need to be aware of the pros and cons of selecting qualitative or quantitative methods of data collection. How do you go about analysing quantitative and qualitative data differently? (*AQA Geography A Level & AS: Human Geography*, 2.10, 2.11.) may give you some ideas.

### Statistical skills

At AS, your study on the use of statistical techniques could include:

• measures of central tendency (mean, mode, median)

• measures of dispersion (range, interquartile range and standard deviation)

• inferential and relational statistical techniques, to include Spearman's rank correlation and application of significance tests.

Think also about your secondary sources. How would you justify your choice of sources? How useful and how reliable are they?

### If your study 'doesn't work'

Remember, there is always room for improvement in any investigation – so reflect on and recognise this when evaluating your methods or analysing your results.

## A Level Geography fieldwork investigation

### Stage 1: Identifying an appropriate question or issue

The specification requires that you undertake:

*'an independent investigation [with] a significant element of fieldwork ... based on either human or physical aspects of geography, or a combination of both.'*

During your A Level course you will be introduced to a range of different fieldwork techniques. Your time 'in the field' will cover techniques that relate to processes in both physical and human geography and should help you to make a decision about what you want to find out more about in your independent investigation.

Once you have chosen a topic, you will need to define your key question, or identify the specific issue that your investigation will focus on and, within that, a number of hypotheses you want to test. This will require some further reading on your part. Take a look at more than one textbook, so go to the library and also look online.

### Choosing your title

In choosing the title of your independent investigation you should consider the following points.

- Your title must be geographical, linked to the specification and written by you.

- It is likely to be a question you would like to answer, supported by (as mentioned above) two or more hypotheses (positive statements) you want to test.

- There must be one or more clear connections to sound geographical theory, concept or process (such as the theory of plant succession (1.9 and 6.9). You'll need to set your key question in the context of relevant literature in your write-up.

- You will need to incorporate primary data (field data collected first-hand through observations or measurements) and/or good-quality secondary data (evidence from other people's field investigations) – either way you will have to demonstrate an understanding of methods used in the field

- The investigation must be based on a small, manageable area of study such as a stretch of coastline or an area of woodland. Consider that a maximum of 1 to 2 days should be needed for collecting the primary data.

- The title must lend itself to the full development of a geographical investigation, prompting the collection of a range of data for qualitative and/or quantitative analysis.

- You must be able to conduct the investigation safely – no abseiling, wading up to your waist in fast-flowing rivers, crossing motorways, etc.

### Some examples of appropriate titles

*Do sand dune characteristics change with distance from the sea at Braunton Burrows?*

**How would the clearance of Sibley's Coppice affect local carbon sequestration?**

The characteristics of fluvioglacial and glacial sediments differ on the Anglesey coast.

# Stage 2: Planning

Many of the common problems students find when conducting an independent investigation come down to lack of proper planning. Thorough, careful planning will reduce problems later on and help you to achieve a successful outcome. If you have time, consider a small pilot study (a few hours of fieldwork to try things out) or take time to reflect on past fieldwork experiences; what worked and what didn't?

When planning your investigation you need to consider the following, presented in Figures **3** and **4**, with regard to your data collection:

⯆ **Figure 3** *Planning your data collection*

| Data collection | |
|---|---|
| **Where are you going to collect your data and when?**<br>• The choice of site(s) is very important and must be justified.<br>• You should also consider when to carry out your data collection: early morning or late afternoon; on a weekday or weekend? Figure **4** lists some factors you should consider when conducting a coastal or an urban study.<br><br>**What kind of data are you going to collect, how and why?**<br>• You need to carefully consider what data you need to fully test your hypotheses and answer your question.<br>• Would qualitative or quantitative data be more useful, and what is the most appropriate technique to employ? (E.g. questionnaire or interviews, see *AQA Geography A Level & AS: Human Geography*, 2.11)<br>• What sort of secondary data would be helpful, in addition to primary data?<br>• Be careful not to collect so much data that you simply become overwhelmed.<br><br>**What sampling strategies are you going to employ and why?**<br>• Make sure you choose an appropriate strategy for your question or issue.<br>• Consider whether you need to use *point*, *line* (transect) or *area* (e.g. quadrat) sampling.<br>• Also consider what kind of sampling you will use – *random* (chance), *systematic* (at regular intervals, e.g. every 10th person or every 10 metres) or *stratified* (biased, e.g. 25 per cent sample points from each of four areas). If sampling the population of a place, you might question a predetermined percentage of people from three or four different age groups, to ensure your sample reflects the age makeup of the community. Check the population data from the 2011 Census. | **What equipment is needed and how does it work?**<br>• You need to select appropriate equipment for an A Level investigation. For example, the use of a rudimentary float to measure river velocity might be fine in Year 7 but at A Level you should ideally use a more scientific instrument, such as a flow meter.<br>• The equipment should be reasonably easy to use, not too expensive and should be of good enough quality to provide you with accurate and reliable results.<br>• Take time to practise using the equipment before you start your investigation.<br>• Consider having 'Plan B' equipment in case your phone or camera does not work (e.g. a pad and pencil) and carrying spare batteries if necessary. |

⯆ **Figure 4** *The type of environment in which you collect your data will affect your planning*

| Considerations | Coast | Urban |
|---|---|---|
| Where? | • Is the length/area of coastline an appropriate size to enable me to collect the data I need?<br>• How can I get there?<br>• Does it have the appropriate characteristics for my study, such as exhibiting marked changes east to west, or with distance from the foreshore?<br>• Can I gain safe (and legal!) access to an appropriate number of sites (e.g. about 8 to 10 if conducting a transect across coastal dunes)? | • Where do I go to collect my data and why?<br>• Do I need to get permission to talk to people in certain places? How do I get it?<br>• Is the settlement big enough? Is it too big?<br>• How safe is the urban area? Are some areas too risky for me to go to alone? |
| When? | • Should I collect data at high tide or low tide? Or both? Why?<br>• Is access available year-round?<br>• Will the time of year affect biological indicators, such as vegetation?<br>• Will the weather be ok? | • Will the time of year affect my results (e.g. tourism)?<br>• Will the time of day (e.g. rush hour) affect my results?<br>• Is the day of the week important (e.g. Sunday compared with Monday)? Do I need to be there on both days?<br>• Will the weather affect footfall or visitor numbers? |

## Risk assessment

It is a legal requirement to carry out a risk assessment. A risk assessment aims to identify the potential risks of your fieldwork investigation and then minimise their likelihood of occurring, particularly for those risks that are potentially most likely or the impacts most severe.

To start, think about the risks that could occur on the day you conduct your fieldwork investigation (ideally, you should pay a preliminary visit to your study site to identify any risks that may be likely). Then, identify ways of minimising the risks, for example, wearing the correct footwear – Figure **5** below shows part of a typical risk assessment of a fieldwork site.

It is also important that you consider what to do if something does happen, for example, if you get a tick bite or the weather turns wet and foggy.

There is a huge amount of helpful advice on the internet, as well as numerous websites providing information about the place you are going to that you can check from your phone, for example, tide times and the weather forecast. Your school or college will be able to help too.

### Steep slopes

*Risk:* slip and fall hazards

*Solution:* take care walking, stick to path, use appropriate footwear

### Vegetation

*Risk:* dense ground vegetation makes it difficult to see the path; possible tick bites

*Solution:* take care walking, avoid trip hazards; wear long trousers, check for ticks

### Weather

*Risk:* warm weather increases the risk of dehydration and sunburn

*Solution:* bring plenty of water, sunscreen, hat, etc.

### Equipment

*Risk:* equipment gets damaged, or is inaccurate or faulty

*Solution:* have spares or alternatives

▲ **Figure 5** *The potential risks and some possible solutions for a fieldwork investigation*

## Data presentation and analysis: before you go

It might seem strange to consider how you are going to present and analyse your data at the start of your investigation, but it is very important that you do so. In order for you to present your data in different diagrams and interpret it using statistical techniques, you need to have the appropriate type and amount of data. For example, if you plan to use Spearman's rank correlation test (see 6.9), you will need to collect two sets of data that can be ranked, such as depth and age of peat, ideally with a minimum of 15 readings of each.

## Stage 3: Data collection

### Be methodical

Be methodical when collecting data in the field, you need to try to ensure that the results you obtain are accurate and reliable. For example, make sure that you take several readings at a particular site, and check the equipment regularly. On the day of the data collection, you may make small amendments to the method you use. The need for these tweaks may only be evident at your first site, once you are knee-deep in the river! As far as possible, keep your method consistent between different sites.

### Keep a record

Remember to make a note of any changes made to your methods so that you can describe and justify your approach in your report. You need to be clear on what you did and why you did it, so talk to other members of your team if you are collecting data in a group.

Do not forget to take photos and draw field sketches. These provide a good record of what you did and also help to put your fieldwork sites in geographical context. For example, the photograph in Figure **5** could be used to help explain certain trends or anomalies in the results of a river study.

### Other people's data

Try to include secondary data if possible. This might include maps, extracts from books, articles in magazines or online, official statistics or other students' work. Having another study to compare with your own is an excellent strategy for your write-up.

## Stage 4: Data presentation

Raw data is of very little value when it comes to data interpretation. It needs to be processed and displayed so that it makes sense and reveals any trends and patterns. Look online or in any fieldwork textbook for presentation methods available to you, such as line, bar and scattergraphs, pie charts, triangular graphs and dispersion diagrams – many are included in this book. At A Level you need to select your method of presentation with care, make sure they are drawn accurately and are appropriate to your study. For example:

- If you carry out a transect in a town, data should be displayed in a linear manner along a scaled line as opposed to a single pie chart.

- Points on a scattergraph should not be joined up to form a line graph but should have a best-fit line to show the trend (drawing the line is a form of analysis).

- Spatial data (e.g. pH readings at a number of sites on sand dunes) is most effectively shown by an overlay. You might use colour to distinguish between different zones within your study area, creating a choropleth map.

Whether drawn by hand or on computer, make sure that all diagrams, are complete. Do not forget to include:

- a clear and focused title
- north arrow and scale bar on a map
- a key, if appropriate
- labelled axes on graphs
- annotations on field sketches and photos including what you observed on the ground.

Remember that any annotation of figures, content of textboxes and other text linked to presentation techniques all count towards the total word count of your final 3000–4000 word report.

Try to present your various datasets clearly and logically (e.g. by date of collection, sub-theme, or location) to guide the reader through your report.

Ensure you have included a range of presentation techniques – demonstrate the breadth of your abilities.

**Which presentation technique?** Ask yourself, is your quantitative data continuous, meaning it can take any value within a range (e.g. temperature) or discrete, meaning it can only take certain values (e.g. the six values on a six-sided dice) or categories that may be unrelated to each other (such as total footfall on different days of the week). When presenting the frequency of phenomena studied, for discrete data a bar graph might be useful, whereas for continuous data you might try a histogram.

Qualitative data gives you the opportunity to be a bit more creative, but shouldn't lack rigour. Does your word cloud or quotation bank help the reader more easily make sense of your findings, or not?

## Stage 5: Interpretation and analysis

In this section of your investigation you need to discuss, describe and, where possible, explain your results. It is also important to critically examine your data. To do this effectively you should:

- make use of appropriate statistical techniques to analyse your data. Consider using measures of central tendency (e.g. mean) or dispersion. If you are examining a relationship between two variables then you should consider using Spearman's rank correlation test (6.9). If you are examining associations or looking to identify significant differences between data sets, you should consider using the Chi-squared or Mann–Whitney U tests (see below).

- consider patterns and trends rather than simply quoting lots of numbers. But be sure to use numbers to support any points you make. Translate raw data into percentages, describe the range.

- look for any exceptions (anomalies) to the general trends and try to explain why they have occurred.

- attempt to explain the patterns and trends you have observed using geographical terminology and referring to concepts and processes as appropriate.

- try to make statements that link your data sets together.

- make connections back to your title.

## Significance levels – don't forget!

With all statistical tests, the calculated result is only as good as the interpretation. Significance tables should be used to assess the probability that the result was due to chance. For most geographical investigations a significance level of 0.5 per cent is reasonable, i.e. there is a 95 per cent likelihood that the result was not caused by chance.

Any test of association or correlation (e.g. Spearman's rank correlation test) must be based on a sound geographical concept or theory.

▲ **Figure 6** *Data analysis*

## Chi-squared test

The Chi-squared test is used to test the difference between grouped data sets. The two sets of data can be either two sets of raw data or a comparison between actual observed values and theoretical 'expected' values representing equal chance.

There are a number of important aspects of this test:

- Data must be in the form of frequencies, grouped or categorised.

- The total number of observations must exceed 20.

- The expected frequency for any one group must exceed 5.

- The categories should not have a directly causal link.

## Mann–Whitney U test

The Mann–Whitney U test is used to test the difference between two sets of data. It is often carried out after drawing a dispersion graph that might suggest a possible difference between two sets of data, for example, the sizes of pebbles collected on two beaches.

There are a number of important aspects of this test:

- It measures the difference around median values.

- Raw individual data must be used rather than grouped data.

- There should be between 5 and 20 values, although it is possible to use more than 20.

- The two data sets do not need to have an equal number of values.

## Qualitative data

Not all data sets are suitable for statistical analysis. Old photographs, your own photographs, field sketches and observation notes all provide valid data for analysis, and should be included, but these qualitative sources may require a different approach. Use overlays or digital annotation that demonstrates your interpretation of these sources. You might wish to present your images on a map ('geolocate' them) or organise them by the date they were taken, to allow the reader to observe the patterns or changes you describe. If you have a lot of this form of data you will need to be selective, presenting those images or sources that best illustrate physical phenomena or the interaction between people and the environment.

## Relating your results to the wider context

Within your report you are expected to set your key question(s) for investigation in context. This means that early on you need to outline the theory or concept that prompted you to design your investigation as you did, demonstrating further reading on your part. Later on, as part of your analysis, you should attempt to explain any patterns and trends, linking back to relevant geographical theory.

## Extending geographical understanding

The experience of completing your independent investigation will deepen your understanding of the theme you choose – you might suggest how and in what ways, as part of your conclusion or evaluation. Furthermore, in the local context, your study will provide colour and detail that overarching theories from textbooks cannot. You can comment on how your specific study develops the theory or process(es) as they manifest themselves in your local or study area. Just like any other researcher, your work will have 'pushed back the boundaries' of geographical understanding – congratulations!

## Stages 6 and 7: Drawing conclusions and your evaluation

Your conclusion should pull together all of your results so that you can address the question or hypothesis topic in your title and any hypotheses you had.

On reflection, you should also be realistic about how your investigation went. Consider what you would do differently if you began again, knowing what you know now. To evaluate your investigation you should:

- consider whether there were any inaccuracies in the data collection
- suggest how data collection could have been improved (e.g. more sites, more data, different types of equipment, etc.)
- consider how representative your sample was and suggest whether or not you would take a different approach to sampling next time
- discuss the accuracy of your results, bearing in mind any problems with the data collection
- assess the validity of the conclusion, i.e. would a different person doing the same investigation get exactly the same results as you did?

Your written report should make reference to fieldwork ethics (below) and how they applied to your investigation, either in the justification of your methods, your evaluation, or both.

## Fieldwork ethics

Over the course of all geographical investigations you should strive to maintain the highest standards of scientific rigour, to protect the environment (Figure **7**) and treat people that you meet with respect.

The Code of Practice for researchers applying for funding from the Royal Geographical Society (with the Institute of British Geographers) is a useful guide on fieldwork ethics. It requires:

- accurate reporting of findings, and a commitment to enabling others to replicate results where possible

- fair dealing in respect of other researchers and their intellectual property

- confidentiality of information supplied by research subjects and anonymity of respondents (unless otherwise agreed with research subjects and respondents)

- independence and impartiality of researchers to the subject of the research.

## It is science, be it physical or human

Do not exaggerate your results or leave data out that does not fit the wider trend or suit your hypothesis. You can discuss any anomalies ('outliers'), exceptions and possible inaccuracies in your evaluation.

In addition, you should give proper credit to other researchers. If you have used their research work to inform your understanding of a place or theme and you have included their data, you must cite this. You usually do this at the foot of the page or in a bibliography, as you would with any other source of secondary data.

## 'Shut the gate!'

Remembering to follow the countryside code is not just for Scouts. The impact of your group on the countryside will be greater if you fail to consider other people who use the place, be they holidaymakers, farmers or local residents. And if you are not careful you might, inadvertently, shape the physical environment for other researchers, leading to studies like: 'Hypothesis 1: the degree of trampling of dune vegetation will decline with distance from the fieldstudy centre'. Stick to established paths where possible!

If you need to collect sediment or soil samples, consider carefully what you will do with the material once your measurements are complete. Can you return it to where you found it? Do you really need to pick any plants? Photography is a simple way of keeping a record of what you found without carrying it all back to the classroom. Areas of interest to you are also of interest to many others and some places will be protected by law, such as Sites of Special Scientific Interest (SSSIs). You may need to find out who owns the land, ask their permission to access it and check what is and is not allowed – follow the directions given by signs you come across on site (Figure **9**).

⬆ **Figure 7** *Beach sign on the Gower Coast, South Wales*

⬆ **Figure 8** *Psammosere study*

⬆ **Figure 9** *A notice on the South Downs Way*

## The role of researcher

If you are investigating the impact on people of an issue or major change in the environment, ensure you approach the subject with sensitivity – it may be something that keeps them up at night! Take care with question design and avoid leading questions that might suggest your views and/or introduce bias to your data. Your role is to remain as neutral as possible and record what is said.

When recruiting people to answer your questionnaire or take part in a group discussion or interview, briefly explain the purpose of your research and how the results will be used. You will need to get their permission to record what they say or permission from their parents, perhaps, if they are not old enough to give their consent. And thank people for their time afterwards; being polite will always pay off. For a full discussion of your role as a researcher in the interview process, see *AQA Geography A Level & AS: Human Geography*, 2.11.

## Naming names: don't do it!

The results of any questionnaire or survey should make no mention of real names and the reader should not be able to identify individual participants from the data you present. Giving your interviewees the guarantee of anonymity allows them to talk freely, so you are much less likely to hear what they think you want to hear, and more likely to get their real views.

## What does a successful independent investigation look like?

▲ **Figure 10** *Interviewing in Zermatt, Switzerland*

▶ **Figure 11** *Your role as a researcher is to remain neutral and record what is said*

▼ **Figure 12** *The Long Mynd, Shropshire*

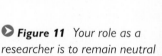

### 3.3.4.2 Methods of field investigation
*Level 4: 12–15 marks*

- Detailed use of a range of appropriate observational, recording and other data collection approaches including sampling.

- Thorough and well-reasoned justification of data collection approaches.

- Detailed demonstration of practical knowledge and understanding of field methodologies appropriate to the investigation of human and physical processes.

- Detailed implementation of chosen methodologies to collect data/information of good quality and relevant to the topic under investigation.

▲ *Figure 13* West Bay, Bridport, Dorset

### 3.3.4.3 Methods of critical analysis
*Level 4: 15–20 marks*

- Effective demonstration of knowledge and understanding of the techniques appropriate for analysing field data and information and for representing results.

- Thorough ability to select suitable quantitative or qualitative approaches and to apply them.

- Thorough ability to interrogate and critically examine field data in order to comment on its accuracy and/or the extent to which it is representative.

- Complete use of the experience to extend geographical understanding.

- Effective application of existing knowledge, theory and concepts to order and understand field observations.

▲ *Figure 14* Honister Pass, the Lake District

### 3.3.4.4 Conclusions, evaluation and presentation
*Level 4: 12–15 marks*

- Thorough ability to write up field results clearly and logically, using a range of presentation methods.

- Effective evaluation and reflection on the fieldwork investigation.

- Complete explanation of how the results relate to the wider context(s).

- Thorough understanding of the ethical dimensions of field research.

- Thorough ability to write a coherent analysis of fieldwork findings in order to answer a specific geographical question.

- Draws effectively on evidence and theory to make a well argued case.

## ACTIVITIES

Study Figures **12–14**. Pick one location and discuss the following with a partner.

**a** The risks of conducting fieldwork in this environment and how you would minimise these risks.

**b** Two possible hypotheses or key questions to investigate.

**c** Any ethical dilemmas that might arise as part of your proposed investigation in this place.

Your AS or A Level course is a lively and interesting one, full of contemporary topics, such as Changing places as well as old favourites like Hazards. It is important to prepare for your assessment well, and you should take the time to understand the structure and assessment criteria of your course.

## What does the AS specification include?

Remember that the AS Geography qualification does not count towards an A Level. The topic content is designed to provide you with a real knowledge of geographical themes and skills, along with a real enthusiasm for the subject.

The AS specification consists of six topics, of which you must study three plus a fieldwork investigation. They are grouped into two components: Physical geography and people and the environment; Human geography and fieldwork.

| Component 1: Physical geography and people and the environment | Component 2: Human geography and geography fieldwork investigation |
|---|---|
| • Section A: you must study one of either Water and carbon cycles (Chapter 1 in this textbook) or Coastal systems and landscapes (Chapter 3) or Glacial systems and landscapes (Chapter 4)<br><br>• Section B: you must study either Hazards (Chapter 5) or Contemporary urban environments (Chapter 3 in *AQA Geography A Level & AS: Human Geography*) | • Section A: Changing places (Chapter 2 in *AQA Geography A Level & AS: Human Geography* ). This topic is compulsory.<br><br>• Section B: Geography fieldwork investigation (Chapter 7) and geographical skills. You will have to answer one question on each of these topics. |

Remember, there is no coursework at AS, but questions will be included in the exam about the fieldwork you complete as part of your AS course.

## What does the A Level specification include?

The A Level specification consists of eleven topics of which you must study six plus a fieldwork investigation. They are grouped into three components: Physical geography; Human geography; Geography fieldwork investigation.

| Component 1: Physical geography | Component 2: Human geography | Component 3: Geography fieldwork investigation |
|---|---|---|
| • Section A: Water and carbon cycles (Chapter 1 in this textbook). This topic is compulsory.<br><br>• Section B: you must study one of either Hot desert systems and landscapes (Chapter 2) or Coastal systems and landscapes (Chapter 3) or Glacial systems and landscapes (Chapter 4)<br><br>• Section C: you must study either Hazards (Chapter 5) or Ecosystems under stress (Chapter 6) | • Section A: Global systems and global governance (Chapter 1 in *AQA Geography A Level & AS: Human Geography* ). This topic is compulsory.<br><br>• Section B: Changing places (Chapter 2 in *AAQA Geography A Level & AS: Human Geography* ). This topic is compulsory.<br><br>• Section C: you must study either Contemporary urban environments (Chapter 3 in *AQA Geography A Level & AS: Human Geography*) or Population and the environment (Chapter 4 in *AQA Geography A Level & AS: Human Geography*) or Resource security (Chapter 5 in *AQA Geography A Level & AS: Human Geography*) | • You must complete an independent investigation which must include data collected in the field.<br><br>• Your investigation must be based on a question or issue defined and developed by you relating to any part of the specification content |

## How will you be assessed?

### AS Level

There are two exams for AS Geography. Unlike A Level there is no coursework – but you must complete two days of AS fieldwork. The AS exams will involve:

#### Component 1: Physical geography and people and the environment

There is a total of 80 marks available, worth 50% of the AS qualification. The paper consists of multiple-choice, short answer, levels of response and includes 9- and 20-mark extended prose questions. You may use a calculator.

You will have 1 hour 30 minutes.

- Section A: you must answer either question 1 (Water and carbon cycles) or question 2 (Coastal systems and landscapes) or question 3 (Glacial systems and landscapes) – **40 marks**

- Section B: you must answer either question 4 (Hazards) or question 5 (Contemporary urban environments) – **40 marks**

#### Component 2: Human geography and geography fieldwork investigation

There is a total of 80 marks available, worth 50% of the AS qualification. The paper consists of multiple-choice, short answer, levels of response and includes 9- and 20-mark extended prose questions. You may use a calculator.

You will have 1 hour 30 minutes.

- Section A: you must answer all questions (Changing places) – **40 marks**

- Section B: this will test fieldwork skills. You will be asked questions on your own fieldwork and the use of fieldwork skills in different situations – **40 marks**

### A Level

There are two exams for A Level Geography and one individual fieldwork investigation – you must complete at least four days of A Level fieldwork. The A Level assessment will involve:

#### Component 1: Physical geography

There is a total of 120 marks available, worth 40% of the A Level qualification. The paper consists of multiple-choice, short answer, levels of response and includes 20-mark extended prose questions. You may use a calculator.

You will have 2 hours 30 minutes.

- Section A: you must answer all questions (Water and carbon cycles) – **36 marks**

- Section B: you must answer either question 2 (Hot desert systems and landscapes) or question 3 (Coastal systems and landscapes) or question 4 (Glacial systems and landscapes) – **36 marks**

- Section C: you must answer either question 5 (Hazards) or question 6 (Ecosystems under stress) – **48 marks**

#### Component 2: Human geography

There is a total of 120 marks available, worth 40% of the A Level qualification. The paper consists of multiple-choice, short answer, levels of response and includes 20-mark extended prose questions. You may use a calculator.

You will have 2 hour 30 minutes.

- Section A: you must answer all questions (Global systems and global governance) – **36 marks**

- Section B: you must answer all questions (Changing places) – **36 marks**

- Section C: you must answer either question 3 (Contemporary urban environments) or question 4 (Population and the environment) or question 5 (Resource security) – **48 marks**

## Component 3: Geography fieldwork investigation

There is a total of 60 marks available for this independent investigation, worth 20% of the A Level qualification. You must complete an independent investigation based on a question or issue defined and developed by you relating to any part of the specification content.

You are advised to write between 3000 and 4000 words. Your investigation will be marked by your teachers and moderated by AQA.

## How are the exam papers marked?

Examiners have to know what it is that they are assessing you on, so they use Assessment Objectives, or AOs for short. There are three AOs for AS and A Level.

- **AO1: Demonstrate knowledge and understanding** of places, environments, concepts, processes, interactions and change, at a variety of scales (30–40%).

- **AO2: Apply knowledge and understanding** in different contexts to interpret, analyse and evaluate geographical information and issues (30–40%).

- **AO3: Use a variety of relevant quantitative, qualitative and fieldwork skills** to:

  - investigate geographical questions and issues

  - interpret, analyse and evaluate data and evidence

  - construct arguments and draw conclusions (20–30%).

## Understanding the mark schemes

There are two types of mark scheme, which depend on the number of marks allocated for shorter questions (up to 4 marks) and those for extended written answers (6 marks or more).

### Shorter answers (worth up to 4 marks)

Questions carrying up to 4 marks are point marked. For every correct point that you make, you earn a mark. Sometimes these are single marks for 1-mark question. Others require the development of a point for a second mark, for example, if you are asked to describe one feature of something for 3 marks. *Outline the role of wind in affecting coastal energy. [3 marks]*

There are three marks for this question and you have to outline the role of wind. You would receive one mark per valid point (e.g. 'wind is responsible for the generation of waves as friction occurs at the surface of the water') with additional credit for development (e.g. 'this has a direct bearing upon the potential for longshore drift depending upon the angle that the waves hit the coastline').

### Extended answers (worth 6 marks or more)

Questions carrying 6 marks or more are *level-marked*. The examiner reads your whole answer and then uses a set of criteria – known as levels – to judge its qualities. There are two levels for questions carrying 6 marks, three levels for 9 marks, and four levels for 20 marks.

Consider this question:

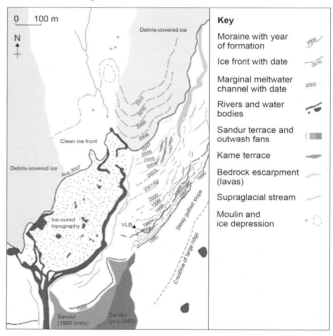

▲ **Figure 1** *A geomorphological map of an Icelandic glacier, Virkisjökull-Falljökull*

*With reference to Figure 1, interpret the evidence that this glacier is changing. [6 marks]*

Examiners would mark this question using the mark scheme on page 361, which consists of two levels. The mark scheme is the same for all 6-mark questions. This question tests interpretation of the quantitative evidence provided, so it is an AO3 question.

| Level | Mark | Descriptor |
|-------|------|------------|
| Level 2 | 4–6 | • Clear interpretation of the quantitative evidence provided, which makes appropriate use of data in support.<br><br>• Clear connection(s) between different aspects of the data and evidence. |
| Level 1 | 1–3 | • Basic interpretation of the quantitative evidence provided, which makes limited use of data and evidence in support.<br><br>• Basic connection(s) between different aspects of the data and evidence. |

Interpretation makes clear links between glacial retreat and evidence provided in Figure **1**. Use and understanding of the map evidence are clear and accurate. Your answer must fit the Level **2** criteria.

## 9- and 20-mark answers

9-mark questions have three levels in their mark schemes. These are identical for each of the exam papers. The balance of marks is split between 4 marks for AO1 and 5 marks for AO2. You should aim to:

- demonstrate accurate knowledge and understanding throughout (AO1)
- apply your knowledge and understanding (AO2)
- produce a full interpretation that is relevant and supported by evidence (AO2)
- make supported judgements in a balanced and coherent argument (AO2).

20-mark questions have four levels in their mark schemes. The balance of marks is split evenly with 10 marks for AO1 and 10 marks for AO2. In addition to the qualities for Level 3 you should aim to:

- reach a detailed evaluative conclusion that is rational and firmly based on knowledge and understanding which is applied to the context of the question (AO2)
- show a detailed, coherent and relevant analysis and evaluation in the application of knowledge and understanding throughout (AO2)
- show full evidence of links between knowledge and understanding and the application of knowledge and understanding in different contexts (AO2).

## Command words

Command words – that is, those words that tell you what you must do – are listed here.

**Analyse** Examine and break down the data provided in a figure, in order to make sense of it. Look for trends or patterns as well as any anomalies. Use evidence to help your description but don't rely on direct lifts. You will need to manipulate the data. Finally look for links between data sets. This generally assesses AO3 and does not require you to explain. Occasionally **Analyse** is the command word used in an AO2 question, if used without a figure.

**Annotate** Add to a diagram, image or graphic a number of words that describe features, rather than just identify them (which is labelling). This assesses AO3.

**Assess** Consider several options or arguments and weigh them up so as to come to a conclusion about their effectiveness or validity. This assesses AO1 (as you have to know something) but mainly AO2 (as you are asked to apply your understanding and make a judgement).

**Compare** Describe the similarities and differences of at least two phenomena. This generally assesses AO1 or AO3.

**Contrast** Point out the differences between at least two phenomena. This generally assesses AO1 or AO3.

**Critically** Often occurs before 'Assess' or 'Evaluate' inviting an examination of an issue from the point of view of a critic with a particular focus on the strengths and weaknesses of the points of view being expressed. This assesses AO1 (as you have to know something) but mainly AO2 (as you are asked to apply your understanding and make a judgement).

**Define..., What is meant by...** State the precise meaning of an idea or concept. This assesses AO1.

**Describe** Give an account in words of a phenomenon which may be an entity, an event, a feature, a pattern, a distribution or a process. This assesses AO1.

**Distinguish between** Give the meaning of two (or more) phenomena and make it clear how they are different from each other. This assesses AO3.

**Evaluate** Consider several options, ideas or arguments and form a view based on evidence about their importance/validity/merit/utility. This assesses AO1 (as you have to know something) but mainly AO2 (as you are asked to apply your understanding and make a judgement).

**Examine** Consider carefully and provide a detailed account of the indicated topic. This assesses AO1.

**Explain… Why… Suggest reasons for…** Set out the causes of a phenomenon and/or the factors which influence its form/nature. This usually requires an understanding of processes. This assesses AO1 and occasionally AO2.

**Interpret** Ascribe meaning to geographical information and issues. This assesses AO3.

**Justify** Give reasons for the validity of a view or idea or why some action should be undertaken. This might reasonably involve discussing and discounting alternative views or actions. This assesses AO1

(as you have to know something) but mainly AO2 (as you are asked to apply your understanding and make a judgement).

**Outline…, Summarise…** Provide a brief account of relevant information. This assesses AO1.

**To what extent…** Form and express a view as to the merit or validity of a view or statement after examining the evidence available and/or different sides of an argument. This assesses AO1 (as you have to know something) but mainly AO2 (as you are asked to apply your understanding and make a judgement).

## Handy hints

It can be tough to keep cool under pressure. However, these hints should help you perform much better.

### Dissect the question

Look at the example to the right. For this question you should aim to make two well-developed points.

Note the command word – 'outline'! So provide a brief account of relevant info.

Outline the impact of temperature variation on weathering processes in hot deserts. [4 marks]

The focus of the question – the impact of temperature variation.

Relate everything you say to weathering processes – and why they're happening in the way they do.

### Choose examples and case studies

Case studies are in-depth examples of particular places, used to illustrate big ideas at localised scales. There are plenty in these books; some examples are a paragraph, while others run to several pages. You should use these to answer questions.

Few questions actually ask for examples but including examples can help you better demonstrate a point. The following is a question about Hazards (Chapter 5), where good examples could produce a better answer:
*To what extent do you agree that seismic events will always generate more widespread and severe impacts than volcanic events? [9 marks]*

In this example question, specific examples of seismic and volcanic hazard would enhance your answer – they could be events that you have studied in this book (see sections 5.8 and 5.13). The important thing is that named examples make your answer precise – and you could find that three explained examples could help you better demonstrate and apply knowledge and understanding of the relative impacts of seismic and volcanic events.

But be selective – no question at this level will ever ask you to write everything you know about a case study.

### Plan answers

Students who plan their answers before writing them usually score higher marks than those who don't. This is because planning:

- stops the student from going off-track
- prevents 'memory blanks'.

Your plans need not to be lengthy – allow 5–10 % of total exam time for planning. Also, your plans should not be elaborate – brief notes will keep your answer on track.

### Keep to time

Many students lose track of time in exams. The exam questions are designed to be completed in the time allowed, so the following suggestion may help you to achieve this. BEFORE the exam:

- work out how long each section should take to complete
- work out 5–10% total planning time, 80–85% writing time, and 10% checking time
- practise timed answers – including shorter answers, as well as longer ones.

# Glossary

**accretion** Coastal sediment being deposited on a beach making it wider

**active layer** The surface layer that is thawed seasonally in periglacial environments

**adaptation** Fitting into the natural system and local ways of life, which may involve changing behaviour

**alluvial fan** Triangular fan-shaped alluvial deposit formed at the edge of a mountain front at the outlet of wadis and canyons

**aridity index** A measure of aridity – it is the ratio between mean annual precipitation (P) and mean annual potential evapotranspiration (PET)

**atmosphere** The air that surrounds the Earth

**bajada** Extensive apron of alluvium formed by the coalescence (merging) of alluvial fans

**barchan** Crescent-shaped sand dune formed at right angles to the prevailing winds

**basal sliding** Large-scale and often quite sudden movement of a portion of ice in a glacier, usually lubricated by subglacial meltwater

**berm** A ridge or plateau on the beach formed by the deposition of beach material by wave action

**biodiversity** The range (diversity) of living organisms within a species, between species and in an ecosystem

**biome** A large-scale ecosystem (e.g. tundra, savanna)

**block separation** Rocks with a clear pattern of joints and bedding plains break apart in the form of blocks

**blockfield** Rock-strewn landscape caused by frost action

**burial and compaction** Where organic matter becomes buried and is then compressed by the overlying sediment

**butte** Isolated pillar of horizontally bedded rock, a relic of an eroded mountain landscape

**carbon budget** A way of using data to describe the amount of carbon that is stored and transferred within the carbon cycle

**carbon cycle** The recycling of carbon between the main carbon stores – the atmosphere, the lithosphere, the hydrosphere and the biosphere

**carbon sequestration** An umbrella term used to describe the long-term storage of carbon in plants, soils, rock formations and oceans

**carbon sink** Anything that absorbs more carbon than it releases

**carbon source** Anything that releases more carbon than it absorbs

**channel flash flooding** Flooding that happens after a storm when the water makes its way into channels in the landscape

**chemical weathering** The breakdown or decay of rocks involving a chemical change

**climatic climax vegetation** The vegetation that would evolve in a climate region if the seral progression is not interrupted by human activity, tectonic processes etc.

**closed system** A system with no inputs or outputs

**coastal morphology** The origin and evolution of a coast

**cold-based glacier** Glacier where the base temperature is too low to enable liquid water to be present so the glacier freezes to the ground

**combustion** The process where carbon is burned in the presence of oxygen and converted to energy, carbon dioxide and water

**compressional flow** 'Piling-up' or thickening of glacier ice due to a decrease in the long profile valley floor gradient

**constructive wave** A powerful wave with a strong swash that surges up a beach usually forming a berm

**continentality** The influence of a large land mass on weather and climate

**cryosphere** The frozen parts of the Earth's surface including ice caps, frozen oceans, glaciers and snow cover

**cusp** Crescent-shaped beach formations with graded sediment; coarse material collects at the 'horns' and finer material collects in the 'bay' area

**dalmatian coast** A submergent landscape of ridges and valleys running parallel to the coast

**decomposition** The process where carbon from the bodies of dead organisms is returned to the air as carbon dioxide

**deflation** The process of wind erosion that involves the removal of loose material from the desert floor, often resulting in the exposure of the underlying bedrock

**desert pavement** Stony desert surface often resembling a cobbled street

**desertification** Turning marginal land into a desert by destroying its biological potential

**destructive wave** A wave formed by a local storm that crashes onto a beach and has a powerful backwash

**drift-aligned beach** Formed when beach deposits (sand and pebbles) are transferred along a coastline by longshore drift, and accumulate to form a wide beach at a headland where the lateral drift is interrupted

**dynamic equilibrium** A state of balance where inputs equal outputs in a system that is constantly changing

**ecosystem** A system in which organisms interact with each other and with their environment

**endoreic river** River that terminates in a desert region usually in a lake

**ephemeral river** River that flows intermittently in a desert region

**episodic flash flooding** Infrequent high intensity rainfall event that results in either sheet flash flooding or channel flash flooding

**esker** Sinuous (winding) ridge found on the floor of a glacial trough, formed by fluvioglacial deposition in a meandering subglacial river

**eustatic change** Variations in relative sea level resulting from changes in the amount of liquid water entering the oceans (e.g. glacial meltwater at the end of an ice age)

**evapotranspiration** the combined losses of moisture through transpiration and evaporation

**exfoliation** The peeling or flaking of the outer skin of rocks due to intense heating and cooling

**exogenous river** River that flows continuously through a desert and has its source in mountains outside the desert

**extensional flow** Stretching or thinning of ice (glacier) in response to an increase in gradient

**fjord** Created when a rise in sea level floods a deep glacial trough

**flood hydrograph** A graph that plots river discharge against time

**flows/transfers** The process of moving water or carbon from one store to another

**frost heave** Small-scale upwards displacement of soil particles as a result from the freezing and expansion of water just below the ground surface

**frost shattering** Repeated freezing and thawing of water trapped within a rock, causing it to shatter

**glacial budget** The balance between inputs and outputs of a glacier

**global atmospheric circulation system** The large-scale circulation of the atmosphere involving three distinct but interconnected circulation cells

**granular disintegration** The crumbling and breaking down of rocks made of grains (such as granite and sandstone into grains of sand

**gravitational sliding** The movement of tectonic plates as a result of gravity

**halosere** Vegetation succession that originated in an area of saline water

**high-energy environments** Coastline with powerful waves where rates of erosion exceeds rates of deposition

**hydrosere** Vegetation succession that originated in an area of fresh water

**hydrosphere** All of the water on or surrounding the Earth, including oceans, seas, lakes, rivers, ice, groundwater in soil or rock, and the water in the atmosphere

**ice wedge** V-shaped ice-filled features formed by the enlargement of a surface cracks by frost action. In time the cracks will become infilled with sediment.

**inselberg** Rounded isolated outcrop of rock, a relic of an eroded upland

**insolation** The amount of heat (short-wave radiation) that reaches the ground surface

**internal deformation** Small-scale inter- and intra-granular movement or deformation of ice crystals in response to gravity and mass

**isostatic change** Rising or falling of a land mass relative to the sea resulting from the release of the weight of ice after the last ice age or by the weight of sediment being deposited

**kame** Mound or hillock found on the floor of a glacial trough formed by fluvioglacial deposition

**lahar** Mudflow composed mainly of volcanic ash mixed with water from a crater lake, snowmelt, glacier melt or prolonged torrential rain

**liquefaction** The jelly-like state of silts and clays resulting from intense ground shaking. This may result in subsidence and collapse of buildings following an earthquake.

**lithosere** A vegetation succession that originated on a bare rocky surface

**lithosphere** The outermost solid layer of the Earth, approximately 100 km thick, comprising the crust and upper mantle

**low-energy environments** Coastline with waves of relatively low power where rates of deposition exceeds rates of erosion

**magma plume** A rising column of hot rock usually at plate a margin (but can also burn through a plate) creating a hot spot

**meltwater channel** Often narrow and steep-sided valleys formed by torrents of meltwater at the end of a glacial period

**mesa** Table-like relic landform formed in horizontally bedded rocks

**mitigation** Reducing or alleviating the impacts or severity of adverse conditions or events

**negative feedback** A cyclical sequence of events that damps down or neutralises the effects of a system

**nivation** Snow-related processes, such as weathering and mass movement, that operate collectively to form shallow hollows in the landscape

**nutrient cycle** Recycling of nutrients between living organisms and the environment

**offshore bars** Submerged (or partly exposed) ridges of sand or coarse sediment created by waves offshore from the coast

**open system** A system with inputs from and outputs to other systems

**outwash plain** An extensive, gently sloping area of sands and gravels formed in front of a glacier

**patterned ground** Concentration of large stones on the ground surface, usually associated with polygonal patterns of ice wedges

**pediment** Gently sloping, usually concave rock surface at the foot of a mountain front

**permafrost** Permanently frozen soil and rock, a key characteristic of a periglacial environment

**photosynthesis** The process whereby plants use the light energy from the sun to produce carbohydrates in the form of glucose

**pingo** Ice-cored mound formed by the freezing of subsurface water bodies and subsequent swelling of the ground surface

**pioneer community** The flora and fauna that first colonise a habitat

**pioneer species** The first plants that colonise an area, usually with special adaptations

**plagioclimax community** The vegetation succession that results from human influence

**playa** Salt lake formed on flat clay deposits on a desert plain characterised by high levels of salinity

**positive feedback** A cyclical sequence of events that amplifies or increases change

**psammosere** Vegetation succession that originated in a coastal sand dune area

**raised beach** The result of isostatic recovery which raises wave-cut platforms and their beaches above the present sea level

**resilience** The psychological quality of strength of character, of being able to respond positively to adversity

**respiration** A chemical process that happens in all cells, which converts glucose into energy

**ria** A sheltered winding inlets with irregular shorelines

**ridge push** The higher elevation at a mid-ocean ridge causes gravity to push the lithosphere that is further from the ridge

**runnel** The dips in the foreshore area of a beach between ridges. They are drained down the beach by channels that break the ridges.

**salt crystallisation** Growth of salt crystals within a rock that can cause it to break apart

**saltation** Rocks or sand that is moved in a series of leaps across a river or sea bed or the desert floor

**saltmarsh** Coastal ecosystem formed on mudflats (e.g. in a river estuary) largely comprising salt-tolerant plants

**sediment budget** An attempt to quantify the various stores and transfers associated with sediment movement

**sediment cell** A conceptual way of describing sediment movement from a source, through various transfers to a sink or output. The movement is usually cyclical.

**seif dune** Linear sand dune formed parallel to the prevailing winds

**seismicity** The frequency and distribution of earthquakes in an area

**seral stage** A stage within the sere

**sere** A complete vegetation succession

**shattering** The breakdown of rocks that do not have separate grains or a clear pattern of joints to form angular fragments

**sheet flooding** Flooding that happens after a storm when water flows across the landscape rather than being diverted into valleys and gullies

**slab pull** Following subduction, the lithosphere sinks into the mantle under its own weight, helping to 'pull' the rest of the plate with it

**soil water** Water that is stored in soil

**soil water budget** The seasonal pattern of water availability for plant growth

**stemflow** Water flowing down the stems of plants or the trunks of trees

**store** An accumulation or quantity of water or carbon

**subduction** Occurs when one tectonic plate slides beneath another, moving down into the mantle. This usually involves oceanic crust sliding beneath continental crust.

**submarine volcanoes** Volcanoes formed beneath the sea (either a single-vent or fissure volcano), where lava is emitted along a crack in the Earth's crust (e.g. the Mid-Atlantic Ridge)

**supervolcano** A huge volcano that often takes the form of a caldera and is associated with massive eruptions capable of having a global impact on people

**surface creep** Sand that is transported by being rolled along the desert floor

**suspension** Sand that is whisked up and carried by the wind often over great distances. This also applies to sediments being carried along within a water body.

**swash-aligned beach** A beach formed in a low-energy environments by waves roughly parallel to the shore

**symbiotic** Diverse organisms that exist in the same environment, often depend on this relationship to survive and prosper

**system** An approach that usually takes the form of a diagram representing the different components and their interrelationships or links between them

**tephra** Pyroclastic material that ranges in size from dust to blocks the size of cars

**terracettes** Steps formed on a slope caused by the freezing and thawing of the ground causing particles to move downhill

**thermal fracture** A form of weathering brought about by the expansion and contraction of the outer surface of a rock caused by intense temperature fluctuations

**thermokarst** A typical periglacial landscape of hollows and hummocks

**tombolo** Ridge of beach material that has formed between an island and the mainland

**transform fault** A fault that cuts across a mid-ocean ridge

**trophic level** An organism's position in the food chain

**ventifact** Sharply angled individual rock usually found on a desert pavement and formed by abrasion

**vulcanicity** The process of molten rock and gases extruding onto the Earth's surface or intruding into the Earth's crust

**wadi** Dry river channel, gully or valley formed by periodic water erosion

**warm-based glacier** Glacier where the base temperature is high enough to enable meltwater to exist and therefore basal sliding to occur

**water balance** An equation used to express the relationship between precipitation, runoff, evapotranspiration and storage

**water cycle** The recycling of water between the main water stores – the lithosphere, the hydrosphere, the cryosphere and the atmosphere

**weathering** The breakdown or decay of rocks in their original place at, or close to, the surface. Chemical weathering involves the absorption of carbon dioxide from the atmosphere.

**wildfire** An uncontrolled fire, either natural or human-made, that occurs in open country or wilderness

**yardang** Ridge formed by abrasion of vertical bands of resistant rock

**zeugen** Ridge formed in horizontal rocks, with a clear resistant cap rock

# Index